Pyrolytic Methods
in Organic Chemistry

APPLICATION OF FLOW AND FLASH
VACUUM PYROLYTIC TECHNIQUES

This is Volume 41 of
ORGANIC CHEMISTRY
A series of monographs
Editor: HARRY H. WASSERMAN

A complete list of the books in this series appears at the end of the volume.

Pyrolytic Methods in Organic Chemistry

APPLICATION OF FLOW AND FLASH VACUUM PYROLYTIC TECHNIQUES

ROGER F. C. BROWN

Department of Chemistry
Monash University
Clayton, Victoria
Australia

ACADEMIC PRESS   1980
A Subsidiary of Harcourt Brace Jovanovich, Publishers
New York   London   Toronto   Sydney   San Francisco

ACADEMIC PRESS, INC.
111 Fifth Avenue, New York, New York 10003

United Kingdom Edition published by
ACADEMIC PRESS, INC. (LONDON) LTD.
24/28 Oval Road, London NW1 7DX

Library of Congress Cataloging in Publication Data

Brown, Roger F C
    Pyrolytic methods in organic chemistry.

    (Organic chemistry, a series of monographs ;)
    Includes bibliographical references and index.
    1. Pyrolysis. 2. Chemistry, Organic. I. Title.
II. Series.
QD281.P9B76   547'.308'6   79—52787
ISBN 0—12—138050—5

PRINTED IN THE UNITED STATES OF AMERICA

80 81 82 83   9 8 7 6 5 4 3 2 1

Contents

Preface

During the last 15 years there has been a revival of interest in the use of pyrolytic reactions in organic synthesis, and this volume offers a guide to the literature of this area. Pyrolytic reactions do not form a distinct mechanistic class, and they include many reactions familiar from solution studies at ordinary temperatures. Thus this literature is not, with some exceptions, readily brought together through the use of *Chemical Abstracts* or indexes of reviews. There are some recent short reviews of flash pyrolytic methods and results and some longer reviews of particular pyrolytic topics, but there has been no successor to C. D. Hurd's (1929) classic monograph "The Pyrolysis of Carbon Compounds."

This volume is much more restricted in its scope than that of Hurd. It deals mainly with the use of flow pyrolysis and flash vacuum pyrolysis in preparative organic chemistry, and its intention is to encourage chemists who are unfamiliar with such methods to use them as readily as they now use photochemical reactions. Organic chemists with no experience of pyrolytic methods are often reluctant to use them, probably because they feel that special and complex apparatus is always needed. This volume is unashamedly propagandist and is aimed at overcoming that reluctance and providing examples of the types of reaction for which flow and flash pyrolytic methods are well suited. It is intended mainly for practicing academic and industrial organic chemists and for advanced and graduate students.

The purpose and scope of the book is explained further in Chapter 1; Chapter 2 gives an account of apparatus and experimental methods. In the remaining seven chapters pyrolytic reactions are grouped together partly by the nature of the overall process (e.g., elimination reactions in Chapter 4), partly by the formal structure of the starting material (fragmentation of cyclic structures in Chapter 6), and partly by mechanistic type. The coverage in individual sections is not exhaustive; only a selection of examples is given with references to reviews and leading papers. This should, however, enable the reader to assess the value of a reaction and to obtain more complete coverage without difficulty. There is limited mechanistic discussion of many reactions, but for adequate discussion the reader is usually referred to more specialized accounts. Material for inclusion is selected mainly from the literature to the end of 1977, but there are some references to papers published or in press in 1978.

Flash pyrolytic methods in particular have permitted the straightforward synthesis of many sensitive and highly reactive organic compounds which are not readily obtained from reactions in solution, and all organic chemists should be aware of their preparative power.

Most of this book was written in Oxford during study leave granted by Monash University, and I am grateful to Professor Sir Ewart Jones for the hospitality of the Dyson Perrins Laboratory and to Professor W. A. Waters who made room for me in his office. I am indebted to the following for valuable discussion, for sending me unpublished data or diagrams, or for other assistance: Professor J. A. Berson (Yale), Dr. F. W. Eastwood (Monash), Professor H-D. Martin (Würzburg), Dr. I. D. Rae (Monash), Professor C. W. Rees (Liverpool), Professor R. J. Spangler (Moscow, Idaho), Dr. R. C. Storr (Liverpool), Professor B. M. Trost (Wisconsin), and Professor K. P. C. Vollhardt (Berkeley). All of the line diagrams were drawn or redrawn by Mr. B. A. Baxter and his assistants in the Department of Chemistry, Monash University. My wife, Mary, retyped the whole of the final manuscript and removed many errors and inconsistencies from the text.

Finally I wish to thank particularly my students and research associates both at Monash and at the Department of Chemistry of the Australian National University for the skill and enthusiasm which they have brought to the study of flash pyrolytic reactions.

Roger F. C. Brown

Terminology and Conventions

I. EQUATIONS AND FLOW SCHEMES

The conditions used in pyrolytic experiments are often summarized on the first arrow of the equation according to the following scheme.

$$\text{Starting material} \xrightarrow[\substack{\text{tube diameter} \times \text{tube length} \\ \text{material of the tube} \\ \text{packing in the tube} \\ \text{contact time (CT)}}]{T^{\circ}(C)/P(\text{mm Hg})/\text{carrier gas}} \text{product}$$

Unfortunately much of this information is not reported for many experiments, particularly those in preliminary communications. In a few cases the conditions in flash vacuum pyrolytic experiments are shown merely as FVP, implying a high vacuum and a very short contact time. All pyrolytic *temperatures* are in °C, but °K are used in describing a few very low temperature spectroscopic measurements. All *pressures* are noted in mm Hg. The term N_2 *flow* implies a pressure of 1 atm of nitrogen.

Transient species are sometimes shown within large square brackets to indicate their nature, but this device is used very sparingly for emphasis, and many transients are not so marked. Recovered starting material is occasionally shown as SM.

II. UNITS OF ENERGY

This book is little concerned with closely reasoned use of thermodynamic and other quantitative data, and energies are quoted in the units used by the original authors so that energies in kilocalories, kilojoules, and electron volts may be found within the one chapter. This will distress SI purists, but it may make it a little easier for the reader to track a particular thermodynamic estimate back to its source.

III.   YIELDS

In the literature percentage yields are often quoted after correction for recovery of starting material. As far as possible all yields noted without comment are uncorrected and based on starting material introduced into the apparatus. Some yields have been recalculated on this basis so that the reader can appreciate the practical yield of product after one passage through a pyrolysis tube. There may be a few cases where high (corrected) yields are quoted in the mistaken belief that they are uncorrected.

Chapter 1

The Place of Flow and Flash Vacuum Pyrolytic Methods in Organic Chemistry

I. HISTORICAL INTRODUCTION

The destructive distillation of mineral or organic material was one of the few preparative methods available to the alchemists and the first chemists. That materials could undergo profound changes at high temperatures became a well-recognized principle which was emphasized by such chemical landmarks as the formation of sulfuric acid by heating ferrous sulfate (copperas) described by Basil Valentine in the late fifteenth century, Brandt's discovery of phosphorus in 1669 by destructive distillation of residues from urine, and the isolation of benzene from oil gas by Faraday in 1825 and of pyrrole from bone oil by Runge in 1834. Mineral acids, alkalis, and many other inorganic reagents were available to chemists working with organic compounds in the early nineteenth century, but in the absence of adequate formulas and structural theory their use often added to the volume of descriptive organic chemistry without clarifying structural relationships. It was perhaps partly for this reason that the vigorous pyrolysis of organic compounds, involving no addition of further groups, remained a common technique for the investigation of structural and chemical behavior until about the end of the nineteenth century. It had the advantage that many high-temperature reactions are fragmentations which produce simple products from more complex starting materials. The direct decarboxylation of some carboxylic acids and the similar decomposition of carboxylate salts by heating with lime or soda lime are obvious examples; Mitscherlich[1] prepared benzene as early as 1834 by vigorous distillation of benzoic acid with lime.

More subtle changes were recognized with the use of the modern technique of passage of the vapor of an organic compound through a red-hot porcelain or iron tube. Berthelot[2] first prepared biphenyl from benzene in this way in 1866, and its present large-scale manufacture differs essentially only in scale from that first experiment. In the late nineteenth century, structural theory and separatory techniques had advanced to the point where in 1888–1891 Pictet and Ankersmit[3,4] were able to rationalize the preparation of phenanthridine from benzylideneaniline [Eq. (1.1)], to comment on the relationship between this process and pyrolytic synthesis of acridine from 2-methyldiphenylamine, and to consider further synthetic approaches in the light of Eq. (1.1).

$$\text{(1.1)}$$

glowing iron tube / pumice packing ; $+ \text{H}_2$

Although the pyrolysis of organic compounds continued to attract attention in the first three decades of the twentieth century, it inevitably declined in relative importance as the emphasis of much organic chemical research shifted from the study of simple aliphatic and aromatic compounds to the formidable problems of the structures of natural products. The larger naturally occurring molecules, often extracted and purified only with considerable effort, were not usually suitable for degradation by the sledgehammer methods of hot-tube chemistry; and hydrolytic or oxidative cleavage reactions were used as the major tools of structure determination. Only reduction to a parent hydrocarbon by distillation of an oxygenated compound over hot zinc dust was much used as a high-temperature reaction which could quickly yield significant structural information. The value of this method was first established by the recognition of the anthracene system in alizarin by Graebe and Liebermann.[5]

In 1929 Hurd's classic monograph "The Pyrolysis of Carbon Compounds" appeared.[6] This book, which reviewed all the pyrolytic processes that had been examined critically at that time, has been a valuable source of factual information and an influence on organic chemical research for almost 50 years. It is arranged by classes of compounds and functional groups, and the term *pyrolysis* is interpreted in a catholic manner to include both the decomposition of malonic acid at its melting point of 135° and the decomposition of methane in a porcelain tube at 1100°. That it was written before the impact of electronic theory and quantum mechanics on organic chemistry is in some respects an advantage to the modern reader; the facts are presented without bias arising from theoretical expectations in a way which would be more difficult for a modern author, and they are thus available without any encumbrance of outmoded theory. I am confident that Hurd's book will still be of value 100 years after its publication.

The rise of the petroleum and petrochemical industry brought with it the use of various high-temperature flow processes on a large scale. This also required developmental work on a laboratory scale, but the use of pyrolytic methods in organic chemical laboratories in the period 1920–1950 tended to be restricted to a few specific reactions, mainly of volatile liquid compounds. These included the generation of ketenes from acid anhydrides or ketones (see Chapter 4, Section V), the preparation of alkenes by pyrolysis of esters (Chapter 4, Section IV,A), the preparation of dienes from cyclohexenes by retro-Diels–Alder fragmentation (Chapter 8, Section III,A), and the synthesis of polycyclic hydrocarbons by the Elbs reaction.[7] The Elbs reaction, which forms anthracenes from *o*-methyldiarylketones, requires brief mention here because it is not treated in later chapters. The preparation of 1,2:5,6-dibenzanthracene[8] shown in Eq. (1.2) is an example of an efficient Elbs synthesis, but the reaction proceeds in very poor yield with the simplest substrate, 2-methylbenzophenone. The mechanism of the reaction is not completely understood, but an electrocyclic reaction of an enol such as **1** may well be involved. The reaction has always been run by heating the high-boiling ketone at 400°–450° or under reflux, but an unpublished experiment in the author's laboratory has shown that it may also occur under flash pyrolytic conditions at a higher temperature.

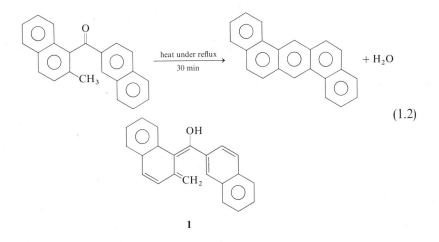

$$(1.2)$$

1

During the period 1920–1950 the study of gas-phase reactions and gas kinetics progressed rapidly in the hands of physical chemists, and the existence and chemical importance of alkyl radicals was established. The early kinetic work tended to be restricted to simple compounds which decomposed cleanly and so could be studied through a single measurement such as that of pressure. Until the widespread introduction of gas chromatography in the mid-1950s it was difficult to analyze complex mixtures of volatile compounds, and many kinetic studies were concerned mainly with the disappearance of the starting material and much less with the nature of any other than the major products.

Since about 1950 interest in the use of pyrolytic methods for the transformation of a much wider range of organic compounds has grown with the explosive growth of organic chemistry itself, encouraged by the same factors including the introduction of spectroscopic methods of analysis and identification, and the development of efficient chromatographic methods of separation. The application of molecular orbital theory to organic chemistry and the enunciation of the Woodward–Hoffmann rules (1965–1969) have focused attention the comparison of photochemical and thermal processes, and since 1950 organic photochemistry has grown into a major division of organic chemistry. The study of thermal reactions has also grown in importance but this growth has been less obvious because thermal reactions are so widespread, and interest has been concentrated on systems with special relevance to the Woodward–Hoffmann rules. The recent period has also been marked by the exploration of the chemistry of transient reactive intermediates such as carbenes, nitrenes, and arynes in addition to that of free radicals. Because of their high reactivity such intermediates are often best generated by photolysis or pyrolysis in the gas phase and can then be studied in the absence of strong interactions with solvents or other reagents.

Gas kineticists became interested in pyrolytic methods which could generate short-lived species with very short times in the reaction zone and times of flight to a detector, and this led to techniques such as the very low pressure pyrolysis (VLPP) method of Golden, Spokes, and Benson (see Chapter 2, Section II,O). A major influence on the thinking of organic chemists was the growth of organic mass spectrometry in the 1960s, which provided a technique for direct detection of molecular ions and for the quantitative measurement of their fragmentation patterns. The mass spectrometer was of immediate value for the detection and analysis of products of pyrolytic reactions, but in the present context its most interesting effect was that it directed attention to the comparison of pyrolytic, photochemical, and mass spectral fragmentations. The multiple fragmentations of molecular ions generated at 70 eV provided a bewildering array of novel and unexpected reactions, some of which were similar to recently studied photochemical reactions, and others to known pyrolytic reactions. Organic chemists thus regained interest in the hot-tube methods typical of the nine-teenth century and applied them to molecules of moderate complexity. The observation of primary processes of high activation energy required that high pyrolytic temperatures be balanced by short contact times at low pressures, and the methods generally referred to as flash vacuum pyrolysis (FVP)[9] and flash vacuum thermolysis (FVT)[10] were developed to satisfy these criteria (see Chapter 2, Section II,D).

This has introduced two very similar terms to describe the same class of experiments. Which should be used? De Mayo's group has consistently used the terms FVT and *thermolysis* on the logical ground[10] that the "Shorter Oxford Dictionary" defines thermolysis as "decomposition or dissociation by heat." In this book the term *pyrolysis* used by Hurd is retained except where it is obviously appropriate to use *thermolysis* to describe experiments in which the FVT apparatus of de Mayo and co-workers was used. Pyrolysis means dissociation "caused or produced by fire," and although its use may thus be somewhat inaccurate it does seem to convey the additional useful meaning of a process occurring well above ordinary laboratory temperatures and approaching the temperature of a fire. The difference between the terms is akin to the difference between the Kelvin and Celsius temperature scales. In principle a thermal process may occur at any temperature above $0°K$, but it is in practice convenient to distinguish broadly between ordinary laboratory processes in which the temperature of a heated reaction is usually measured with a $0°-360°$ mercury-in-glass thermometer, and pyrolyses at higher temperatures. The temperatures of reactions mentioned in this book span a range from about $150°$ to $1200°$, so that some inconsistencies will be found, but the majority involve temperatures of $350°-900°$ which are clearly above those used routinely in most organic laboratories.

II. SCOPE OF THIS BOOK AND OF PREVIOUS REVIEWS

Pyrolytic reactions may be run in either a static apparatus or in a flow system. In static experiments the sample is heated under reflux or in a sealed vessel and consequently the primary products of decomposition may undergo further changes leading to complex mixtures or, on occasion, to intractable tars. Nevertheless, static reactors are widely used in gas kinetics, and re-actions run in this way are mentioned frequently in succeeding chapters because in many cases the literature contains no report referring to a flow system. The purpose of this book is to emphasize that preparative pyrolyses of organic compounds are best run in flow systems in which the primary products are removed from the hot zone as quickly as possible. It is usually these primary or early products of unimolecular decompositions which are required, and the isolation of primary processes with avoidance of secondary and intermolecular reactions is often readily achieved by flash vacuum pyrolysis. A brief account of the major types of apparatus which have been used for flow pyrolytic and flash vacuum pyrolytic work is given in Chapter 2, and later chapters provide examples of the types of reaction which have been reported. Thermal reactions run in solution or in a melt are not included unless they are relevant to other flow experiments, so that much interesting chemistry including, for example, most of the modern work on electrocyclic ring opening is omitted. The Elbs reaction [see Eq. (1.2)] is not strictly within the scope of this book because it is usually run in the liquid phase.

The intention is to provide a guide to the literature and a broad survey of flash and flow pyrolytic reactions, but it is impossible to do this within one volume and to retain the complete coverage characteristic of most chemical reviews. Only a selection of examples is given in each area, but usually the major references cited will help the reader to find a complete set of references to a topic. Mechanistic discussion is also necessarily brief in most cases, and there is no attempt to review as such the scope and mechanism of reactions which have been thoroughly reviewed elsewhere. Many sections begin with a selection of references to major reviews and monographs which will provide a mechanistic background.

Since Hurd's monograph[6] there has been a long gap in the literature reviewing pyrolytic reactions in organic chemistry, broken only by reviews of specific groups of reactions such as that of DePuy and King[11] on pyrolytic cis-eliminations. In the late 1960s and early 1970s, however, the growth of interest in flash pyrolytic methods led to a number of short review articles explaining the techniques and reporting results. These included articles by Hedaya[9] on flash vacuum pyrolysis, by de Mayo[12] on flash vacuum thermolysis, by Hageman and Wiersum[13] on application of flash vacuum thermolysis, and by Golden et al.[14] on applications of very low pressure pyrolysis (VLPP). More recently Seybold[15] has reviewed the principles of flash thermolysis of organic compounds with a description of new apparatus and a survey of applications, and Wentrup has published a lecture[16] which extends certain aspects of his long review[17] of rearrangements of carbenes and nitrenes in the gas phase.

In the introduction to their account of work on the reactions of aromatic compounds at high temperatures Fields and Meyerson[18] wrote of "The pyrolysis of carbon compounds": "No researcher since has tried to write a comprehensive treatise similar to Hurd's; it would be a formidable task indeed." The present book is, as already outlined, much more restricted than the one of Hurd, but it does try to provide a wider coverage than the reviews of Refs. 12–16, some of which concentrate on the immediate research interests of the authors and on the more spectacular examples. It would be impossible (and undesirable) to write a book of this kind without reflecting personal research enthusiasms, but some attempt has been made to keep these in check. It is unashamedly propagandist in intent: organic chemists with no experience of pyrolytic methods are often reluctant to use them, probably because of a feeling that special and complex apparatus is needed. Chapter 2 shows that for many preparative purposes quite simple, reliable, and inexpensive apparatus will suffice, and later chapters provide many examples of the clear preparative advantages of flow and flash pyrolytic methods.

III. SCOPE OF PYROLYTIC EXPERIMENTS

A. Introduction

In Section III the scope of pyrolytic experiments is discussed and a few specific experiments are described to show the types of investigations which have been made, and the kinds of molecules which are most suitably produced by flow and flash pyrolysis rather than by photochemical or conventional solution methods. Some of this material overlaps that of later more detailed chapters, and for this only internal chapter references are given.

B. Study of Complex Pyrolyzates; Formation of Polycyclic Hydrocarbons

When simple aromatic or aliphatic hydrocarbons are exposed to temperatures of 500°–1200° for more than a few seconds, mixtures of polycyclic aromatic hydrocarbons including the highly carcinogenic 3,4-benzopyrene may be formed, almost regardless of the structure of the starting material. 3,4-Benzopyrene was first isolated from coal tar in 1933, and was quickly established as a probable causative agent in a variety of industrial cancers associated with exposure to lubricating oils, coal tar, and other materials which had been heated strongly. In 1958 Badger and co-workers published the introductory paper[19] of a long systematic series on the formation of aromatic and polycyclic aromatic hydrocarbons at high temperatures. This topic was reviewed with the main conclusions from the researches of his group by Badger[20] in 1965, and so will not be discussed in detail here. The work is notable for the early use of a battery of separatory and spectroscopic techniques [gas-liquid chromatography (GLC), adsorption and partition chromatography, infrared (ir), ultraviolet (uv), and fluorescence spectroscopy] for the identification of individual hydrocarbons in the complex tars produced, and for the use of ^{14}C-labeling experiments to assess possible pathways for their formation. Although the present book, being preparatively oriented, is concerned as far as possible with clean pyrolytic reactions which give few products, there are many pyrolyses which give complex mixtures even with quite short contact times. Badger's work on hydrocarbon tars is a model for such investigations, which usually include the following phases: (1) Identification of the products; (2) repyrolysis of each product under the same conditions to see whether some minor products may be plausible intermediates in the formation of the major products; (3) proposal of a reaction pathway usually involving some identified intermediates and some hypothetical intermediates; (4) testing of this hypothesis by further pyrolytic and labeling experiments; and (5) proposal of a reaction pathway modified by the results of (4). The problems are similar to those encountered in the investigation of biosynthetic pathways, although most experimental difficulties are less severe and that of low incorporation of label is absent.

Badger's group investigated the pyrolysis at about 700° of such hydrocarbons as the following: isooctane, decane, and dotriacontane; acetylene, butadiene, 4-vinylcyclohexene, and isoprene; benzene, *n*-alkylbenzenes, styrene and methylstyrenes, and indene; naphthalene, tetralin, and 1-(4-phenylbutyl)naphthalene; anthracene and phenanthrene. The approach used is well illustrated by a single paper[21] on the pyrolysis of [1-^{14}C]naphthalene in which a table of previous results shows that naphthalene can be formed by pyrolysis of acetylene, butadiene, isooctane, decane, styrene, vinylcyclohexene, indene, tetralin, and alkylbenzenes, and is thus a plausible intermediate in the formation of higher aromatic hydrocarbons. Pyrolysis of naphthalene at 700° through a 90 × 2.5 cm inner diameter (ID) silica tube in a stream of nitrogen (60 ml/min) with the naphthalene added dropwise at 5 drops/min gave a tar containing binaphthyls, perylene, 10,11-benzofluoranthene (2), 11,12-benzofluoranthene (3), and a trace of 3,4-benzopyrene. Insufficient material could be isolated to measure the radioactivity of perylene or of 3,4-benzopyrene formed from [1-^{14}C]naphthalene, but the other polycyclic products each contained two labeled carbon atoms per molecule. It was considered that these were formed via naphthyl radicals which added to naphthalene to give binaphthyls; cyclodehydrogenation of the binaphthyls then led to perylene and the benzofluoranthenes. A possible route to 1,2-binaphthyl and the benzofluoranthenes 2 and 3 is shown in Eq. (1.3), and other products were derived from the other two possible binaphthyls. The formation of a trace of 3,4-benzopyrene (4) was attributed to the reduction by free hydrogen of some of the naphthalene to tetralin, followed by the radical fission, addition, and cyclodehydrogenation outlined in Eq. (1.4) in terms of the carbon skeleta only. This interpretation was based on previous experiments with tetralin itself.

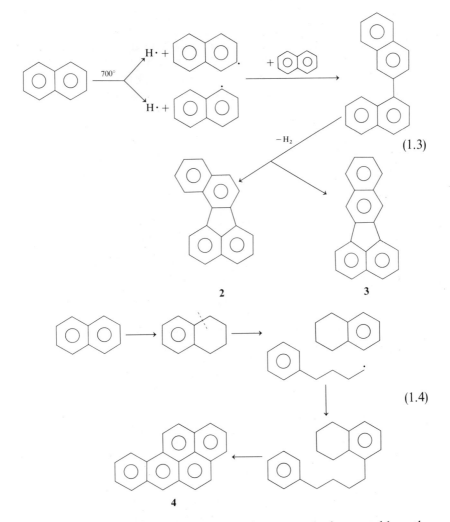

(1.3)

(1.4)

Such work was concerned mainly with the proposal of reasonable pathways for the formation of polycyclic products. The mechanism of individual steps remains uncertain, although radical fission, addition, and cyclodehydrogenation reactions were shown to be important together with additions and eliminations of C_2 and C_4 units. Later work on the high-temperature chemistry of arynes (Chapter 3, Section V), arylcarbenes (Chapter 5, Section III,D), acetylenes (Chapter 5, Section II,D), and tetralins (Chapter 8, Section III,D) has added further mechanistic possibilities, but the schemes proposed by Badger are not dependent on such detail.

C.   Generation of Thermodynamically Unstable Species

Thermodynamically unstable species may sometimes be produced in detectable amounts by equilibration of a system at a high temperature followed by direct physical detection at a high temperature, or by rapid freezing of the equilibrium by immediate collection of the mixture on a very cold surface.

Anet and Squillacote[22] observed spectra of two *conformers* of methylcyclohexane. An equilibrium mixture containing the less stable axial-methyl conformer was produced by passing the vapor of methylcyclohexane through a quartz tube (30 × 2 mm) at 500° and 0.2 mm [Eq. (1.5)] with immediate quenching to −175°. The nuclear magnetic resonance (nmr) spectrum of the condensate in a mixture of $CHClF_2$ and CCl_2F_2 at −160° showed a strong doublet at δ 0.86 due to the equatorial methyl group of the more stable conformer, and a weaker doublet at δ 1.00 due to 10–25% of the less stable conformer with the methyl axial. Similarly, 25% of the unstable cis-conformer of N-methylformamide was produced by flash equilibration at 540° [Eq. (1.6)] followed by quenching at −65°, and kinetic data for the conversion of the cis- to the trans-conformer in 1,2-dichloroethane at −10° to +40° were obtained by nmr spectrometry. The method of flash equilibration is applicable to systems in which a conformer of higher energy is separated from one of lower energy by a barrier of 8 kcal/mole or more.

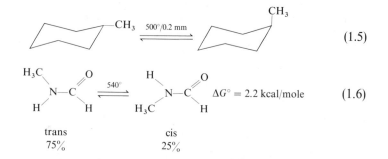

$$\tag{1.5}$$

$$\Delta G° = 2.2 \text{ kcal/mole} \tag{1.6}$$

trans   cis
75%   25%

The barrier to return of the twist boat conformer of cyclohexane to the more stable chair conformer is only 5.3 kcal/mole. Squillacote *et al.*[23] subjected cyclohexane to flash equilibration at 800° with a contact time of 0.01 sec [Eq. (1.7)] and collected the vapor alone or with argon on a CsI plate at 20°K. Infrared spectra of the product in argon matrix showed well-resolved bands due to the twist boat conformer, and the rate of conversion back to the chair conformer could be measured by infrared spectrometry of films of flash equilibrated cyclohexane collected at 20°K and annealed at 72.5°–74°K.

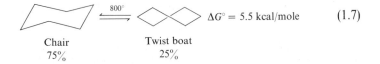

Chair	Twist boat
75%	25%

$$\Delta G° = 5.5 \text{ kcal/mole} \qquad (1.7)$$

The conformers of Eqs. (1.5)–(1.7) differ in energy by less than 6 kcal/mole. Unstable *isomers* differing in energy by 10–40 kcal/mole but with a relatively low barrier to return to the more stable form have also been generated by flash equilibration. The formation of hydrogen isocyanide, HNC, in a stream of hydrogen cyanide flowing through nozzles heated to 730° is noted in Chapter 5, Section II,E; the isocyanide was detected directly by microwave spectrometry. Acetylene appears to undergo similar isomerization to methylenecarbene, $CH_2=C:$, at 850° (Chapter 5, Section II,D). In this case the isomer is proposed as an intermediate in the isomerization of $HC\equiv^{13}CD$ to $DC\equiv^{13}CH$, and this isomerization in substituted acetylenes can lead to products of trapping of the carbene.

Detty and Paquette[24] used flash vacuum pyrolysis in the search for a *transition state* having hexahomobenzene character which would lead to the automerization of the deuterated trishomo-1,4,7-nonatrienes 5 and 6. No automerization was detected by nmr spectrometry in a series of pyrolyses at temperatures up to 500°, above which some decomposition occurred. Although this result was negative [Eq. (1.8), contact time (CT)], the work shows how readily such searches can be carried out using only small amounts of valuable materials. These can be recovered almost quantitatively from runs at temperatures below the onset of reaction.

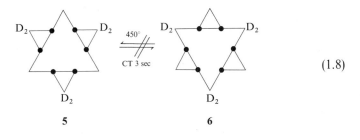

$$(1.8)$$

5	6

D. Generation of Transient and Highly Reactive Molecules

Flash pyrolysis is of considerable value for the generation of transient reactive intermediates such as free radicals (Chapter 3), arynes (Chapter 3 and Chapter 6, Section II,I), carbenes and nitrenes (Chapter 5), sulfenes (Chapter 6, Section III,A), and silaethenes (Chapter 8, Section II,B). Such species have extremely short lifetimes because of the great facility of dimerization or intramolecular rearrangement, and the study of these rearrangements

in the gas phase is an important area of pyrolytic chemistry. Reactive intermediates are often readily generated by the high-temperature extrusion of
very stable small molecules (e.g., N_2, C_2H_4, CO, or CO_2) from precursors
which are ordinarily regarded as thermally stable. The formation of benzyne
from phthalic anhydride is a case in point.

Another group of substances conveniently approached by flash vacuum
pyrolysis includes compounds which are thermodynamically quite stable as
isolated molecules in the gas phase at low pressure, but which dimerize
or decompose readily in the condensed phase at ordinary temperatures.
The distinction between this second group and the first group of reactive
intermediates is not a firm one, but most molecules in the second group
decompose by intermolecular reactions.

An example not treated in a later chapter is the work of Clapp and
Westheimer[25] on the generation of monomeric methyl metaphosphate (8)
by the retro-Diels–Alder reaction of methyl 2-butenylphostonate (7) shown
in Eq (1.9). Monomeric alkyl metaphosphates had been postulated for
many years as intermediates in phosphorylation reactions and in the hydrolysis of phosphoric and pyrophosphoric esters, but the ease of their trimerization and polymerization made them difficult to detect. Monomeric 8 was
identified in the pyrolyzate stream by condensation into a cold trap containing a stirred solution of N-methylaniline and subsequent isolation of
an N-methyl-N-phenylphosphoramidate,[25] and by reaction with dimethyl-
aniline.[26]

$$
\underset{7}{\text{(structure)}} \xrightarrow[\substack{1 \times 8 \text{ cm quartz} \\ \text{tube} \\ \text{CT } 0.02 \text{ sec}}]{600°/0.2 \text{ mm } N_2} \underset{8}{\text{(structure)} + \text{(structure)} }
\tag{1.9}
$$

Typical examples from later chapters include strained ketenes and methyl
eneketenes (Chapter 6, Section II,J), pentalenes (Chapter 8, Section III,B),
isoindoles and isobenzofuran (Chapter 8, Section III,D), azetes and benzazetes (Chapter 6, Section IV,D), and benzthiete (Chapter 6, Section III,C).
Fortunately most of these compounds are quite stable when condensed into
a cold trap cooled with liquid nitrogen ($-196°$), and some are stable in dilute
solution even at room temperature. Some of these molecules contain small
and somewhat strained rings, but it will be obvious that pyrolytic methods
are not generally suitable for the preparation of highly strained systems with
many small rings which are likely to be susceptible to radical cleavage. Such
systems are often best approached photochemically because in many cases
highly strained products absorb light less strongly than their more unsaturated precursors, and the products can thus survive the conditions of

irradiation. Pyrolysis is less discriminating in this respect, and even with short contact times highly strained products will usually undergo further cleavage and rearrangement.

The study of transient and very highly reactive molecules may require the use of flash vacuum pyrolysis equipment directly coupled to a mass spectrometer, microwave spectrometer, or photoelectron spectrometer, and much recent work of this kind has required a new approach to organic chemistry in which the results of trapping and isolation experiments have become secondary to the detailed information available from the spectrometers. The work of Hedaya's group on cyclobutadiene (Chapter 3, Section VI) made extensive use of coupled mass spectrometry. Microwave spectrometry has provided the main tool for determination of the structures in the gas phase of such highly reactive molecules as sulfines (Chapter 8, Section II,D), allene episulfide (Chapter 8, Section III,B, Table 8.1), and cumulenones (Chapter 4, Section V,A) and of less stable tautomers such as hydrogen isocyanide (Chapter 5, Section II,E) and vinyl alcohol (Chapter 4, Section IV, F). The generation and microwave characterization of vinyl alcohol by Saito[26a] has ended a long period in which textbooks (and students) could invoke its existence only as a hypothetical intermediate. Molecules such as sulfine, thioformaldehyde (Chapter 4, Section IV,F) and thioketene (Chapter 4, Section V,A) have been further characterized in pyrolyzate streams by photoelectron spectroscopy.

E. Preparation of Otherwise Less Accessible Compounds by Fragmentation of Readily Available Starting Materials

This category includes such reactions as the preparation of benzocyclobutenes from *o*-methylbenzyl halides (Chapter 4, Section VI,A), of benzocyclobutenone from homophthalic anhydride (Chapter 8, Section III,D), of benzocyclobutenedione from ninhydrin (Chapter 6, Section II,B), and of biphenylenes from phthalic anhydrides (Chapter 6, Section II,I). Because the commercial availability of inexpensive starting materials is largely based on the chance occurrence that some compounds have practical applications and others do not, this category is chemically a mixed bag. It does, however, include many simple and useful preparations of compounds which are tedious to make by conventional methods.

F. Large-Scale Preparations

In principle all flow pyrolytic reactions can be scaled up merely by continuous operation until sufficient pyrolyzate has been accumulated to provide the pure product in the quantity required. This is also the case in practice for readily volatile compounds such as esters of moderate size, which can readily be pyrolyzed to give alkenes on a large scale (Chapter 4, Section IV,A). The thermal rearrangement of β-cyclogeranyl vinyl ether [Chapter 9, Section IV,C, Eq. (9.81)] is an example of a reaction run in a flow system on a substantial scale (0.76 mole, 135 gm) to give an aldehyde in high yield.

The scaling up of flash vacuum pyrolyses of less volatile solids is more difficult, and the most satisfactory preparative scale is usually 0.5–1 gm of starting material. The difficulty lies in the volatilization; vacuum sublimation of more than about 1 gm of many poorly volatile solids leads to difficulties due to premature decomposition of the solid in the sublimation pot. Nevertheless, some apparatus has been designed to deal with as much as 10 gm of starting material (see Chapter 2), and even repeated batchwise pyrolysis may not require excessive working time. As an example of the latter method, 5 gm of the benzofuranol **10** was prepared in three working days by flash vacuum pyrolysis of the Meldrum's acid derivative **9** in 1 gm batches (see Chapter 9, Section II,C and Table 9.2).

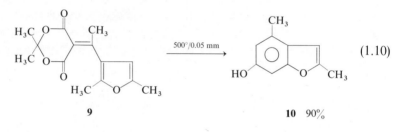

$$(1.10)$$

9   **10**   90%

IV. THERMAL, PHOTOCHEMICAL, AND MASS SPECTRAL PROCESSES AND THEIR CORRELATION

The flash vacuum pyrolysis of organic compounds may be an extremely mild process, as shown by the survival of some *p*-tolyldiazomethane formed in the pyrolysis of 5-(*p*-tolyl)tetrazole at 420° and 0.01 mm,[27] or it may at much higher temperatures lead to deep-seated rearrangements initiated by the breaking of carbon–carbon, carbon–oxygen, or other strong bonds. The chemist new to this field will probably find the high temperature rearrangements of aromatic carbenes and nitrenes discussed in Chapter 5 somewhat overwhelming, but this is an unusual area in which complexity results from the operation of many competing equilibria. In many simple systems the outcome of a fragmentation reaction may be predicted on the basis of the stability of the various possible radicals or other reactive intermediates, and analogy with the ease of generation of such intermediates in solution will be a reasonable basis for prediction. This approach tends to break down for other than simple cases, and it is often found that structurally very closely related systems will fragment in quite different ways—the behavior of *o*-phenylene sulfite and *o*-phenylene carbonate (Chapter 6, Section II,D) is a notable example. In such situations a little more than elementary chemical intuition is needed for successful prediction.

Three different approaches have occasionally been used to rationalize and predict the outcome of pyrolytic fragmentations: identification of the bond of lowest calculated bond order, estimation of the thermochemistry of alternative pathways, and empirical correlation with the mass spectral fragmentation of the same molecule.

The first approach, calculation of bond orders or related properties of the substrate, has been little used. Dougherty and co-workers correlated the thermal loss of CO_2 and the mass spectral loss of NO from 3-substituted sydnones (**11**) with the π-bond orders in the ground-state molecule and the ground-state molecular ion. DeJongh and Thomson[29] used a CNDO/2 program to calculate Mulliken overlap populations, indicative of total bond strengths, for *o*-phenylene sulfite and carbonate (**12**, X = S or C) and found that initial fissions of the ArO-SO and Ar-OCO bonds were correctly predicted. They also used calculated energies of the *products* to rationalize the preferred pathways (see Chapter 6, Section II,D).

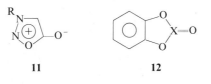

11   **12**

Wentrup[30] has preferred a thermochemical approach to the problem of the behavior of *o*-phenylene sulfite and carbonate, and has estimated heats of formation for the alternative sets of diradicals and final products using the methods of Benson *et al.*[31] Similarly, Wentrup[32] used thermodynamic estimates to construct energy profiles which rationalize the rearrangements of aromatic carbenes and nitrenes discussed in Chapter 5, Section III. Bird *et al.*[33] and Crow and Khan[34] have used estimated heats of formation of intermediate radicals to rationalize the products of pyrolysis of ketazines. Pyrolysis of acetophenone azine at 450° and 0.05 mm gives benzonitrile (96%) but no acetonitrile,[33] and the estimated thermochemistry of fragmentation of the intermediate iminyl radical **13** shows clearly that the formation of the conjugated benzonitrile should be strongly favored [Eq. (1.11)]. The pyrolytic behavior of cyclopentanone azine[34] is dominated by secondary rearrangements of the iminyl radical **14**, and the thermochemical estimates of Eq. (1.12) rationalize the observation that the major products are derived

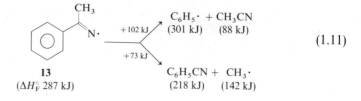

$$
\begin{array}{ll}
& C_6H_5 \cdot \ + \ CH_3CN \\
& (301 \text{ kJ}) \quad (88 \text{ kJ})
\end{array}
$$

$$
\begin{array}{ll}
& C_6H_5CN + \ CH_3 \cdot \\
& (218 \text{ kJ}) \quad (142 \text{ kJ})
\end{array}
$$

(1.11)

13
(ΔH_F° 287 kJ)

from the stabilized radical **16** rather than from the intermediate radical **15**. The pyrolysis of azines is discussed further in Chapter 3, Section III,E. The energetics of pyrolytic and mass spectral reactions of the reverse Diels–Alder type have been discussed by Loudon *et al.*[35,36] These are very simple examples, but in principle most pyrolytic fragmentations with several possible pathways can be analyzed by the thermochemical methods of Benson and his group.

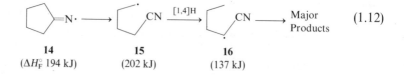

(1.12)

14   **15**   **16**
(ΔH_F° 194 kJ)   (202 kJ)   (137 kJ)

The third approach relies on the fact that mass spectral and pyrolytic fragmentations often appear to be quite closely parallel, so that the pyrolytic behavior of a compound may often be predicted from consideration of its mass spectrum or that of a closely related compound from the literature. The simplicity of this approach and the great wealth of mass spectral data available have been an important influence on the recent growth of interest in flash pyrolysis, but its theoretical justification is uncertain and there are many examples of its failure. The parallelism is very marked in the case of the decarbonylation of simple polycarbonyl compounds such as indanetrione [see Chapter 6, Eqs. (6.8) and (6.9)], but this compound has so few processes open to it that parallelism is scarcely surprising. Early examples (1962–1968) have been reviewed by Bentley and Johnstone;[37] they conclude that, "There is very little direct evidence to support the drawing of 'analogies' between mass spectrometry and pyrolysis because of our poor knowledge of ion structures and mechanisms in mass spectrometry."

Nevertheless, such analogies continue to be mentioned in the literature, and the consideration of published mass spectra has repeatedly been of value in encouraging pyrolytic investigations in my own laboratory. Thus the use of arylidene derivatives (**17**) of Meldrum's acid for the pyrolytic generation of arylideneketenes (Chapter 6, Section II,J) was suggested by their published mass spectra[38] which show intense peaks corresponding in composition to the ion **18**. Similarly, Spangler and co-workers[39] found a strong correlation

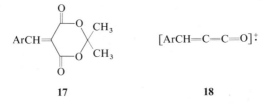

[ArCH=C=C=O]‡

17   **18**

between the yield of a benzocyclobutene obtained on pyrolysis of an iso-chromanone and the intensity of the $[M - CO_2]^{\ddagger}$ peak in its mass spectrum.

The analogy between pyrolytic and mass spectral processes has been noted for a variety of reverse Diels–Alder reactions and for the fragmentation of carbonyl compounds, anhydrides, lactones and lactams and their thio derivatives, carbonates and sulfites, nitro compounds, and thionylanilines. Specific examples are mentioned in later chapters, and the reviews by Bentley and Johnstone[37] and by Dougherty[40] give detailed accounts of many cases. One of the earliest and most explicit comparisons of electron-impact and pyrolytic fragmentations was due to Cava and co-workers[41] who studied aromatic and heteroaromatic anhydrides. As a single example, the pyrolysis of pyridine-2,3-dicarboxylic anhydride, discussed further in Chapter 6, Section II,I, gave CO_2, CO, HC≡CCH=CHCN (MW 77), and much HCN by a process which appears very similar to the major mass spectral pathway of Eq. (1.13).

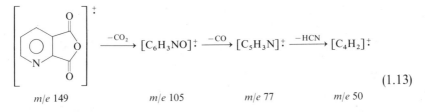

$$\xrightarrow{-CO_2} [C_6H_3NO]^{\ddagger} \xrightarrow{-CO} [C_5H_3N]^{\ddagger} \xrightarrow{-HCN} [C_4H_2]^{\ddagger}$$

(1.13)

m/e 149   *m/e* 105   *m/e* 77   *m/e* 50

In 1968 Dougherty[42] published an important theoretical paper on a perturbation molecular orbital approach to the interpretation of mass spectra and the relationship between mass spectrometric, thermolytic, and photolytic reactions. He concluded that reactions of ions in low lying doublet or singlet states (Class I reactions) should be analogous to thermal processes and should be associated with significant metastable peaks; reactions occurring from excited electronic states ions and from doubly ionized molecules (Class II reactions) should not show significant metastable peaks, and should correlate with the corresponding photochemical processes. In a second paper Dougherty[43] discussed the reverse Diels–Alder reaction, the McLafferty and related rearrangements, and the mass spectrum of hexa-helicene. The mass spectrum of this hydrocarbon showed an intense peak for the loss of C_2H_4 and a strong metastable peak for this transition; the remainder of the mass spectrum of hexahelicene was identical with that of coronene. Dougherty predicted that the pyrolytic cyclization of hexahelicene to coronene would occur by loss of C_2H_4 rather than of $C_2H_2 + H_2$, and the formation of ethylene and coronene in an evacuated sealed tube at 485° was confirmed [Eq. (1.14)]. Dougherty has followed this work with a major review[40] of analogies between mass spectral, pyrolytic, and photolytic reactions.

$$\xrightarrow{485°} \quad + C_2H_4 \qquad (1.14)$$

Dougherty's approach has been criticized by Bentley and Johnstone[37] on the grounds that his criterion of the observation of a significant metastable peak may not be reliable as an indicator of the energy level of the fragmenting ion. Johnstone and Ward[44] have examined the mass spectrometric cyclization of the diphenylmethyl cation and of the molecular ions of stilbene and isoelectronic systems in relation to the Woodward–Hoffmann rules for electrocyclic reactions. They concluded that these systems, which showed strong metastable peaks for the final loss of a hydrogen molecule or atoms, involved excited ions which followed *photochemical* rules for the stereochemistry of cyclization. Bishop and Fleming,[45] however, have pointed out that it is unreliable to compare even-electron and odd-electron ions in this way, and they provide examples of disrotatory ring closure in the mass spectrometer of ions with either $4n$ or with $4n + 2$ π electrons. Bishop and Fleming[45] support Dougherty's argument[42,43] that only those ions in the electronic ground state will have sufficient lifetime to give rise to significant metastable peaks, but they warn that the Woodward–Hoffmann rules cannot safely be used to interpret the behavior of ions in the mass spectrometer and that both "allowed" and "disallowed" processes are commonly observed.

The preceding four papers[42–45] are of course more directly concerned with mass spectrometry than with pyrolytic processes, but they provide a valuable introduction to discussion of analogies between mass spectral and and other processes. My own present view is that attempts to provide general theoretical justification for pyrolytic-mass spectral correlations are bound to fail, but that such correlations should not be ignored because they have provided, and will continue to provide, a useful rule-of-thumb stimulus to the discovery of interesting new pyrolytic reactions.

REFERENCES

1. Mitscherlich, E. (1834). *Justus Liebigs Ann. Chem.* **9**, 39.
2. Berthelot, M. (1866). "Jahresbericht über die Fortschritte der Chemie" "Beilstein's Handbuch der Organische Chemie," Vol. 5, p. 185. Springer-Verlag, Berlin and New York.
3. Pictet, A., and Ankersmit, H. J. (1899). *Ber. Dtsch. Chem. Ges.* **22**, 3339.
4. Pictet, A., and Ankersmit, H. J. (1891). *Justus Liebigs Ann. Chem.* **266**, 138.
5. Graebe, C., and Liebermann, C. (1868). *Ber. Dtsch. Chem. Ges.* **1**, 49.
6. Hurd, C. D. (1929). "The Pyrolysis of Carbon Compounds." Chem. Catalog Co., New York.
7. Fieser, L. F. (1942). *In* "Organic Reactions" (R. Adams, ed.), Vol. 1, p. 129. Wiley, New York.
8. Fieser, L. F., and Dietz, E. M. (1929). *Ber. Dtsch. Chem. Ges.* **62**, 1827.
9. Hedaya, E. (1969). *Acc. Chem. Res.* **2**, 367.
10. King, J. F., de Mayo, P., McIntosh, C. L., Piers, K., and Smith, D. J. H. (1970). *Can. J. Chem.* **48**, 3704.
11. DePuy, C. H., and King, R. W. (1960). *Chem. Rev.* **60**, 431.
12. De Mayo, P. (1972). *Endeavour* **31**, 135.
13. Hageman, H. J., and Wiersum, U. E. (1973). *Chem. Br.* **9**, 206.
14. Golden, D. M., Spokes, G. N., and Benson, S. W. (1973). *Angew. Chem. Int. Ed. Engl.* **12**, 534.
15. Seybold, G. (1977). *Angew. Chem. Int. Ed. Engl.* **16**, 365.
16. Wentrup, C. (1977). *Chimia* **31**, 258.
17. Wentrup, C. (1976). *Top. Current Chem.* **62**, 173.
18. Fields, E. K., and Meyerson, S. (1969). *Acc. Chem. Res.* **2**, 273.
19. Badger, G. M., Buttery, R. G., Kimber, R. W. L., Lewis, G. E., Moritz., A. G., and Napier, I. M. (1958). *J. Chem. Soc.*, p. 2449.
20. Badger, G. M. (1965). *In* "Progress in Physical Organic Chemistry" (S. G. Cohen, A. Streitwieser, and R. W. Taft, eds.), Vol. 3, p. 1. Wiley (Interscience), New York.
21. Badger, G. M., Jolad, S. D., and Spotswood, T. M. (1964). *Aust. J. Chem.* **17**, 771.
22. Anet, F. A. L., and Squillacote, M. (1975). *J. Am. Chem. Soc.* **97**, 3243.
23. Squillacote, M., Sheridan, R. S., Chapman, O. L., and Anet, F. A. L. (1975). *J. Am. Chem. Soc.* **97**, 3244.
24. Detty, M. R., and Paquette, L. A. (1977). *Tetrahedron Lett.*, p. 347.
25. Clapp, C. H., and Westheimer, F. H. (1974). *J. Am. Chem. Soc.* **96**, 6710.
26. Clapp, C. H., Satterthwait, A., and Westheimer, F. H. (1975). *J. Am. Chem. Soc.* **97**, 6873.
26a. Saito, S. (1976). *Chem. Phys. Lett.* **42**, 399.
27. Gleiter, R., Rettig, W., and Wentrup, C. (1974). *Helv. Chim. Acta* **57**, 2111.
28. Dougherty, R. C., Foltz, R. L., and Kier, L. B. (1970). *Tetrahedron* **26**, 1989.
29. DeJongh, D. C., and Thomson, M. L. (1972). *J. Org. Chem.* **37**, 1135.
30. Wentrup, C. (1973). *Tetrahedron Lett.*, p. 2919.
31. Benson, S. W., Cruickshank, F. R., Golden, D. M., Haugen, G. R., O'Neal, H. E., Rodgers, A. S., Shaw, R., and Walsh, R. (1969). *Chem. Rev.* **69**, 279.
32. Wentrup, C. (1974). *Tetrahedron* **30**, 1301.
33. Bird, K. J., Chan, A. W. K., and Crow, W. D. (1976). *Aust. J. Chem.* **29**, 2281.
34. Crow, W. D., and Khan, A. N. (1976). *Aust. J. Chem.* **29**, 2289.
35. Loudon, A. G., Maccoll, A., and Wong, S. K. (1970). *J. Chem. Soc. (B)*, p. 1727.
36. Loudon, A. G., Maccoll, A., and Wong, S. K. (1970). *J. Chem. Soc. (B)*, p. 1733.
37. Bentley, T. W., and Johnstone, R. A. W. (1970). *In* "Advances in Physical Organic Chemistry" (V. Gold, ed.), Vol. 8, p. 151. Academic Press, New York.

38. Egger, H. (1967). *Monatsh. Chem.* **98**, 1245.
39. Spangler, R. J., Beckmann, B. G., and Kim, J. H. (1977). *J. Org. Chem.* **42**, 2989.
40. Dougherty, R. C. (1974). *Top. Current Chem.* **45**, 93.
41. Cava, M. P., Mitchell, M. J., DeJongh, D. C., and Van Fossen, R. Y. (1966). *Tetrahedron Lett.*, p. 2947.
42. Dougherty, R. C. (1968). *J. Am. Chem. Soc.* **90**, 5780.
43. Dougherty, R. C. (1968). *J. Am. Chem. Soc.* **90**, 5788.
44. Johnstone, R. A. W., and Ward, S. D. (1968). *J. Chem. Soc. (C)*, p. 1805.
45. Bishop, M. J., and Fleming, I. (1969). *J. Chem. Soc. (C)*, p. 1712.

Chapter 2

Apparatus and Methods

I. INTRODUCTION

All flow pyrolytic experiments involve most of the operations and variables shown in the block diagram of Fig. 2.1, but the apparatus used to control these varies enormously in complexity and cost.

We shall be concerned mainly with the pyrolysis of liquid and solid compounds covering a wide range of volatility, and we may roughly distinguish four kinds of experiments which use somewhat different apparatus. (1) Pyrolysis of (mainly liquid) compounds at atmospheric pressure. (2) Pyrolysis of (less volatile) compounds at reduced pressure (1–30 mm). (3) Flash pyrolysis of liquids or solids in a moderate vacuum (0.01–1 mm), usually with a short contact time (0.1–1 sec). (4) True flash vacuum pyrolysis (FVP;[1] FVT;[2] very low pressure pyrolysis, VLPP[3]) at pressures of 10^{-3} mm or less, with contact times of the order of milliseconds, and usually with a short distance and time of flight before analysis or collection.

These distinctions are not sharp, and experiments of types (1), (2), and (3) are closely related. Type (4) is rather different because of the very low pressure involved, which changes the mode of thermal activation of molecules and may change the fate of products. Many experiments in the literature involve pressures at the lower end of the range for type (3) or apparatus which produces conditions having some but not all of the characteristics of true flash vacuum pyrolysis and so lies in a gray area between types (3) and (4). Another distinction evident from Fig. 2.1 is that between preparative pyrolysis, in which the collected products are subjected to static analysis by chromatographic, chemical, or spectroscopic methods under conditions in which they are stable, and pyrolysis with dynamic analysis in which the emergent gas stream is analyzed by a sensitive technique (e.g., mass spectrometry or microwave spectrometry) for species which may be quite short-lived.

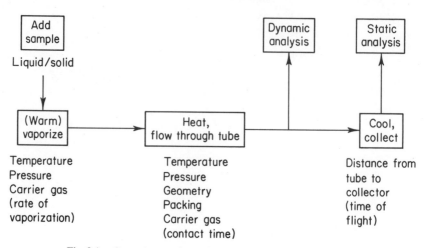

Fig. 2.1.   Operations and variables in flow pyrolytic experiments.

II. APPARATUS

A. Pyrolysis at Atmospheric Pressure

Several forms of apparatus suitable for the pyrolysis of liquid compounds at atmospheric pressure are illustrated or described in "Organic Syntheses." The simplest form is perhaps that shown by Hurd[4] for the preparation of ketene from acetone. Acetone is easily vaporized by dripping it into an empty flask heated by boiling water, and the vapors are led into a horizontal glass tube packed with broken porcelain. The tube is heated to about 650° with a gas furnace, and excess acetone is condensed from the gas stream containing ketene and methane by passage through a vertical water-cooled condenser. Conia and Le Perchec[5] illustrate a similar apparatus for less volatile compounds which are vaporized by passing a stream of nitrogen through the heated liquid and into a vertical heated tube; this apparatus has been used for the thermal cyclization of unsaturated carbonyl compounds (Chapter 9, Section II,E)

More commonly the starting material is added dropwise in a slow stream of nitrogen at the top of a vertical packed tube as shown in generalized form in Fig. 2.2. The liquid is introduced slowly from a long-stemmed funnel (as used for the pyrolysis of ethyl acrylate, bp 100°, to acrylic acid[6]) or from a pressure-equalizing funnel preferably fitted with a needle valve or a Hershberg wire valve.[7] Vaporization occurs at the top of the packing, which is usually of short lengths (5–20 mm) of Pyrex or silica tubing of 5–10 mm outer diameter (od.) or of Pyrex rings or helices. The tube is of Pyrex, Vycor

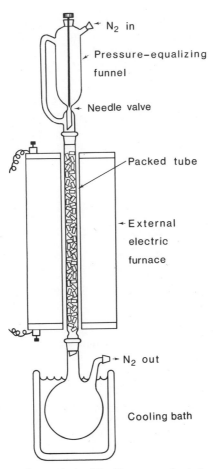

Fig. 2.2. Apparatus for pyrolysis of liquid compounds at atmospheric pressure.

or silica heated with an external electric furnace, and a thermocouple may well be incorporated within the tube to permit measurement of the internal temperature. Detailed descriptions of such apparatus are given by Ratchford,[6] Andreades and Carlson[8] (pyrolysis of diketene to ketene at 550°), and Benson and McKusick[7] (pyrolysis of 1,5-pentanediol diacetate to 1,4-pentadiene at 575°).

B. Pyrolysis at Reduced Pressure

In some preparative pyrolyses carbonization of the tube and packing, and tar formation, can be avoided by working under reduced pressure. Rice and co-workers[9] describe and illustrate a simple apparatus for the cracking of cyclohexene to butadiene and ethylene at 700°–790° and 10–15 mm. The cyclohexene is vaporized from a conical flask heated only on the upper surfaces so that vaporization occurs from the liquid surface and bumping is avoided.

The pyrolysis of a high-boiling liquid often leads to more volatile products, and high overall yields can sometimes be achieved by using an apparatus which recycles the high boiling starting material at reduced pressure. In this way the pyrolysis can be carried out at a lower temperature, with incomplete

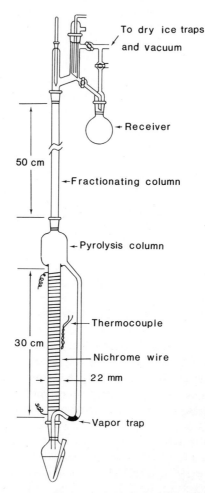

Fig. 2.3. Recycling apparatus for pyrolysis of 2-acetoxycyclohexanone. (Ref. 10; reproduced by permission.)

conversion in each cycle, but with a diminished tendency to carbonization of the hot tube and formation of unwanted secondary products. Williamson and co-workers[10] have used the apparatus shown in Fig. 2.3 for the preparation of cyclohexenone by β-elimination of acetic acid from 2-acetoxycyclohexanone. The yield of cyclohexenone from this reaction was poor when a conventional downflow apparatus similar to that of Fig. 2.2 was used, and cyclohexenone was accompanied by cyclopentene [Eq. 2.1)] in yields which increased with increasing carbonization of the tube and packing. Use of the recycling apparatus gave cyclohexenone in 91% yield (after purification) in a much shorter time [Eq. (2.2)]. In this case the volatility of starting material and product are not markedly different, and a fractionating column is required on top of the pyrolysis column.

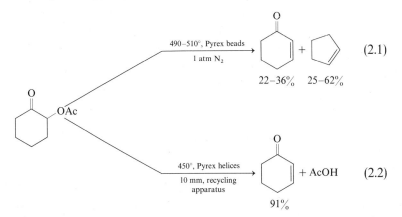

A recycling apparatus along similar lines, shown by Wiersum and Nieuwenhuis,[11] has been used for the preparation of the very volatile hydrocarbon fulvenallene from phthalide by pyrolysis through an empty silica tube at 700°–750° and 0.1 mm (see Chapter 5, Section III,B). A much simpler arrangement with a spiral condenser kept at 75° sufficed to return phthalide (mp 73°) to the reactor and allow the fulvenallene to pass to a cold trap.

The conventional ketene lamp[12,13] used for the pyrolysis of acetone to methane and ketene employs a glowing Nichrome filament operating in acetone vapor at atmospheric pressure. Deposition of carbon on the filament and part of the apparatus near the filament may occasionally be a problem. Devices which use a stream of nitrogen at reduced pressure (2–25 mm) to sweep the vapor of a solid from a sublimation chamber into a pyrolysis chamber containing a heated Nichrome filament have proved very effective for flash pyrolytic work, and have been used extensively by groups led by M. P. Cava, D. C. DeJongh and R. J. Spangler. Full details and a diagram (Fig. 2.4) are given by Spangler and co-workers[14] for a form of the apparatus

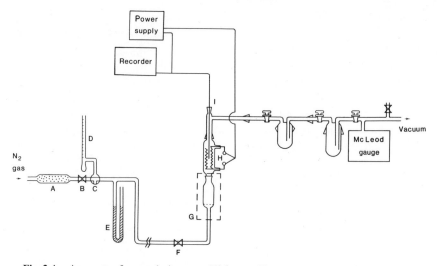

Fig. 2.4. Apparatus for pyrolysis over a Nichrome filament. A, drying tube; B, F, needle valves; D, bubble flowmeter; E, pressure gauge; G, sublimation chamber; H, pyrolysis chamber; and I, thermocouple. (Ref. 14; reproduced by permission.)

suitable for preparative work on a 1 gm scale, and an account of a slightly modified apparatus has been submitted to "Organic Syntheses." The solid, supported on a porous glass disk, is heated in the sublimation chamber and the vapor is swept by the flow of nitrogen into the pyrolysis chamber, where the temperature is recorded with a thermocouple. In this paper[14] the apparatus was used for the preparation of liquid or low melting benzocyclobutenes which were collected in the first trap. Pyrolysis of 6,7-methylenedioxyiso-chroman-3-one at 500° gave the methylenedioxybenzocyclobutene in 90% yield [Eq. (2.3)].

$$\text{(structure)} \xrightarrow[\text{Nichrome spiral}]{500°,\ 3\ mm\ N_2} \text{(structure)} + CO_2 \qquad (2.3)$$

This form of apparatus is particularly suitable for the trapping of short-lived or highly reactive species with added reagents, and it has been used frequently for the trapping of pyrolytically generated ketenes with alcohols. The elaboration of the apparatus required for trapping experiments with methanol is shown in Fig. 2.5, taken from the paper of Spangler and co-workers[15] on the trapping of the ketene generated by pyrolysis of 2-diazoindan-1,3-dione (see Chapter 5, Section II,A). This arrangement makes it possible to deliver a flood of reactant vapor right at the end of the hot zone, and respectable

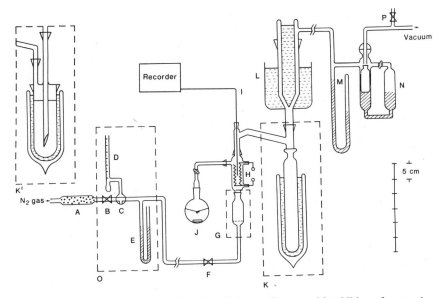

Fig. 2.5. Apparatus for pyrolysis over a Nichrome filament with addition of a trapping reagent such as methanol. A, drying tube; B, F, needle valves; D, bubble flowmeter; E, pressure gauge; G, sublimation chamber; H, pyrolysis chamber; I, thermocouple; J, flask containing trapping reagent; K, cold trap; L, cold finger; M, manometer; and N, manostat. K' is an alternative form of cold trap used for experiments with added *t*-butanol. (Ref. 15; reproduced by permission.)

yields of trapping products can be obtained. A very detailed description of the construction and operation of this apparatus is given.[15]

C.   Flash Pyrolysis in a Moderate Vacuum

In the most frequently used form of preparative apparatus the starting material is sublimed at a pressure of less than 1 mm from a flask or test tube into a horizontal silica tube heated with an external electric furnace and the emergent vapors are collected in cold traps or on a cold finger cooled with liquid nitrogen. Seybold[16] gives a diagram of such a simple apparatus with provision for distillation of solvent into the cold trap. Descriptions of components and assembly are given by Crow and Solly,[17] Gleiter and co-workers,[18] and by Trahanovsky and co-workers.[19] The author's own bias has been toward the use of silica and glass components with tapered joints immediately compatible with other glassware in an organic chemical laboratory, because this enables other workers less committed to pyrolytic work to use, adapt, or duplicate equipment with little difficulty. This involves some problems with the instability of silicone grease in joints close to the furnace in high-temperature operation (>700°), and O-ring joints[19] are probably much more satisfactory for such experiments. A diagram of the apparatus used by Brown and co-workers[20] for the generation of methyleneketenes is shown in Fig. 2.6. With this form of apparatus, poorly volatile products condense or crystallize in the exit elbow and volatile liquid products appear in the cold trap. Figure 2.7 shows details of the apparatus mounted on a trolley.

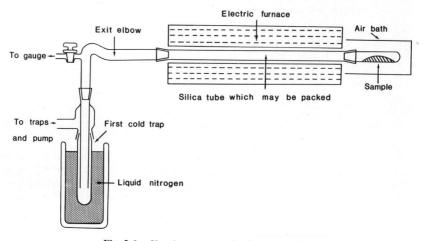

Fig. 2.6.   Simple apparatus for flash pyrolysis.

The sample may be sublimed by warming with an external air bath,[20] with heating tapes,[21] or, on a small scale, with a hot air blower.[22] It is desirable to vaporize the sample by sublimation rather than by distillation from a melt, but this not always practicable with low melting compounds. Molten compounds may undergo unwanted decompositions which lead to the formation of involatile dark polymeric residues. Such compounds are difficult to sublime on other than a very small scale. Brent and co-workers[23] have described difficulties experienced in the pyrolysis of 2-pyridone on a 15 gm scale; the sublimation rate was satisfactory at first but the sample became coated at 170–210° with a film of polymeric material which retarded sublimation, even though decomposition of the sample was otherwise scarcely detectable. In some cases rapid sublimation of small samples can be achieved by spreading the compound as a thin film over the whole inner surface of the sublimation vessel by evaporation of a solution in a volatile solvent.[22]

The pyrolysis tube is usually of silica or Vycor, although Pyrex tubes can be used up to about 520°. Packing may be introduced to prevent streamlined

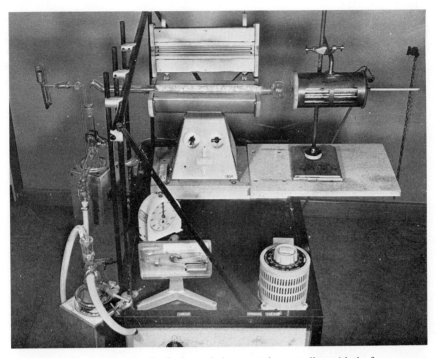

Fig. 2.7. Simple apparatus for flash pyrolysis mounted on a trolley, with the furnace open to show the pyrolysis tube.

viscous flow;[17] in general the effect of a packing of short lengths of silica tubing is to reduce the minimum temperature required for complete decomposition of a compound in one pass by 50°–100°, but it is not always clear whether this is due to effects on contact time and thermal equilibration or whether more specific surface effects of the packing are involved. Pyrolysis tubes may be permanently wound with resistance wire, heated with a close fitting wire-wound silica sleeve,[24] or heated with a stainless steel sleeve and a large current at low voltage.[25] Many workers find it convenient to use a commercial tube furnace such as the 800-W Lindberg Hevi-Duty "Mini Mite"[19] (Lindberg Hevi-Duty Division of Sola Basic Industries, Watertown, Wisconsin), or a Heraeus oven Type ROK 3/30[18] (W. C. Heraeus GmbH, D-6450 Hanau, Germany). In the author's laboratory, Lindberg Hevi-Duty furnaces Type 70-I-S (750 W, maximum temperature 1000°) have given maintenance-free service for over 10 years. These hinged furnaces are particularly useful for ease of setting up and aligning silica and glass apparatus.

Tube temperatures may be measured internally using a thermocouple in an internal well, or on the outside of the tube using a thermocouple placed on

the outer wall. Current Lindberg Hevi-Duty furnaces have a built-in pyro-
meter in addition to regulating controls. For most preparative purposes
temperature control to $\pm 10°-20°$ over the range $400°-900°$ measured at the
center of the heated zone provides adequate reproducibility.

Pressures are usually measured with a Pirani or McLeod gauge either
fairly close to the hot zone (Fig. 2.6) or beyond a cold trap or cold finger, and
both the local pressure in the hot zone during reaction and any pressure
difference across the length of the tube are usually unknown. A few pyrolyses,
such as the unusual and almost explosive decomposition of phenyl azide,[26]
produce uncontrollable fluctuations in pressure but in most cases a two-stage
oil pump can cope with gaseous products of reaction and little or no change
in pressure is observed as sublimation of the starting material into the hot
zone begins. Two-stage Dynavac 8 (rated pumping speed 2.4 liters/sec at
0.1 mm) and Edwards Speedivac 2SC20A pumps are used in the author's
laboratory.

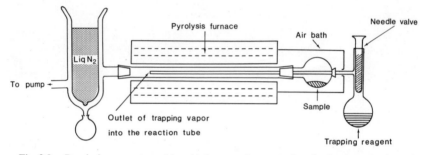

Fig. 2.8. Pyrolysis apparatus with cold finger and concentric tube for introduction of a
stable trapping reagent.

The apparatus of Fig. 2.6 suffers from the disadvantage that the cold trap
may be as much as 20–40 cm from the end of the hot zone, and it may be
desirable to collect some highly reactive products as close as possible to the
furnace. The gap may be reduced to about 10 cm by using a cold finger of
the type shown in Fig. 2.8, cooled with liquid nitrogen, Figure 2.8 also shows
an arrangement for delivery of the heated vapor of a stable trapping reagent
(e.g., aniline or methanol) at a predetermined point close to the end of the hot
zone. With some unstable species such as methyleneketene, $CH_2{=}C{=}C{=}O$,
trapping can be achieved with this arrangement[27] but introduction of the
reagent vapor into the exit tube 5–10 cm beyond the hot zone is quite inefficient. Bonnett and co-workers[22] show a cold finger (Fig. 2.9) arranged for
the collection of isoindole, a substance stable only in cold, dilute, and oxygen-
free solutions, for nmr spectral measurements.

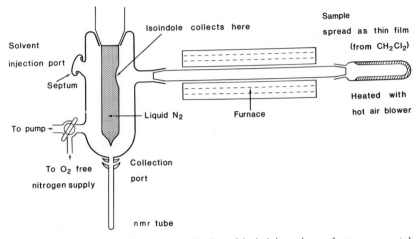

Fig. 2.9. Cold finger arranged for collection of isoindole and transfer to an nmr tube. (Ref. 22; reproduced by permission.)

A useful form of vertical pyrolysis tube and integral cold cavity condenser has been developed in the laboratory of C. W. Rees. The first form of the apparatus was described with diagrams by Anderson *et al.*,[25] but the improved versions shown in Fig. 2.10 are those used in more recent work on the chemistry of benzazete (see Chapter 6, Section IV,D). The following account of the use of this equipment is condensed from notes kindly provided by P. W. Manley, C. W. Rees, and R. C. Storr of the University of Liverpool, and Fig. 2.11 shows the apparatus in use.

The material to be pyrolyzed is spread as a thin film on the walls of a 50 cm^3 round bottomed flask, usually by rotary evaporation of a solution in dichloromethane, to facilitate sublimation. The flask is heated with an external electric oven and the whole apparatus is evacuated with a high-capacity vacuum pump to a pressure of less than 0.01 mm. The cold cavity condenser is cooled to $-78°$ with acetone-solid carbon dioxide, and the silica pyrolysis tube is heated with a close-fitting electric furnace constructed from stainless steel tube (see below). When the pyrolysis is complete an unstable pyrolyzate can be put under dry nitrogen and then washed off the cold finger with cold solvent (apparatus of Fig. 2.10). An air-stable pyrolyzate can be allowed to warm to room temperature and the apparatus (Fig. 2.10b) opened at the flanged point.

The furnace tubes are made from stainless steel tubes (2.54 cm, ID; 2.86 cm, OD), the length of which depends on the desired contact time. The steel tube is first electrically insulated by a thin layer of Triton Kaowool Mastic, which after hardening is sanded down. Copper wire terminals are then wired onto

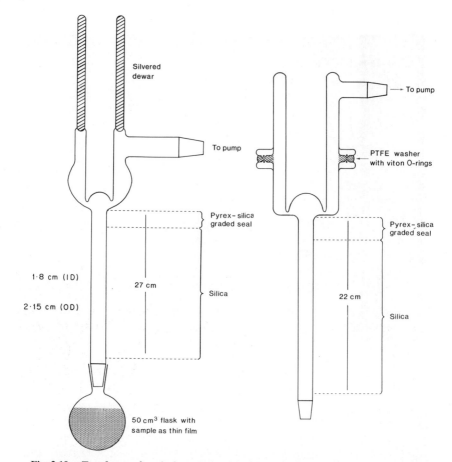

Fig. 2.10.   Two forms of vertical pyrolysis tube with cold cavity condenser: (a) with integral condenser as used for the collection of unstable pyrolyzates, and (b) with detachable condenser for the collection of more stable products. (From drawings by Mr. P. W. Manley, University of Liverpool; reproduced by permission.)

each end of the tube, over the Mastic. The heating element is wound between the terminals with Nichrome wire (resistance 8 ohms/m) to a resistance of about 70 ohms. The coils of the element are wound more closely together at each end so as to keep the temperature more uniform throughout the length of the furnace. Finally, the terminals are silver soldered and a coating of Mastic is put over the element and allowed to dry. Such a furnace can be used to 900°.

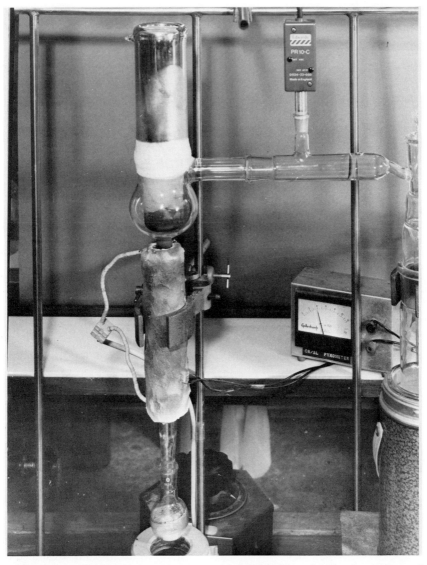

Fig. 2.11. Vertical pyrolysis apparatus assembled with the integral cold cavity condenser of Fig. 2.10a. (Courtesy of C. W. Rees, University of Liverpool.)

Rosen and Weber[28] have optimized the conditions for thermal cyclization of 1-(ω-pyridyl)-1,3-butadienes to give dihydroquinolines without secondary dehydrogenation (see Chapter 9, Section II,C, and Table 9.1). They used a quartz tube (8 mm, ID; 9 mm, OD) 250 cm long wound in a helical spiral of 30 turns and contained within the heated zone (30 × 3.5 cm) of a vertical tube furnace. The compounds were pyrolyzed at 650° and 0.1 mm on a 1 gm scale using upward flow and a contact time of approximately 0.1 sec.

D. True Flash Vacuum Pyrolysis

Hedaya[1] has reviewed the revival of interest in flash pyrolytic methods arising from the long development of the gas-phase chemistry of free radicals and carbenes in the hands of Paneth and Rice, Lossing, Benson, Skell, and their co-workers. Direct study of these short-lived reactive species by mass spectrometry or matrix isolation requires short contact times in the pyrolysis step, a short time of flight before analysis, and consequently a greater investment in high vacuum technology as compared with the simple apparatus of Section II,C. Hedaya's review[1] gives diagrams showing both the coupling of a small pyrolysis oven to the inlet system of a mass spectrometer and an apparatus for preparative flash vacuum pyrolysis; both systems were used in the study of cyclopentadienyl radicals generated from nickelocene. In such apparatus[1,29] the substantial pyrolysis tubes of Section II,C are reduced to small heated sections of smaller tubes contained *within* the evacuated system, and the gap between the hot zone and the analyzing system or a cold collecting surface can be reduced to a few centimeters.

The very low pressure pyrolysis (VLPP) method for examining the kinetics of decomposition and subsequent reactions of a range of organic molecules at pressures of about 10^{-3} mm is reviewed with diagrams of the apparatus by Golden *et al.*[3] In the VLPP technique molecules decompose within a Knudsen cell reactor with a residence time $t_{res} = 4V/\bar{c}A_h$ dependent only on the volume of the cell V, the area of the escape aperture A_h, and the mean unimolecular speed of the molecules $\bar{c}$. The emergent species are detected by quadrupole mass spectrometry. Under these conditions intermolecular collisions are rare and molecules are activated by collision with the hot wall of the reactor. The induction collision number Z_i is defined as the number of collisions needed for a molecule to acquire an initial energy in excess of E_a for decomposition. Golden *et al.*[3] estimate Z_i to be about 40 collisions for a number of small organic molecules, but show that it is difficult to arrange conditions to approach this minimum number of collisions, and in real experiments collision numbers are of the order of 10^2–10^4. It will be evident that preparative pyrolyses, even with some of the sophisticated equipment discussed below, will not always approach the low pressures in the reaction zone used in the VLPP technique.

Reactive species isolated in a frozen matrix of inert gas atoms can be studied by various spectroscopic methods; both simple[30] and detailed[31] accounts of this matrix isolation technique are available. As an example of the method applied to an unusual heterogeneous flow reaction we may note the measurement by Chapman and co-workers[32] of the ultraviolet and infrared spectra of the unstable hydrocarbon benzocyclobutadiene in an argon matrix at 8°K. Benzocyclobutadiene was generated from *cis*-1,2-diiodobenzocyclobutene by passing it at 10^{-6} mm over zinc at 230°, condensing out less volatile products at 5°, and condensing the product and argon atoms on a plate at 8°K. A diagram is given[32] showing the arrangement of the reaction tube, argon inlet, and vacuum shroud with quartz and potassium bromide windows.

King and co-workers[2] give diagrams (see Fig. 2.12) and details of construction of their flash vacuum thermolysis (FVT) apparatus in which the sublimation and pyrolysis ovens are formed by winding separate 10 cm sections of a 13 mm (OD) ceramic tube with resistance wire, and both are contained within the evacuated system. Pressures can be monitored at several points with Hastings gauges, and an oil diffusion pump and backing pump are used. In this paper pyrolyses of thietane and thietanone 1,1-dioxides were carried out at 930°–960° with pressures of 10^{-1} mm or less in the hot tube and 10^{-3} mm at the pump, and with contact times of the order of 1–5 msec.

Similar apparatus has been used by Chapman and McIntosh[33] both for preparative work and for experiments in which pyrolytic products are collected for infrared spectroscopy on sodium chloride plates conductively cooled by liquid nitrogen. De Mayo[34] gives diagrams of the arrangement of the FVT apparatus for spectroscopic studies using such a cold plate and cryostat device, and Lewars and Morrison[35] give a clear diagram (Fig. 2.13) of a simple infrared cryostat and details of its construction.

A somewhat simplified version of the FVT apparatus (Fig. 2.14) has been developed by Seybold and Jersak.[16,36] In this version interchangeable pyrolysis tubes are used so that the geometry of the hot zone can be changed in a defined way. At low pressures (10^{-3} mm or less) the residence or contact time in the hot zone is independent of pressure and for a particular compound depends only on tube geometry and temperature,[3,16,37] so that in a series of experiments contact times (10^{-3}–10^{-1} sec) can be changed by changing pyrolysis tubes, and can be calculated.[16] Full details of the construction and operation of this apparatus and of the associated high-vacuum pumping system (80 liter/sec) are given,[16] and it is now available commercially from Otto Fritz GmbH (Normag), Feldstrasse 1, D-6238, Hofheim, Germany.

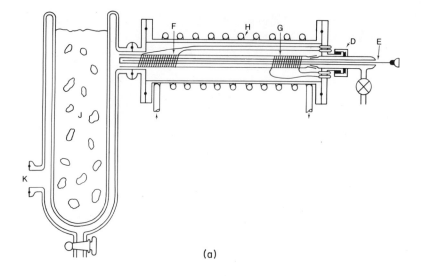

(a)

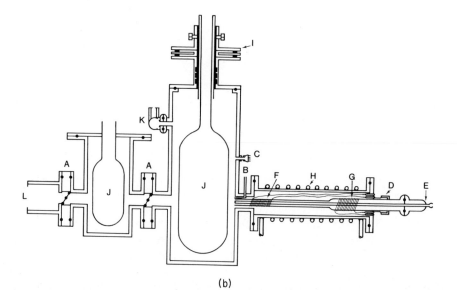

(b)

Fig. 2.12. Flash vacuum thermolysis apparatus: (a) with conventional Dewar cold finger, (b) with high-efficiency cold finger. A, quarter-swing butterfly valves; B, trapping agent inlet; C, vacuum gauge (the apparatus in Fig. 2.12a also has a Hastings gauge located adjacent to the vacuum seal coupling D and not shown on the drawing); D, vacuum seal coupling; E, thermocouple well; F, main oven; G, sublimation oven; H, water-cooling coil; I, ball bearing race; J, liquid nitrogen cold finger; K, connection leading to the sample-handling manifold; and L, connection to main pump. (Ref. 50; reproduced by permission.)

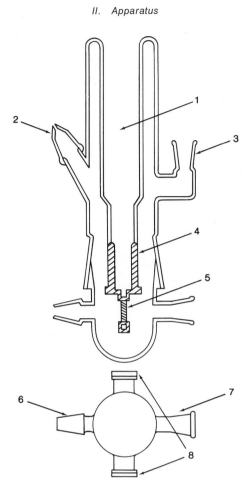

Fig. 2.13. Infrared cryostat. 1, liquid nitrogen reservoir; 2, head carrying thermocouple wires (wires sealed at orifice with epoxy resin); 3, joint for receiving Pirani gauge; 4, metal collar (attached to reservoir through glass-to-metal seal); 5, NaCl disk; 6, to vacuum system; 7, to thermolysis tube; 8, optical windows. Silicone grease was used on all ground glass joints. (Ref. 51; reproduced by permission.)

It is suitable for preparative pyrolyses of up to 10 gm of liquid or solid compounds.

Much of the work with the FVP and FVT apparatus described above has been concerned with molecules of modest molecular weight (100–300) or of reasonable volatility under vacuum. Seybold and Heibl[38] show a modified sublimation system for use with substances of low volatility. The pyrolysis of poorly volatile and high melting macrocyclic disulphones

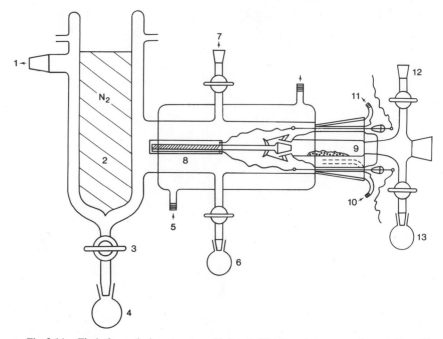

Fig. 2.14. Flash thermolysis apparatus with internal heating. 1, to pump; 2, rotatable cold finger; 3, coolable tap; 4, receiver; 5, cooling water inlet; 6, solvent supply vessel; 7, to manometer; 8, electrically heated quartz tube; 9, sublimation tube; 10, connection for thermostat; 11, electric leads; 12, inert-gas inlet; and 13, storage vessel for volatile substances. The compounds to be pyrolyzed can optionally be introduced at 9 or 13. (Ref. 16; reproduced by permission of Verlag Chemie GMBH.)

(mp range 230°–360°) has been achieved by Grütze and Vögtle[39] using a plain silica tube (25 cm long; 4 mm, ID) heated by three movable coaxial ring-ovens with separate temperature controls. Small samples (50–100 mg) were pyrolyzed at 400°–1080° and 0.01–0.5 mm by subliming the sample from the closed end of the tube with ring-oven I and using ring-ovens II and III to provide a hot zone about 4 cm long which was moved toward the sublimation zone during the experiment.

Finally it is worth noting the technique of Curie-point pyrolysis which, although mainly an analytical tool outside the scope of this book, has been used by Schaden[40] to survey the pyrolytic decomposition of quinones and carbonyl compounds in conjunction with preparative flash pyrolysis. In Curie-point pyrolysis[41,42,43] a microgram sample of a relatively involatile compound spread on the surface of a ferromagnetic wire is heated by a high-frequency induction field from room temperature to 300°–900° in less than 0.1 sec, and the pyrolysis products are analyzed by gas chromatography or combined gas chromatography–mass spectrometry (GC-MS). The upper temperature is set by the Curie point of the metal of the wire and is thus readily reproducible. Various analytical techniques which use pyrolysis in conjunction with gas chromatography have been reviewed by McKinney[43] with diagrams of pyrolytic devices.

III.   CHOICE OF PYROLYTIC CONDITIONS

Pyrolysis at atmospheric pressure with a high partial pressure of reactant and a slow flow of carrier gas (Section II,A) leads to long contact times of the order of 20–30 sec in most conventional apparatus, and is suitable mainly for a few reactions, such as the pyrolytic β-elimination of esters, which happen to be considerably faster at temperatures of $300°$–$600°$ than the range of possible secondary reactions of the products. At higher temperatures secondary reactions ·become inevitable, and above $800°$ tars containing polycyclic aromatic compounds are produced by complex sequences of the type reviewed by Badger.[44] This form of pyrolysis is also useful for the generation of highly reactive intermediates in the presence of a large excess of a trapping reagent. The generation of benzyne by pyrolysis of solutions of phthalic anhydride (precursor) in benzene (trapping reagent) is an example in which the high partial pressure of benzene leads to efficient trapping (see Chapter 3, Section V,B).

The aim of preparative flash pyrolytic methods is to isolate the primary unimolecular decomposition or an early short sequence of unimolecular reactions. This is achieved by working at high temperatures so that a short contact time will suffice for the primary reaction, at low pressures to inhibit secondary bimolecular reactions, and with reactors and flows of carrier gas arranged to give short contact times and times of flight to cold collectors in which all reactions stop.

These criteria may be adequately met for the preparation of thermodynamically stable products ʼof reasonable kinetic stability by pyrolysis under reduced pressure (Section II,B) or in a moderate vacuum (Section II,C) with or without a carrier gas. The compound passes through the hot zone under conditions of viscous flow, with a mean free path which is small compared to the dimensions of the tube. The retention time depends on the geometry of the tube, the viscosity of the carrier gas, the pressure drop across the length of the tube, and the temperature.[37] Under these conditions molecules are activated by intermolecular collision, decompose, and the products rapidly reach thermal equilibrium with the emergent gas stream. In pyrolysis over a Nichrome filament[14,15] most molecules are probably activated by collision with nitrogen molecules.

True flash vacuum pyrolytic conditions (Section II,D) are essential when the products are thermodynamically stable as isolated molecules, but are kinetically highly unstable with respect to dimerization and other bimolecular reactions; the geometry of the apparatus must provide the shortest possible path to the cold surface or analyzing device. Both Golden *et al.*[3] and Seybold[16] discuss design and rate-of-throughput criteria for the observation of reactive species which can be destroyed by fast molecule–molecule, radical–molecule, or radical–radical reactions. It is difficult to obtain direct evidence for species which rearrange rapidly by a unimolecular mechanism to thermodynamically more stable products; the difficulty of observing intermediate isonitriles in high-temperature reactions (Chapter 5, Section II,E) is a case in point.

At very low pressures ($<10^{-3}$ mm) under conditions of molecular flow, molecules are activated by collision with the reactor walls, and intermolecular collisions within the hot zone and in the emergent gas stream are rare. This means that products of an exothermic reaction may not be in true thermal equilibrium, and effects of chemical activation may appear. These effects can be particularly important in the common case where a product is formed in an exothermic reaction by expulsion of a small stable fragment (e.g., CO_2 or N_2) and the excess energy is retained largely as vibrational energy of the product. This excess energy cannot be lost by collisional deactivation at the low pressure, and further decomposition or isomerization may occur before collection. A related situation, reviewed by Frey,[45] arises in the formation of cyclopropanes by attack of methylene, $:CH_2$, on alkenes in static gas-phase systems. The primary products are formed with excess energy and at low pressures they isomerize completely. At high pressures collisional deactivation occurs, and some primary products survive.

Chemical activation may complicate attempts to assess the plausibility of proposed intermediates in a pyrolytic sequence at low pressure. If products B and C are isolated from the pyrolysis of A and consecutive reactions are suspected [Eq. 2.4)], it is usual to repyrolyze B under conditions as nearly identical as possible.

$$A \xrightarrow{\text{FVP}} B \longrightarrow C \tag{2.4}$$

If B is formed in a chemically activated state from A then the further reaction to give C may occur much more extensively than in the pyrolysis of B at the same temperature. By raising the total pressure with a few millimeters of carrier gas it may be possible to depress further reaction of chemically activated B by collisional deactivation.

The effects of chemical activation have been invoked by Wentrup[46,47] in the area of arylcarbene and arylnitrene rearrangements (Chapter 5, Section III), and Wentrup's group has often used the comparison of products from low-pressure experiments and from experiments with added carrier gas in the investigation of mechanisms. For example, phenylnitrene formed from 5-(2-pyridyl)tetrazole by rearrangement of 2-pyridylcarbene is considered to be chemically activated. In a low-pressure experiment most of the phenyl-nitrene retains sufficient energy to rearrange further to cyanocyclopentadiene, whereas addition of 1 mm of nitrogen effects some collisional deactivation, and the yield of unrearranged dimer, azobenzene, increases[47] [Eq. (2.5)].

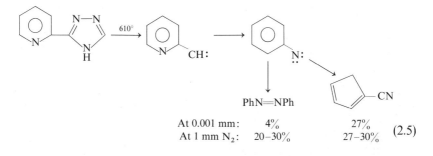

$$
\begin{array}{llll}
\text{At 0.001 mm:} & 4\% & 27\% & \\
\text{At 1 mm N}_2: & 20\text{--}30\% & 27\text{--}30\% & (2.5)
\end{array}
$$

Contact times have been mentioned frequently in this chapter without comment on their reliability in most preparative experiments. The estimation of true contact times at pressures above 0.1 mm is complex, particularly when reactions involving changes in the number of gas molecules are occurring.[48,49] Most workers use instead a simple approach based on expansion of m moles of the sample to a volume of gas at the pressure P mm and absolute temperature T of the hot zone, and passage of that volume during t sec through the hot zone of volume V liters, which leads to a formula of the type of Eq. (2.6).

$$
\text{Contact time (CT)} = \frac{273}{T} \times \frac{V}{22.4} \times \frac{P}{760} \times \frac{t}{m} \qquad (2.6)
$$

Such expressions ignore changes in the number of molecules during reaction and must frequently be used with unrealistically low pressure readings made far from the hot zone. These apparent contact times are clearly of dubious significance, though relative contact times within a series of experiments may be of value.

REFERENCES

1. Hedaya, E. (1969). *Acc. Chem. Res.* **2**, 367.
2. King, J. F., de Mayo, P., McIntosh, C. L., Piers, K., and Smith, D. J. H. (1970). *Can. J. Chem.* **48**, 3704.

3. Golden, D. M., Spokes, G. N., and Benson, S. W. (1973). *Angew. Chem. Int. Ed. Engl.* **12**, 534.
4. Hurd, C. D. (1941). *In* "Organic Syntheses" (H. Gilman, ed.), Coll. Vol. 1, p. 330. Wiley, New York.
5. Conia, J. M., and Le Perchec, P. (1975). *Synthesis*, p. 1.
6. Ratchford, W. P. (1955). *In* "Organic Syntheses" (E. C. Horning, ed.), Coll. Vol. 3, p. 30. Wiley, New York.
7. Benson, R. E., and McKusick, B. C. (1963). *In* "Organic Syntheses" (N. Rabjohn, ed.), Coll. Vol. 4, p. 746. Wiley, New York.
8. Andreades, S., and Carlson, H. D. (1973). *In* "Organic Syntheses" (H. E. Baumgarten, ed.), Coll. Vol. 5, p. 679. Wiley, New York.
9. Rice, F. O., Ruoff, P. M., and Rodowskas, E. L. (1938). *J. Am. Chem. Soc.* **60**, 955.
10. Williamson, K. L., Keller, R. T., Fonken, G. S., Szmuszkovicz, J., and Johnson, W. S. (1962). *J. Org. Chem.* **27**, 1612.
11. Wiersum, U. E., and Nieuwenhuis, T. (1973). *Tetrahedron Lett.*, p. 2581.
12. Williams, J. W., and Hurd, C. D. (1940). *J. Org. Chem.* **5**, 122.
13. Fieser, L. F., and Fieser, M. (1967). "Reagents for Organic Synthesis" Vol. 1, p. 528. Wiley, New York.
14. Spangler, R. J., Beckmann, B. G., and Kim, J. H. (1977). *J. Org. Chem.* **42**, 2989.
15. Spangler, R. J., Kim, J. H., and Cava, M. P. (1977). *J. Org. Chem.* **42**, 1697.
16. Seybold, G. (1977). *Angew. Chem. Int. Ed. Engl.* **16**, 365.
17. Crow, W. D., and Solly, R. K. (1966). *Aust. J. Chem.* **19**, 2119.
18. Gleiter, R., Rettig, W., and Wentrup, C. (1974). *Helv. Chim. Acta* **57**, 2111.
19. Trahanovsky, W. S., Ong, C. C., Pataky, J. G., Weitl, F. L., Mullen, P. W., Clardy, J., and Hansen, R. S. (1971). *J. Org. Chem.* **36**, 3575.
20. Brown, R. F. C., Eastwood, F. W., and Harrington, K. J. (1974). *Aust. J. Chem.* **27**, 2373.
21. Trahanovsky, W. S., Ong, C. C., and Lawson, J. A. (1968). *J. Am. Chem. Soc.* **90**, 2839.
22. Bonnett, R., Brown, R. F. C., and Smith, R. G. (1973). *J. Chem. Soc. Perkin Trans. 1*, p. 1432.
23. Brent, D. A., Hriber, J. D., and DeJongh, D. C. (1970). *J. Org. Chem.* **35**, 135.
24. Badger, G. M., Buttery, R. G., Kimber, R. W. L., Lewis, G. E., Moritz, A. G., and Napier, I. M. (1958). *J. Chem. Soc.*, p. 2449.
25. Anderson, D. J., Horwell, D. C., Stanton, E., Gilchrist, T. L., and Rees, C. W. (1972). *J. Chem. Soc. Perkin Trans. 1*, p. 1317.
26. Crow, W. D., and Wentrup, C. (1967). *Tetrahedron Lett.*, p. 4379.
27. Brown, R. F. C., Eastwood, F. W., and McMullen, G. L. (1977). *Aust. J. Chem.* **30**, 179.
28. Rosen, B. I., and Weber, W. P. (1977). *J. Org. Chem.* **42**, 47.
29. Farmer, J. B., and Lossing, F. P. (1955). *Can. J. Chem.* **33**, 861.
30. Cradock, S., and Hinchcliffe, A. J. (1975). "Matrix Isolation." Cambridge Univ. Press, London and New York.
31. Hallam, H. E., ed. (1973). "Vibrational Spectroscopy of Trapped Species" Wiley, New York.
32. Chapman, O. L., Chang, C. C., and Rosenquist, N. R. (1976). *J. Am. Chem. Soc.* **98**, 261.
33. Chapman, O. L., and McIntosh, C. L. (1971). *Chem. Commun.*, p. 770.
34. de Mayo, P. (1972). *Endeavour* **31**, 135.
35. Lewars, E. G. and Morrison, G. (1977). *Can. J. Chem.* **55**, 966.
36. Seybold, G., and Jersak, U. (1977). *Chem. Ber.* **110**, 1239.
37. Dushman, S., and Lafferty, J. M. (1962). "Scientific Foundations of Vacuum Technique," 2nd Ed. Wiley, New York.
38. Seybold, G., and Heibl, C. (1977). *Chem. Ber.* **110**, 1225.
39. Grütze, J., and Vögtle, F. (1977). *Chem. Ber.* **110**, 1978.

40. Schaden, G. (1977). *Angew. Chem. Int. Edit. Engl.* **16**, 50.
41. Giacobbo, H., and Simon, W. (1964). *Pharm. Acta Helv.* **39**, 162.
42. Schmid, P. P., and Simon, W. (1977). *Analyt. Chim. Acta* **89**, 1.
43. McKinney, R. W. (1969). *In* "Ancillary Techniques of Gas Chromatography" (L. S. Ettre and W. H. McFadden, eds.), pp. 55–87. Wiley (Interscience), New York.
44. Badger, G. M. (1965). *In* "Progress in Physical Organic Chemistry" (S. G. Cohen, A. Streitwieser, and R. W. Taft, eds.), Vol. 3, p. 1. Wiley (Interscience), New York.
45. Frey, H. M. (1964). *In* "Carbene Chemistry" (W. Kirmse, ed.), p. 217. Academic Press, New York.
46. Wentrup, C. (1974). *Tetrahedron* **30**, 1301.
47. Wentrup, C. (1976). *Topics Curr. Chem.* **62**, 173.
48. Benson, S. W. (1960). "The Foundations of Chemical Kinetics." McGraw-Hill, New York.
49. Mulcahy, M. F. R. (1973). "Gas Kinetics." Nelson, London.
50. National Research Council of Canada (1970). *Can. J. Chem.* **48**, 3704.
51. National Research Council of Canada (1977). *Can. J. Chem.* **55**, 966.

Chapter 3

Pyrolytic Generation and Reactions of Free Radicals, Arynes, and Cyclobutadienes

I. INTRODUCTION

The three groups of species which are the subject of this chapter are related by their common very high reactivity rather than by similarity of generation or of structure. All three dimerize rapidly so that their lifetimes at ordinary temperatures and pressures are extremely short, and their direct detection requires sophisticated spectroscopic methods.

Radical and diradical intermediates are involved in a great many of the high-temperature reactions discussed in later chapters. This is scarcely surprising; free radical chemistry is a major division of organic chemistry, and the development of the study of reactions in the gas phase has been closely connected with growing interest in radical pathways. This chapter largely ignores the production of radicals from conventional initiators such as aliphatic peroxides and azo compounds and instead concentrates mainly on the generation and trapping of radicals from sources such as nitroarenes and oxalic esters which decompose only at high temperatures.

The section on arynes similarly deals with only one aspect of aryne chemistry, namely, the formation of arynes from phthalic anhydrides and related compounds and reactions involving the trapping of these arynes by other aromatic compounds. The pyrolysis of a wider range of cyclic anhydrides is discussed in Chapter 6; in the present chapter many of the reactions considered are those of benzyne formed from phthalic anhydride.

Pyrolytic methods have contributed relatively little to the study of cyclobutadiene and its derivatives; the topic is introduced here mainly because of the high reactivity of cyclobutadienes and because some of the experimental methods used in the area are related to those used for radicals and arynes.

II. PHYSICAL DETECTION, IDENTIFICATION, AND SPECTROSCOPIC STUDY OF FREE RADICALS

The use of a pyrolysis oven coupled to an electron bombardment chamber and mass spectrometer can provide a wealth of information about pyrolytic processes and about short-lived species including free radicals. Hedaya[1] gives a comprehensive list of references to the development by Lossing, Benson, and others of mass spectral methods for the study of radicals at low pressures. The course of a pyrolytic reaction in a flow system can sometimes be established from the appearance and disappearance of species of different m/e as the pyrolytic temperature is varied; and the measurement of appearance potentials and ionization potentials can assist in the identification of short-lived species. Some pyrolytic reactions which have been employed recently in the measurement of ionization potentials (IP) of common radicals are shown in Eqs. (3.1)–(3.6), all of which refer to flash vacuum pyrolysis through short silica tubes with millisecond contact times. The cyanoisopropyl radical [Eq. (3.1)] was studied by HeI photoelectron spectroscopy.[2]

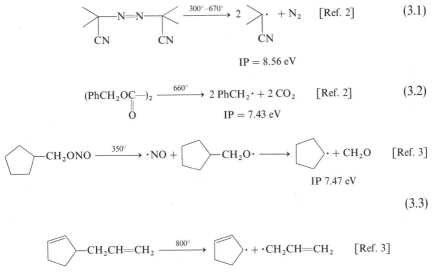

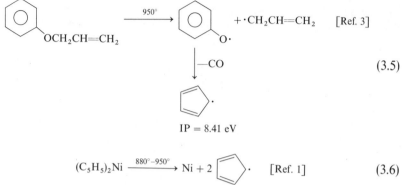

$$\text{IP} = 8.41 \text{ eV}$$

(3.5)

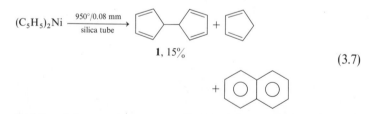

The pyrolysis of nickelocene [Eqs. (3.6) and (3.7)] recalls that the experiments of Paneth and Hofeditz[4] on the generation of methyl radicals by pyrolysis of tetramethyllead are a landmark in the history of free radical chemistry, and the work of Hedaya and co-workers[1,5] on nickelocene demonstrates the continued value of organometallic compounds as precursors of radicals in the gas phase. By using the technique of neon matrix isolation[6] electron spin resonance (esr) spectra of cyclopentadienyl radicals from nickelocene have been measured at 4°–12°K, as have the infrared and ultraviolet spectra. Preparative pyrolysis of nickelocene at 950° gives a carbon-containing deposit of nickel on the walls of the pyrolysis tube[7] and affords a surprisingly high yield of 9,10-dihydrofulvalene (**1**). Hedaya and co-workers[5] found that collection of cyclopentadienyl in a xenon matrix at 50°K, or annealing of neon or argon matrices formed at 4°K or 20°K, led to a metastable intermediate which on further annealing gave 9,10-dihydrofulvalene. This metastable intermediate was tentatively identified as "nullocene," a sandwich association of two cyclopentadienyl radicals lacking the metal filling.

Another example of the generation of a radical for esr spectroscopy in a rare gas matrix is given by the pyrolysis of phenacyl iodide studied by Kasai and co-workers.[8] The benzyl radicals ultimately produced [Eq. (3.8)] were collected in a neon matrix on a sapphire rod at 4°K. Ultraviolet irradiation

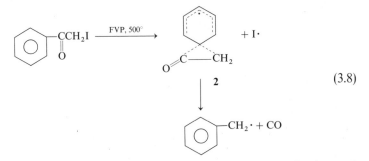

$$(3.8)$$

of the precursor phenacyl iodide in an argon matrix at 4°K gave only phenacyl radicals, $PhCOCH_2\cdot$, whereas in the pyrolytic system rearrangement and decarbonylation to the benzyl radical occurs, probably via **2**.

III.   PYROLYTIC GENERATION OF SOME O-, S-, AND N-CENTERED RADICALS

A.   From Peroxides

The chemistry of oxygen radicals and diradicals has been reviewed in some detail by Kochi.[9] The generation of acyloxy and alkoxy radicals from peroxides has usually been accomplished by heating the peroxide in dilute solution, and few gas-phase reactions of preparative interest have been reported, presumably because of the danger of explosion during vaporization. Adam and Arce[10] have found formation of polymer and of more than 30 volatile products in the decomposition of the unusually stable peroxide 2-ethylidene-3,3,5,5-tetramethyl-1,2-dioxolane (**3**) in a static system at 315° and 0.01 mm for 5 hr. The major early processes are shown in Eq. (3.9);

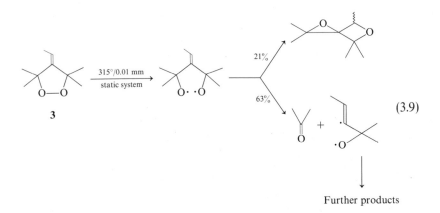

$$(3.9)$$

Further products

this appears to be a reaction which could profitably be reexamined by flash pyrolytic methods. Trimethylenemalonyl peroxide (4) on flash vacuum pyrolysis at 360° gave mainly cyclobutanone, formed by decarbonylation of the intermediate α-lactone 5;[11] some loss of carbon dioxide from 5 to form a carbene which subsequently rearranged to cyclobutene was also observed [Eq. (3.10)]. Moore[12] first showed that fragmentation of dihydro-ascaridole 6 at 240° under reflux gives the 1,4-dione 7 and ethylene, but the reaction is not clean and much polymer is produced. More recently Haynes and co-workers[13] have demonstrated the similar but smooth fragmentation of the peroxide 8 on flash pyrolysis to give mainly ethylene and benzophenone [Eq. (3.12)].

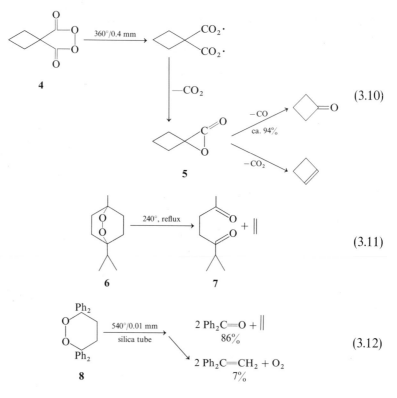

B.   From Nitrite Esters

The gas-phase pyrolysis of alkyl nitrites is a general method for the generation of alkoxy radicals $RO \cdot$ and nitric oxide by homolytic cleavage of the $RO-NO$ bond. The reaction has been reviewed by Steacie[14] and by Gray and Williams,[15] and we have already seen an application of this reaction in

Section II, Eq. (3.3). The photochemical decomposition of nitrite esters in solution also involves RO—NO fission and the reaction with intramolecular hydrogen transfer within the alkoxy radical (the Barton reaction) is of considerable synthetic importance.[16] Barton and co-workers[17,18] originally considered that photochemically generated alkoxy radicals were activated with respect to thermally produced radicals, but later work[19] showed that similar hydrogen transfers can occur in thermal reactions in the gas phase. In the liquid phase, ionic reactions catalyzed by impurities lead to disproportionation of primary nitrites to aldehydes, alcohols, and nitric oxide. Under flash pyrolytic conditions, however, recapture of nitric oxide is less significant than in photochemical experiments in solution, and the major pyrolytic products are formed by dimerization of alkyl radicals produced by intramolecular hydrogen transfer. The gas-phase pyrolysis of nitrites is thus of much lesser synthetic interest. Two flash pyrolytic reactions studied by Barton *et al.*[19] are shown in Eqs. (3.13) and (3.14). Photochemical decomposition of 4-phenylbutylnitrite (**9**) in solution gave the nitroso alcohol as a nitroso dimer in 60% yield.

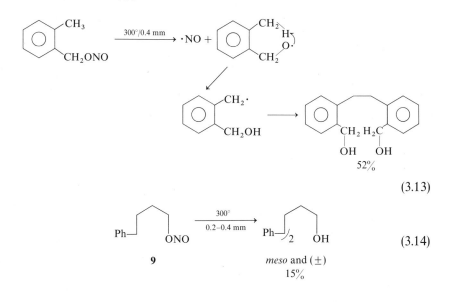

$$(3.13)$$

$$(3.14)$$

9   *meso* and ($\pm$)
            15%

C. From Aromatic Ethers

Flash vacuum pyrolysis of anisole at about 950° has been shown by Harrison *et al.*[20] to give first methyl and phenoxy radicals and then cyclopentadienyl radicals, all detected by mass spectrometry [Eq. (3.15)]. 7-Methoxycycloheptatriene decomposed by a similar pathway at 400°–550°. The decomposition of anisole in a preparative pyrolysis at atmospheric

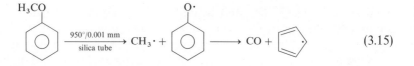

$$\text{(3.15)}$$

pressure is illustrated by an experiment due to Freidlin and co-workers[21,21a] [Eq. (3.16)] which gave 52% of phenol at 640°. Thus the methoxy group is not a thermally stable substituent at very high temperatures or at high temperatures with a long contact time as in Eq. (3.16).

The formation of phenoxy and allyl radicals from phenyl allyl ether on flash vacuum pyrolysis is shown in Section II, Eq. (3.5), and the relationship between this fission and the normal thermal Claisen rearrangement is discussed in Chapter 9, Section V,C. Another example of fission of an allyl aryl ether occurs in the pyrolysis of the o-methoxy compound 10, which affords a substantial yield of salicylaldehyde as outlined in Eq. (3.17) and reported by Marty and de Mayo.[22] The aryloxy radical 11 is also formed in low yield on pyrolysis of o-nitroanisole[22] because nitroarenes tend to fragment in two modes:[22,23,24] first by Ar–NO$_2$ homolysis, and second by rearrangement and fission as ArO–NO (see Section IV,C).

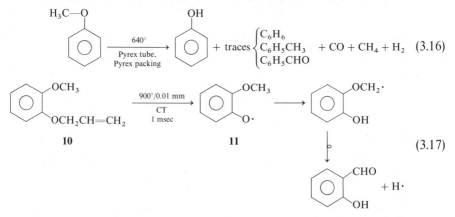

$$\text{(3.16)}$$

$$\text{(3.17)}$$

Similar homolytic fission of an enol ether has been proposed by Schirmer and Conia[25] to explain the cyclization of the enol ether 12 in the liquid phase to the cyclopentyl ketone 13, a reaction which has all the characteristics of an intermolecular radical transfer.

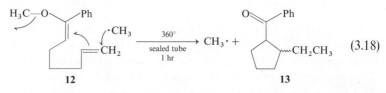

$$\text{(3.18)}$$

D. From Disulfides, Sulfides, and Sulfenic Acids

The chemistry of sulfur radicals has been reviewed by Kice.[26] In the present section we consider only a few reactions of styrylthiyl and related radicals which lead to ring closures. Ando and co-workers[27] found that the sulfoxides **14** ($R^1 = C_2H_5$ or $i - C_3H_7$; $R^2 = H$ or CH_3) on pyrolysis at 580° gave 3-phenylbenzothiophenes via the corresponding sulfenic acids $Ph_2C{=}CR^2SOH$ and thiyl radicals $Ph_2C{=}CR^2S\cdot$ [see Chapter 4, Eq. (4.64)]. A benzene solution of the disulfide **15** gave 85% of 3-phenyl-benzothiophene on flow pyrolysis at 580° [Eq. (3.19)] and a series of thioethers of the type **16** ($R^1 = CH_3$, C_2H_5, or $i - C_3H_7$) gave benzothiophenes in 50–70% yield.

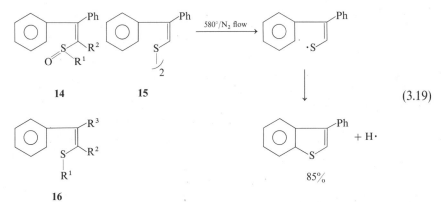

14   **15**   (3.19)

16   85%

A related synthetically useful pyrolysis is that of benzyl 2-(4-pyridyl)-ethylsulfide (**17**), which over Pyrex helices at 600° gave thieno[2,3-c]pyridine in good yield [Eq. (3.20)]. Klemm and co-workers[28] also found that the corresponding 2-pyridyl compound gave the isomeric thieno[3,2-b]pyridine in 28% yield.

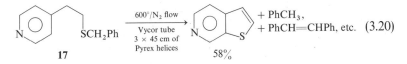

17   58%

E. Iminyl Radicals from Azines and Oxime Esters

The melt decomposition of azines derived from aromatic carbonyl compounds tends to give many products through the operation of several mechanisms, but recent work from Crow's laboratory[29,30,31] on the gas-phase decomposition of azines and related compounds has produced some interesting and synthetically useful results. The central N—N bond in an azine R_2C=N—N=CR_2 is estimated[29] to require a dissociation energy as low as 220 kJ/mole, so that pyrolytic dissociation to iminyl radicals R_2C=N · is a favored primary process.

Flash vacuum pyrolysis of aryl aldehyde azines gives mixtures of aryl nitriles and alkenes ArCH=CHAr. Formation of the alkenes is favored by higher pressures, and crossover and other experiments[29] have revealed that an intermolecular reaction is involved, but that free arylcarbene is not an intermediate. The major product is usually the nitrile, formed by ejection of one substituent of the iminyl radical [hydrogen in the case of Eq. (3.21)].

$$PhCH=N-N=CHPh \xrightarrow[0.15\ mm]{600°} \begin{array}{l} \nearrow PhCH=CHPh \\ \quad\ 5\% \\ \\ \searrow PhCH=N\cdot \longrightarrow PhCN + H\cdot \\ \qquad\qquad\qquad\quad 81\% \end{array} \qquad (3.21)$$

The azines of aryl ketones give high yields of aryl nitriles on flash pyrolysis at 450°; Bird *et al.*[29] discuss the thermochemistry of such reactions, and they conclude that fragmentation of the radical $PhC(CH_3)$=N · from aceto-phenone azine [Eq. (3.22), R=CH_3] to PhCN + CH_3 · is favored by 30 kJ/mole with respect to the alternative fragmentation to $CH_3CN + Ph$ ·.

$$\begin{array}{c} Ar \\ \diagdown \\ \quad\ C=N-N=C \\ R\diagup \qquad\qquad \diagdown R \end{array} \xrightarrow{450°/0.05-0.1\ mm} \begin{array}{c} Ar \\ \diagdown \\ \quad\ C=N\cdot \longrightarrow ArCN + R\cdot \\ R\diagup \qquad\qquad 80-96\% \end{array} \quad (3.22)$$

Ring cleavage of the iminyl radical **18** derived from cyclopentanone azine leads first to the 4-cyanobutyl radical (**19**) which then undergoes fragmentation to acrylonitrile and ethylene, rearrangement to 1-cyanobutyl radical (**20**) and fragmentation to form ethyl radical, and formation of methyl and propyl radicals. The pyrolyzate formed at 450° [Eq. (3.23)] contains ethylene, butane, acrylonitrile (the major product), methacrylonitrile, ethacrylonitrile,

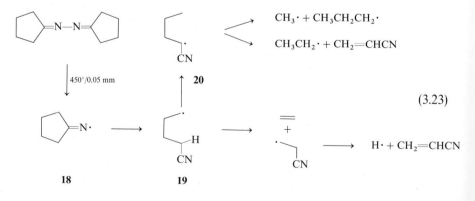

$$(3.23)$$

butanenitrile, 2-methylpentanenitrile, hexanenitrile, heptanenitrile, 2-ethyl-pentanenitrile, and 2-propylpentanenitrile. These products can all be derived from fragmentation or recombination of the radicals $CH_3 \cdot$, $C_2H_5 \cdot$, $C_3H_7 \cdot$, **19**, and **20**. The products of pyrolysis of cyclohexanone azine could be rationalized in a similar fashion.

Similar results were obtained by flash pyrolysis of oxime esters **21** at 440° and 5 mm. Extension of this approach to the 1,2-oxazine derivative **22** led to a useful synthesis of a terminally unsaturated nitrile [Eq. (3.24)].[31]

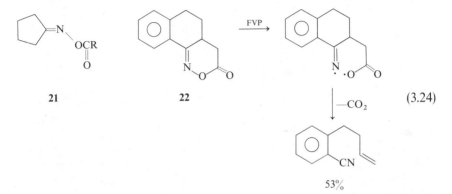

$$\text{(3.24)}$$

      21   **22**

53%

IV. PYROLYTIC FORMATION OF CARBON RADICALS FROM RELATIVELY STABLE PRECURSORS

A. From Azo Compounds and Sulfones

The decomposition of azomethane, which at high partial pressures may be explosive, occurs readily above 300° in a static system.[32] More highly substituted azoalkanes tend to decompose at lower temperatures; thus such compounds do not fall strictly within the scope of this section. Their thermal and photochemical decomposition has been reviewed by Muller[33] and by Mackenzie.[34] The synthetically valuable decomposition of cyclic azo compounds[34] is also covered in a recent review of the gas-phase decomposition of heterocyclic compounds by Braslavsky and Heicklen.[35] Five-membered systems such as 1-pyrazolines decompose at rates suitable for kinetic study in the temperature range 130°–240° to give, initially, 1,3-diradicals.

It seems appropriate, however, to mention here the spectacular *failure* of the cyclic azo compound diazabasketene (**23**) to give the 1,4-diradical required for cyclization to cubane. McNeil *et al.*[36] found that on flash vacuum pyrolysis it gave instead azocine (**24**) and hydrogen cyanide by a sequence of pericyclic reactions summarized in Eq. (3.25).

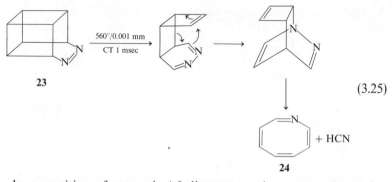

(3.25)

23

24

The decomposition of aromatic 1,2-diaza aromatic compounds to give products of ring fragmentation or ring contraction is discussed in Chapter 6, Section IV,C.

Aliphatic sulfones decompose over the temperature range 360–640° to give radicals and sulfur dioxide;[37] the lower end of the range refers to allylic and benzylic compounds, and the upper end to dimethyl sulfone. Arrhenius activation energies similarly range from 47.7 kcal/mole $(CH_2=CHCH_2SO_2CH_3)$ to 60.6 kcal/mole $(CH_3SO_2CH_3)$. The pyrolysis of sulfones will not be mentioned much further here, but the flash pyrolysis of cyclic sulfones is a very important method for the preparation of certain cyclic and macrocyclic hydrocarbons and is discussed in Chapter 6, Sections III,D, and III,E.

An example of pyrolytic reactions in which the use of a cyclic azo compound or a cyclic sulfone is equally satisfactory is provided by the generation of the trimethylenemethane **26** from the azo compound **25**[38] or the sulfone **27**.[39] Bushby and Pollard[38,39] found that each gave the diene **28** on flash vacuum pyrolysis [Eq. (3.26)], probably by hydrogen migration within the diradical **26**.

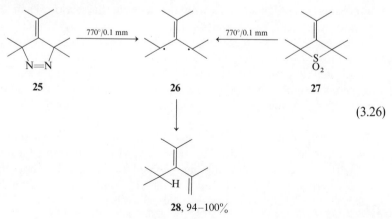

25   **26**   **27**

(3.26)

28, 94–100%

B. From Oxalic Esters

The pyrolytic decomposition of dialkyl oxalates RO_2CCO_2R can lead initially to an alkyl radical, carbon dioxide, and an oxycarbonyl radical $\cdot CO_2R$; the major products usually include R—R and R—CO_2R, the latter being particularly favored in the condensed phase (radical cage reaction).[40] Diallyl oxalate decomposes more readily than allyl acetate or benzoate,[41] but the reaction is not controlled only by thermolysis of the central C—C bond. Dimethyl oxalate is essentially stable to flow pyrolysis at 490° with a contact time of 2.3 min, consistent with a fairly high dissociation energy of 64 kcal/mole or more[42] for this bond. Thus the pyrolytic decomposition of diallyl or allyl alkyl oxalates proceeds mainly according to Eq. (3.27), with initial breaking of two bonds and formation of the stabilized allyl radical.

$$CH_2{=}CHCH_2{\dagger}OC{\dagger}COR \longrightarrow CH_2{=}CHCH_2\cdot + CO_2 + \cdot CO_2R \qquad (3.27)$$
$$\overset{\|}{O} \;\; \overset{\|}{O}$$

Flash vacuum pyrolysis of dibenzyl oxalates consequently leads to good yields of bibenzyls[43] by further decarboxylation of the oxycarbonyl radical and statistical recombination of benzyl radicals [Eq. (3.28)]. Trahanovsky and co-workers[43] comment that the yields of bibenzyls (44–76% for the benzyl and eight substituted benzyl compounds) are strikingly better by flash vacuum pyrolysis than by decomposition in the condensed phase or by flow pyrolysis at 1 atm. Trahanovsky and Ong[44] found that diallyl oxalate similarly gives biallyl [Eq. (3.29)] and dipropargyl oxalate gives benzene, presumably via 1,5-hexadiyne [see Chapter 9, Section V,F, Eqs. (9.92) and (9.93)]. Cinnamyl radicals produced on pyrolysis of di-E-cinnamyl oxalate, however, underwent cyclization rather than coupling; the major products from pyrolysis of 1 mole of the oxalate ($=2$ moles of cinnamyl radicals) were indene (0.82 mole), styrene (0.19), E-β-methylstyrene (0.11), allylbenzene

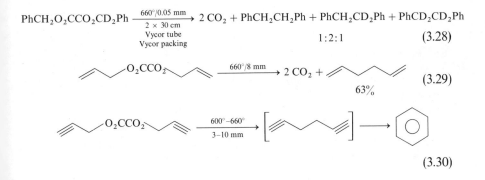

$$PhCH_2O_2CCO_2CD_2Ph \xrightarrow[\substack{2 \times 30\ cm \\ \text{Vycor tube} \\ \text{Vycor packing}}]{660°/0.05\ mm} 2\ CO_2 + PhCH_2CH_2Ph + PhCH_2CD_2Ph + PhCD_2CD_2Ph$$
$$1:2:1 \qquad (3.28)$$

$$\xrightarrow{660°/8\ mm} 2\ CO_2 + \qquad (3.29)$$
$$63\%$$

$$\xrightarrow[3-10\ mm]{600°-660°} \left[\quad\right] \longrightarrow \qquad (3.30)$$

(0.01), and toluene (0.01). Thus, indene must be formed by $E \rightarrow Z$ isomerization of the cinnamyl radical, cyclization, and loss of a hydrogen atom [Eq. (3.31)]. Cinnamyl formate and acetate similarly gave indene, in 82% and 72% yield, respectively, on pyrolysis at 650°. Syntheses of fluorene, xanthene, and dihydroanthracene have also been achieved by the related pyrolytic cyclization of appropriate o-substituted benzyl oxalates.[44] Lehr and Wilson[45] have noted that α-elimination to give a diarylcarbene occurs to a minor extent in the flash pyrolysis of bis(4-methylbenzhydryl) oxalate.

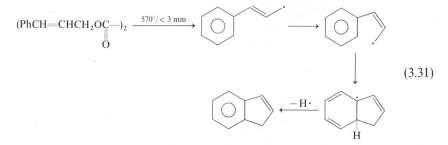

$$(3.31)$$

A further interesting example of the use of oxalate pyrolysis is the generation of trimethylenemethane and the subsequent isolation of its dimer, 1,4-dimethylenecyclohexane (5%), by pyrolysis of the cyclic oxalate **29** [Eq. (3.32)]. No methylenecyclopropane was detected among the several other products of this reaction. Schirmann and Weiss[46] suggest that the diradical is generated under these conditions as the ground-state triplet and thus does not cyclize to methylenecyclopropane. By contrast, Crawford and Cameron[47] have shown that the cyclic azocompound 4-methylene-1-pyrazoline gives only methylenecyclopropane on decomposition at 165°–188° and 100–200 mm.

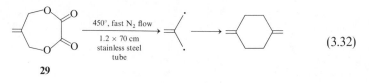

29

$$(3.32)$$

C.   From Nitroarenes

In 1967 Fields and Meyerson discovered that the homolytic decomposition of nitromethane[48] and nitrobenzene[49] occurred readily in a flow system at 400°–600° and that the radicals $CH_3\cdot$, $Ph\cdot$, and $\cdot NO_2$ thus formed could be intercepted by reaction with an excess of another aromatic compound such as benzene. The reaction involving nitromethane is not of much further interest because it appears to form nitrobenzene by attack of $\cdot NO_2$ on benzene, and the major product, biphenyl, is probably then produced by decomposition of this nitrobenzene in the presence of excess of benzene. The decomposition of nitrobenzene itself and of other nitroarenes is, however, of considerable synthetic interest because it provides a new method of radical arylation of aromatic compounds as an alternative to the decomposition of aroyl peroxides or the photolysis of aryl iodides. The gas-phase decomposition of nitrobenzene in Pyrex at 455°–515° had previously been studied by Smith;[50] the activation energy was determined as 51 kcal/mole, but the nongaseous products were not investigated.

In the standard procedure of Fields and Meyerson[23] the nitroarene is dissolved in benzene (5 or 25 moles) and the solution is passed under nitrogen through a Vycor tube packed with Vycor beads and heated to 600°. Contact times are about 5–25 sec. At a $PhNO_2/PhH$ ratio of 1:25 the products and relative concentrations were biphenyl (100), terphenyl (23), quaterphenyl (2), phenol (1), diphenyl ether and hydroxybiphenyl (3), and dibenzofuran (1). The yield of biphenyl and terphenyl per mole of nitrobenzene was 170% at this ratio of reactants, which suggested that more than one phenyl radical is produced for each mole of nitrobenzene decomposed. This result can be rationalized through the reactions of Eqs. (3.33)–(3.35); a second phenyl radical is considered to be formed from $\cdot NO_2$ and benzene, and this is supported by the appropriate direct experiment with $\cdot NO_2$, which also gave biphenyl and terphenyl.

$$
\text{(3.33)}
$$

$$
\text{(3.34)}
$$

$$
\text{(3.35)}
$$

The oxygenated compounds are thought to be formed by an alternative minor pathway which leads to a phenoxy radical and nitric oxide by rearrangement to phenyl nitrite and homolytic decomposition of this nitrite [Eq. (3.36)].

$$
\text{(3.36)}
$$

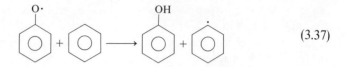

(3.37)

    At higher $PhNO_2/PhH$ ratios, various other minor products mainly formed by further phenylation are obtained. These two modes of fragmentation of nitrobenzene parallel those observed in the mass spectrum of nitrobenzene. Thus $[PhNO_2]^{+}$ decomposes both by loss of $\cdot NO_2$ and of $\cdot NO$ to form Ph^+ and PhO^+, although in both pyrolysis and mass spectrometry the formation of the phenyl species predominates. The mechanistic pathways of Eqs. (3.33)–(3.38) are largely supported by the results of experiments with $C_6H_5NO_2$ and C_6D_6,[23] and the phenylation of mono-, di-, and hexafluorobenzenes has also been studied.[51] There is some evidence for minor formation of benzyne by overall elimination of HNO_2 from nitrobenzene at 600°.[23]

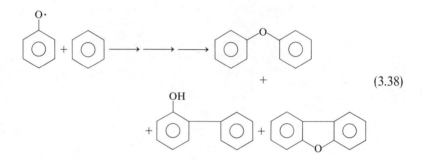

+

(3.38)

    This arylation has been extended[49] to the use of 1,3-dinitrobenzene with benzene (1:10) which gave 37% of m-terphenyl at 600°, and to 1,3,5-trinitrobenzene which gave mixtures of triphenylbenzene, nitroterphenyl, and dinitrobiphenyl. Benzene has been replaced as a trap for phenyl radicals by pyridine, and by thiophene; the ratios of isomeric phenylation products were fairly similar to those for radical phenylation with benzoyl peroxide in solution.

    Reactions with toluene[52] and with methyl benzoate[53] have also been reported. The latter gives, in addition to a mixture of phenylated benzoic esters, a substantial yield of phthalide by abstraction of hydrogen from the methoxyl group followed by cyclization [Eq. (3.39)]. Arylation of benzene with m-nitrotoluene or p-nitrotoluene occurs normally, but o-nitrotoluene showed very different behavior. It failed to arylate benzene satisfactorily at 600°, and on pyrolysis with methanol it gave a mixture of o-toluidine and N-methyltoluidine, o-cresol, and aniline and methyl anthranilate, the last

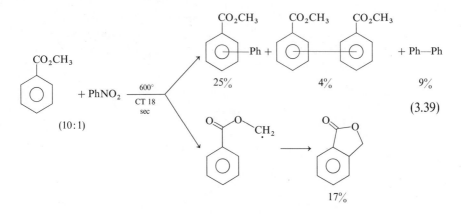

$$(3.39)$$

two being the major products. The detailed mechanism of this reaction is uncertain, but Fields and Meyerson[54] consider that catalysis by a surface acid–base site may be involved. The first species formed are probably **30** and **31**, or closely related charged species.

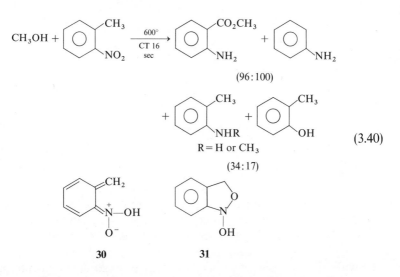

$$(3.40)$$

Other *o*-substituted nitroarenes also show interesting rearrangements. *o*-Dinitrobenzene on flash pyrolysis at 900° undergoes loss of both ·NO$_2$ and ·NO to give a diradical or oxocarbene **32** which then undergoes ring contraction to a ketene; Grützmacher and Hübner[24] were able to trap this with methanol [Eq. (3.41)]. Marty and de Mayo[22] found two pathways of decomposition of *o*-nitroanisole [Eq. (3.42)]. Loss of ·NO led to the phenoxy radical **33** and thus to the minor product, salicylaldehyde [compare Eq.

(3.17)], whereas loss of $\cdot NO_2$ gave the o-methoxyphenyl radical which rearranged to form benzaldehyde, the major product, via a postulated bicyclic radical **34**. Phenol, o-nitrophenol, and cyclopentadienone were also formed.

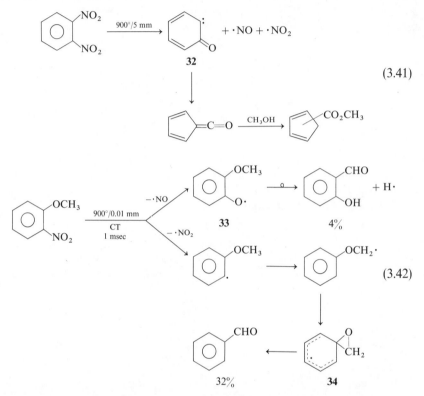

(3.41)

(3.42)

Flash vacuum pyrolysis of β-nitrostyrene and derivatives has been investigated by Kinstle and Stam.[55] β-Nitrostyrene itself forms benzaldehyde, polymeric material, and hydrogen cyanide at 600°, but the course of the reaction has been more fully examined in the case of β-methyl-β-nitrostyrene [Eq. (3.43)]. This gives benzaldehyde and other products which are consistent with the operation of a major pathway through the cyclic intermediate **35**. The pyrolysis of nitrotriphenylethylene at 600° is similar in that benzophenone, benzene, benzonitrile and phenyl isocyanate are obtained, in parallel with the pathways of Eq. (3.43), but 2,3-diphenylbenzofuran and an unidentified hydrocarbon $C_{20}H_{14}$ (probably 9-phenylphenanthrene) are also formed. These last two products probably result from the operation of two radical pathways [Eq. (3.44)] of the type already noted for the pyrolysis of nitroarenes.

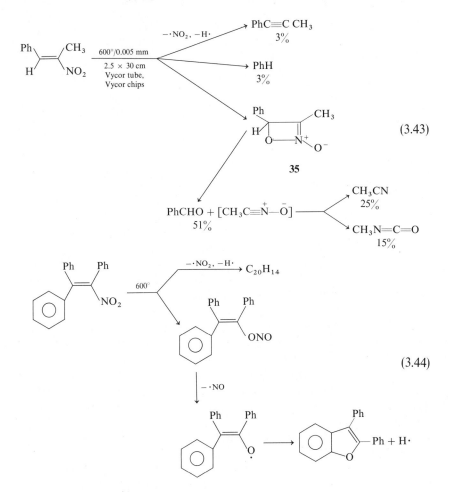

(3.43)

(3.44)

V. TRAPPING OF ARYNES FORMED AT HIGH TEMPERATURES

A. Formation of Benzyne from Indantrione and Phthalic Anhydride

Arynes can be generated in the gas phase at high temperatures from a variety of ordinarily stable precursors of the general structure **36**, in which X, Y, and Z represent groups which can lead to thermodynamically highly stable fragments such as CO, CO_2, N_2, and SO_2. The range of structures suitable for aryne generation is considered in Chapter 6; in this section we are mainly concerned with compounds such as indantrione and phthalic

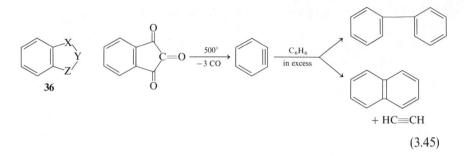

$$(3.45)$$

anhydride which decompose in flow systems at atmospheric pressure and with relatively long contact times at about 500° and 700°, respectively. The benzyne so formed reacts immediately with a large excess of an aromatic compound to produce products of apparent insertion, cycloaddition, and further transformation. Such experiments involve dropwise addition of a dilute (0.1 M) or very dilute solution of the benzyne precursor in the aromatic compound to a vertical packed pyrolysis tube under nitrogen, followed by analysis of the solution of products of trapping. Contact times are usually about 5–20 sec.

The first reported experiment of this kind was due to Brown and Solly[56] who obtained biphenyl, naphthalene, and acetylene from the pyrolysis of a very dilute solution of indantrione in benzene over Pyrex helices [Eq. (3.45)]. They confirmed the origin of naphthalene by replacing the benzene with chlorobenzene, which then gave 1-chloronaphthalene (13%) and 2-chloronaphthalene (39%). In the absence of benzene, pyrolysis of indantrione at 600° and 0.7 mm gave biphenylene (23%) and other products.[57] Almost simultaneously Fields and Meyerson[58] reported the first of a long series of experiments on the pyrolysis of solutions of aromatic anhydrides in aromatic hydrocarbons and heterocycles. A 0.1 M solution of phthalic anhydride in benzene pyrolyzed at 690° over Vycor beads gave biphenyl, naphthalene, and acetylene, with trace amounts of biphenylene, triphenylene, and fluorenone. Both groups were guided to the use of these precursors through consideration of their mass spectral fragmentations, and Fields and Meyerson have made extensive use of mass spectrometry and of coupled GC-MS in the analysis of the complex mixtures of products which are sometimes produced.

Further development of such work has depended entirely on the use of the more accessible aromatic anhydrides, and Fields and Meyerson have produced several valuable reviews dealing with the origins of the work and with the trapping of arynes and aromatic radicals in flow experiments,[59] with the formation and reactions of arynes at high temperatures,[60] and with the reactions of arynes with thiophene at high temperatures.[61] These reviews should be consulted for a detailed account of work in this area; the following sections merely outline a few of the key reactions and the more surprising results.

B.   Reactions of Benzyne with Benzene and Derivatives

In their 1965 communication Fields and Meyerson[58] showed that pyrolysis of solutions of phthalic anhydride at 690° gave benzyne, but that the reaction was possibly stepwise rather than concerted [Eq. (3.46)]. Their proposal of a dipolar intermediate **37a** or **37b** was based on the isolation of fluorenone and on an unexpectedly high ratio of biphenyl to naphthalene among the hydrocarbon products. The major pathway to biphenyl is probably by overall insertion of benzyne into a C—H bond of benzene; and naphthalene is formed by either 2 + 2 or 2 + 4 cycloaddition of benzyne to benzene, followed by elimination of acetylene [Eq. (3.47)]. It was proposed[58]

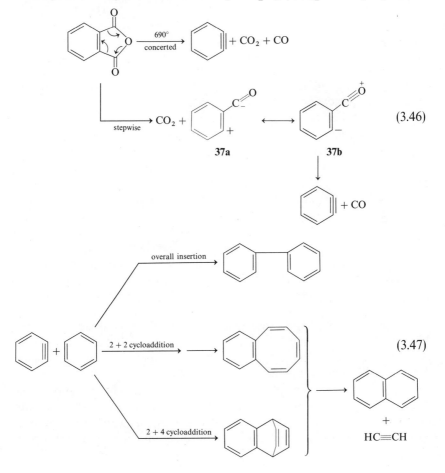

that additional biphenyl resulted from decarbonylation and hydrogen migration in a dipolar intermediate such as **38**, which can also cyclize and lose hydrogen to form fluorenone [Eq. (3.48)]. At 690° there is also some direct pyrolysis of benzene to biphenyl.

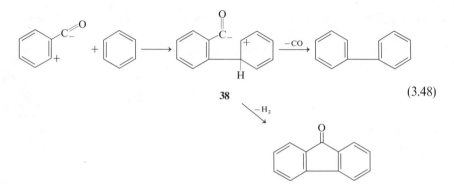

38   (3.48)

These reactions have been explored using benzene-d_1[62] or benzene-d_6[60,63] as the trapping agent, and the pathways of Eq. (3.47) have been substantially confirmed, although the interpretation of the results is complicated by migration and exchange of hydrogen and deuterium. This is serious at 690° with contact times as long as 20 sec; even with a contact time of 4.2 sec the major naphthalene-d_4 formed from benzyne and benzene-d_6 was accompanied by an almost equal amount of naphthalene-d_5 and some naphthalene-d_6, -d_7, and -d_8. Friedman and Lindow[63] reexamined the reaction between phthalic anhydride and benzene, and benzene-d_6, at 690°; they identified the additional products acenaphthene and acenaphthylene, which are thought to be formed from benzocyclooctatetraene. They also discovered conditions in which the cycloaddition of benzyne to benzene-d_6 led almost exclusively to naphthalene-d_4 (86% of the naphthalene fraction). This result was obtained by the use of benzene-d_6 accidentally contaminated with unlabeled tetrahydrofuran, which apparently served as a hydrogen donor to suppress deuterium exchange reactions which were leading to more highly deuterated naphthalenes.

The pyrolysis of phthalic anhydride with hexafluorobenzene is not complicated by such exchange reactions;[64] the only significant fluorinated hydrocarbons obtained were 1,2,3,4-tetrafluoronaphthalene and 2,2′,3,4,5,6-hexafluorobiphenyl in the ratio 1:5 [Eq. (3.49)].

Benzyne derived from phthalic anhydride has also been trapped with various chloro-,[65] dichloro-,[66] and chlorofluorobenzenes.[60] An interesting point which emerged from this investigation is that chlorobenzene itself gives small amounts of naphthalene, chloronaphthalenes, and biphenylene

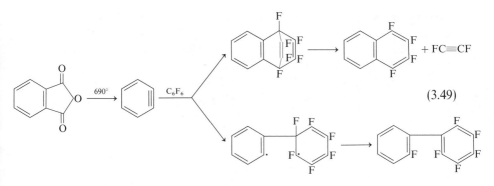

(3.49)

on pyrolysis at 690°,[65] and these appear to be produced by elimination of hydrogen chloride to form benzyne, followed by attack of this benzyne on chlorobenzene, or by its dimerization. Pyrolysis of phthalic anhydride with pyridine (1:10) at 690°[67] gave products exactly analogous to the major products from benzene, and included phenylpyridines, naphthalene, quinoline, isoquinoline, hydrogen cyanide, and acetylene.

C. Probable Formation of Benzyne by Pyrolysis of Acetylene

A remarkable discovery[64] which arose from the investigation of the pyrolysis of phthalic anhydride is that pyrolysis of acetylene at 690° with a contact time of 4–5 sec leads to a mixture of hydrocarbons very similar to that produced under the same conditions from phthalic anhydride. Fields and Meyerson[60] concluded that acetylene on pyrolysis forms benzyne or a closely related "benzynoid" precursor. Pyrolysis of acetylene with hexafluorobenzene gave tetrafluoronaphthalene and hexafluorobiphenyl in the ratio 4:1, whereas benzyne from phthalic anhydride gave the same products in the ratio 1:5 [see Eq. (3.49)]. Acetylene-d_2 with hexafluorobenzene gave tetrafluoronaphthalene-d_4 and hexafluorobiphenyl-d_4 (9:1) together with some tetrafluoroanthracene-d_6, in accordance with reactions expected of benzyne-d_4 and diacetylene-d_2.

The study of the pyrolysis of acetylene to mixtures of aromatic hydrocarbons has a long history which is summarized by Fields and Meyerson.[60] They speculate that acetylene is first converted into diacetylene, and that benzyne is formed by a 2 + 4 cycloaddition of acetylene to diacetylene [Eq. (3.50)]. The formation of tetrafluoroanthracene from acetylene and hexafluorobenzene is attributed to the further addition of diacetylene to

$$3\ HC\equiv CH \xrightarrow{\ 690°\ } H_2 + \quad\quad \longrightarrow \quad\quad \longleftrightarrow \quad\quad \tag{3.50}$$

benzyne, shown in Eq. (3.51). It is obviously extremely difficult to obtain direct evidence for the early stages of these processes under the normal conditions, and several points of uncertainty remain. The more recent discovery[68] that acetylene is in equilibrium with $CH_2{=}C\colon$ at high temperatures suggests further possibilities for this trimerization, but the nature of the dehydrogenation step which ultimately leads to benzyne rather than benzene is then still obscure.

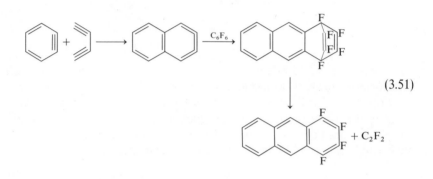

$$(3.51)$$

D.   Reactions of Arynes with Thiophenes

The reactions of benzyne generated from phthalic anhydride with thiophene[60,61] are more complicated than those involving benzene or pyridine, and only the major features will be mentioned here. The pyrolysis of phthalic anhydride with thiophene (1:20) at 690° [61,69] gave naphthalene (the major product), benzothiophene, phenylthiophenes, and bithienyl, the last compound being derived from thiophene alone. These products are probably formed as shown in Eq. (3.52). In addition there were formed minor products such as thiophthene (39), benzothiophthene (40), and naphthothiophene which can best be explained by assuming the formation of 2,3-thiophyne in the pyrolysis with subsequent reaction as summarized in Eq. (3.53).

2,3-Thiophyne is considered to be formed by two routes. The first requires hydrogen transfer to benzyne from thiophene [Eq. (3.54)], and the second direct dehydrogenation of thiophene [Eq. (3.55)]. More recently, Reinecke and Newsom[70] have generated 2,3-thiophyne by flash vacuum pyrolysis of thiophene-2,3-dicarboxylic anhydride in the presence of trapping agents including thiophene and hydrogen. The major products from attack of 2,3-thiophyne on thiophene at 500° were benzothiophene (59%) and sulfur [Eq. (3.56)]; the formation of minor products such as thiophthene (39) under these conditions was not reported. In the presence of hydrogen thiophene was obtained, presumably by the reversal of Eq. (3.55).

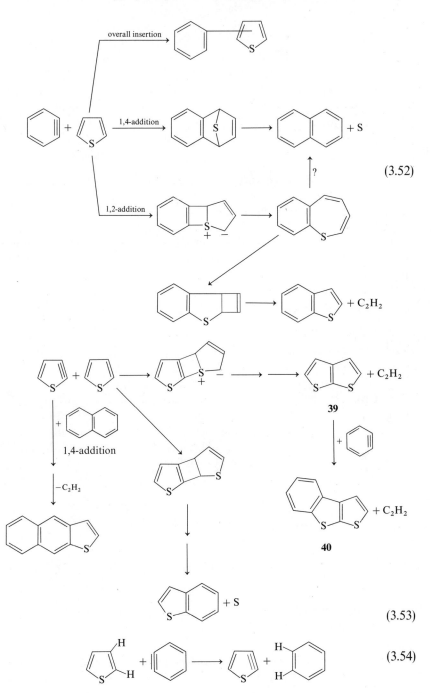

(3.52)

39

40

(3.53)

(3.54)

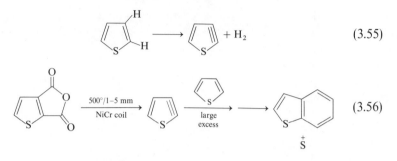

Reactions of thiophene with the arynes or dehydroaromatics generated from tetrachlorophthalic anhydride, benzene-1,2:4,5-tetracarboxylic bisanhydride and its 1,2:3,4 isomer, naphthalene 2,3- and 1,8-dicarboxylic anhydrides, and pyridine-2,3-dicarboxylic anhydride have also been reported,[61] as has the reaction of benzothiophene with benzyne. The products of most of these reactions are consistent with pathways similar to those shown in Eqs. (3.47) and (3.52).

E.   Generation of Arynes from Other Anhydrides

Pyridine-2,3-dicarboxylic anhydride copyrolyzed with benzene at 690° gave 14% of insertion and addition products of 2,3-pyridyne including phenylpyridines and quinoline.[67] Pyrazine-2,3-dicarboxylic anhydride with benzene similarly gave a very low yield of 2-phenylpyrazine, which suggests that pyrazyne itself may have a fleeting existence before rearranging[71] to maleodinitrile and fumarodinitrile [see Chapter 5, Section II,E, Eq. (5.27)].

Fields and Meyerson[72] have suggested that pyrolysis of 4-nitrophthalic anhydride (41) generates the benzynyl radical (42), based on an examination of the products of copyrolysis of 41 with benzene.

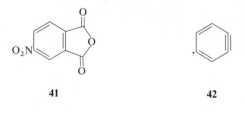

41   42

VI.   PYROLYTIC GENERATION OF CYCLOBUTADIENE AND BENZOCYCLOBUTADIENE

The long struggle to generate and detect free cyclobutadiene, reviewed by Cava and Mitchell,[73] ended with the work of Pettit's group[74] on its generation by oxidation of the iron tricarbonyl complex and with its spectroscopic characterization in an argon matrix by Chapman *et al.*[75,76] and by Lin and Krantz.[77] The latter experiments employed photolysis of photo-α-pyrone (**44**) in argon matrices. The status of the vexed problem of the electronic structure and geometry of ground-state cyclobutadiene has been summarized briefly by Dewar and Komornicki.[78]

Flash vacuum pyrolysis of cyclobutadiene iron tricarbonyl appears not to give significant amounts of free cyclobutadiene. Instead Hedaya and co-workers[79] obtained a complex mixture of rearranged and dimeric products [Eq. (3.57)]. This result can be compared with the similar failure[80] to produce the elusive isomeric species methylenecyclopropene by pyrolysis of Feist's anhydride (**43**) [Eq. (3.58)]. Free cyclobutadiene was, however, produced on flash vacuum pyrolysis of photo-α-pyrone (**44**),[81] and the *syn*-dimer **45** could be isolated from the pyrolyzate, which also contained cycloocta-tetraene, dihydropentalene, benzene, and a trace of furan. Cyclobutadiene was also characterized in the gas phase by titration with methyl or allyl radicals to give new species formed by addition of one and two such radicals; these were detected by mass spectrometry.

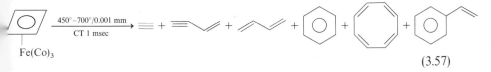

$$(3.57)$$

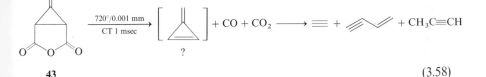

$$(3.58)$$

$$(3.59)$$

Chapman and co-workers[82] have isolated benzocyclobutadiene in an argon matrix at 8°K for spectroscopic study. The benzocyclobutadiene was produced by passing the vapor of *cis*-1,2-diiodobenzocyclobutene at 10^{-6} mm over zinc heated at 230°, with removal of dimeric products and *trans*-diiodo compound by condensation at 5°.

REFERENCES

1. Hedaya, E. (1969). *Acc. Chem. Res.* **2**, 367.
2. Koenig, T., Snell, W., and Chang, J. C. (1976). *Tetrahedron Lett.*, p. 4569.
3. Lossing, F. P., and Traeger, J. C. (1975). *J. Am. Chem. Soc.* **97**, 1579.
4. Paneth, F., and Hofeditz, W. (1929). *Ber. Dtsche. Chem. Ges.* *62B*, 1335.
5. Hedaya, E. (1971). "23rd International Congress of Pure and Applied Chemistry," Vol. 4, "Special Lectures Presented at Boston, U.S.A., 26–30th July, 1971," p. 195. Butterworth, London.
6. Weltner, W. (1967). *Science* **155**, 155.
7. Hedaya, E., McNeil, D. W., Schissel, P., and McAdoo, D. J. (1968). *J. Am. Chem. Soc.* **90**, 5284.
8. Kasai, P. H., McLeod, D., and McBay, H. C. (1974). *J. Am. Chem. Soc.* **96**, 6864.
9. Kochi, J. K. (1973). In "Free Radicals" (J. K. Kochi, ed.), Vol. 2, p. 665. Wiley (Interscience), New York.
10. Adam, W., and Arce, J. (1975). *J. Am. Chem. Soc.* **97**, 926.
11. Martin, M. M., Hammer, F. T., and Zador, E. (1973). *J. Org. Chem.* **38**, 3422.
12. Moore, C. G. (1951). *J. Chem. Soc.*, p. 234.
13. Haynes, R. K., Probert, M. K. S., and Wilmot, I. D. (1978). *Aust. J. Chem.* **31**, 1737.
14. Steacie, E. W. R. (1959). "Atomic and Free Radical Reactions," p. 239. Van Nostrand–Reinhold, Princeton, New Jersey.
15. Gray, P., and Williams, A. (1959). *Chem. Rev.* **59**, 239.
16. Nussbaum, A. L., and Robinson, C. H. (1962). *Tetrahedron* **17**, 35.
17. Barton, D. H. R., Beaton, J. M., Geller, L. E., and Pechet, M. M. (1960). *J. Am. Chem. Soc.* **82**, 2640.
18. Barton, D. H. R., Beaton, J. M., Geller, L. E., and Pechet, M. M. (1961). *J. Am. Chem. Soc.* **83**, 4076.
19. Barton, D. H. R., Ramsay, G. C., and Wege, D. (1967). *J. Chem. Soc. (C)*, p. 1915.
20. Harrison, A. G., Honnen, L. R., Dauben, H. J., and Lossing, F. P. (1960). *J. Am. Chem. Soc.* **82**, 5593.
21. Freidlin, L. Kh., Balandin, A. A., and Nazarova, N. M. (1949). *Izvest. Akad. Nauk. SSSR, Otdel. Khim. Nauk.*, p. 102.
21a. Freidlin, L. Kh., Balandin, A. A., and Nazarova, N. M. (1949). *Chem. Abstr.* **43**, 5758i.
22. Marty, R. A., and de Mayo, P. (1971). *J. Chem. Soc. Chem. Commun.*, p. 127.
23. Fields, E. K., and Meyerson, S. (1967). *J. Am. Chem. Soc.* **89**, 3224.
24. Grützmacher, H. F., and Hübner, J. (1971). *Tetrahedron Lett.*, p. 1455.
25. Schirmer, U., and Conia, J. M. (1974). *Tetrahedron Lett.*, p. 3057.
26. Kice, J. L. (1973). In "Free Radicals" (J. K. Kochi, ed.), Vol. 2, p. 711. Wiley (Interscience), New York.
27. Ando, W., Oikawa, T., Kishi, K., Saiki, T., and Migita, T. (1975). *J. Chem. Soc. Chem. Commun.*, p. 704.
28. Klemm, L. H., Shabtai, J., McCoy, D. R., and Kiang, W. K. T. (1968). *J. Heterocycl. Chem.* **5**, 883.
29. Bird, K. J., Chan, A. W. K., and Crow, W. D. (1976). *Aust. J. Chem.* **29**, 2281.
30. Crow, W. D., and Khan, A. N. (1976). *Aust. J. Chem.* **29**, 2289.
31. Crow, W. D., McNab, H., and Philip, J. M. (1976). *Aust. J. Chem.* **29**, 2299.
32. Allen, A. O., and Rice, R. K. (1935). *J. Am. Chem. Soc.* **57**, 310.
33. Müller, E. (1967). In "Methoden der Organischen Chemie (Houben-Weyl)" (R. Stroh, ed.), Vol. 10, Part 2, p. 790. Thieme, Stuttgart.

34. Mackenzie, K. (1975). *In* "The Chemistry of the Hydrazo, Azo, and Azoxy Groups" (S. Patai, ed.), p. 329. Wiley, New York.
35. Braslavsky, S., and Heicklen, J. (1977). *Chem. Rev.* **77**, 473.
36. McNeil, D. W., Kent, M. E., Hedaya, E., D'Angelo, P. F., and Schissel, P. O. (1971). *J. Am. Chem. Soc.* **93**, 3817.
37. Busfield, W. K., and Ivin, K. J. (1961). *Trans. Farady Soc.* **57**, 1044.
38. Bushby, R. J., and Pollard, M. D. (1977). *Tetrahedron Lett.*, p. 3671.
39. Bushby, R. J. (1975). *J. Chem. Soc. Perkin Trans. 1*, p. 2513.
40. Trahanovsky, W. S., Lawson, J. A., and Zabel, D. E. (1967). *J. Org. Chem.* **32**, 2287.
41. Louw, R. (1971). *Rec. Trav. Chim. Pays-Bas* **90**, 469.
42. Louw, R., van den Brink, M., and Vermeeren, H. P. W. (1973). *J. Chem. Soc. Perkin Trans. 2*, p. 1327.
43. Trahanovsky, W. S., Ong, C. C., and Lawson, J. A. (1968). *J. Am. Chem. Soc.* **90**, 2839.
44. Trahanovsky, W. S., and Ong, C. C. (1970). *J. Am. Chem. Soc.* **92**, 7174.
45. Lehr, R. E., and Wilson, J. M. (1971). *J. Chem. Soc. Chem. Commun.*, p. 666.
46. Schirmann, J. P., and Weiss, F. (1967). *Tetrahedron Lett.*, p. 5163.
47. Crawford, R. J., and Cameron, D. M. (1966). *J. Am. Chem. Soc.* **88**, 2589.
48. Fields, E. K., and Meyerson, S. (1967). *Chem. Commun.*, p. 494.
49. Fields, E. K., and Meyerson, S. (1967). *J. Am. Chem. Soc.* **89**, 724.
50. Smith, R. E. (1940). *Trans. Faraday Soc.* **36**, 983.
51. Fields, E. K., and Meyerson, S. (1967). *J. Org. Chem.* **32**, 3114.
52. Fields, E. K., and Meyerson, S. (1968). *J. Org. Chem.* **33**, 2315.
53. Feinstein, A. I., and Fields, E. K. (1972). *J. Org. Chem.* **37**, 118.
54. Fields, E. K., and Meyerson, S. (1968). *J. Org. Chem.* **33**, 4487.
55. Kinstle, T. H., and Stam, J. G. (1970). *J. Org. Chem.* **35**, 1771.
56. Brown, R. F. C., and Solly, R. K. (1965). *Chem. Ind. (London)*, p. 181.
57. Brown, R. F. C., and Solly, R. K. (1965). *Chem. Ind. (London)*, p. 1462.
58. Fields, E. K., and Meyerson, S. (1965). *Chem. Commun.*, p. 474.
59. Fields, E. K., and Meyerson, S. (1969). *Acc. Chem. Res. 2*, 273.
60. Fields, E. K., and Meyerson, S. (1968). *In* "Advances in Physical Organic Chemistry" (V. Gold, ed.), Vol. 6, p. 1. Academic Press, New York.
61. Fields, E. K., and Meyerson, S. (1967). *In* "Organosulfur Chemistry" (M. J. Janssen, ed.), p. 143. Wiley (Interscience), New York.
62. Fields, E. K., and Meyerson, S. (1966). *J. Am. Chem. Soc.* **88**, 21.
63. Friedman, L., and Lindow, D. F. (1968). *J. Am. Chem. Soc.* **90**, 2329.
64. Fields, E. K., and Meyerson, S. (1967). *Tetrahedron Lett.*, p. 571.
65. Fields, E. K., and Meyerson, S. (1966). *J. Am. Chem. Soc.* **88**, 3388.
66. Meyerson, S., and Fields, E. K. (1966). *Chem. Ind. (London)*, p. 1230.
67. Fields, E. K., and Meyerson, S. (1966). *J. Org. Chem.* **31**, 3307.
68. Brown, R. F. C., Eastwood, F. W., and Jackman, G. P. (1978). *Aust. J. Chem.* **31**, 579.
69. Fields, E. K., and Meyerson, S. (1966). *Chem. Commun*, p. 708.
70. Reinecke, M. G., and Newsom, J. G. (1976). *J. Am. Chem. Soc.* **98**, 3021.
71. Brown, R. F. C., Crow, W. D., and Solly, R. K. (1966). *Chem. Ind. (London)*, p. 343.
72. Fields, E. K., and Meyerson, S. (1971). *Tetrahedron Lett.*, p. 719.
73. Cava, M. P., and Mitchell, M. J. (1967). "Cyclobutadiene and Related Compounds." Academic Press, New York.
74. Watts, L., Fitzpatrick, J. D., and Pettit, R. (1965). *J. Am. Chem. Soc.* **87**, 3253.
75. Chapman, O. L., McIntosh, C. L., and Pacansky, J. (1973). *J. Am. Chem. Soc.* **95**, 614.
76. Chapman, O. L., De La Cruz, D., Roth, R., and Pacansky, J. (1973). *J. Am. Chem. Soc.* **95**, 1337.

77. Lin, C. Y., and Krantz, A. (1972). *J. Chem. Soc. Chem. Commun.*, p. 1111.
78. Dewar, M. J. S., and Komornicki, A. (1977). *J. Am. Chem. Soc.* **99**, 6174.
79. Hedaya, E., Krull, I. S., Miller, R. D., Kent, M. E., D'Angelo, P. F., and Schissel, P. (1969). *J. Am. Chem. Soc.* **91**, 6880.
80. Krull, I. S., D'Angelo, P. F., Arnold, D. R., Hedaya, E., and Schissel, P. O. (1971). *Tetrahedron Lett.*, p. 771.
81. Hedaya, E., Miller, R. D., McNeil, D. W., D'Angelo, P. F., and Schissel, P. (1969). *J. Am. Chem. Soc.* **91**, 1875.
82. Chapman, O. L., Chang, C. C., and Rosenquist, N. R. (1976). *J. Am. Chem. Soc.* **98**, 261.

Chapter 4

Elimination Reactions

I. INTRODUCTION

In this chapter, we consider a variety of reactions which are formally related because they involve elimination of a small fragment molecule X—X or X—Y from a larger molecular framework. These reactions include thermal dehydrogenations with elimination of H—H and dealkylations with elimination of H—R, as well as reactions involving loss of H—OR, H—Hal, and related species which are conventionally regarded as eliminations. Although some reactions with similar mechanisms are grouped together, the arrangement is primarily by formal reaction type rather than by mechanism, so that radical and radical chain reactions appear beside concerted molecular reactions. The coverage of particular types of reaction may seem to be, perversely, in proportion inverse to their synthetic and general importance. The reason for this will be obvious in the case of ester pyrolysis, for which several very detailed and comprehensive reviews are already available, and which is firmly incorporated into the mainstream of organic chemistry in most advanced textbooks.

II. DEHYDROGENATIONS AND DEALKYLATIONS

A. Dehydrogenations

Dehydrogenations are quite common in high-temperature pyrolyses, particularly as the final step in a sequence leading to the formation of an aromatic system, and many examples without special comment will be found throughout this book. The mechanism of dehydrogenation and the precise structure of the penultimate species have not in most cases been established.

Two typical mechanistic possibilities are shown by the kinetic work of Ellis and Frey[1] on the decomposition of the 1,3- and 1,4-cyclohexadienes.

1,3-Cyclohexadiene decomposes at 440° by a radical chain mechanism to give benzene and some cyclohexene; the initial step is considered to be homolytic fission to give a cyclohexadienyl radical [Eq. (4.1)], and the reaction is partly inhibited by added propene. By contrast 1,4-cyclohexadiene decomposes smoothly at much lower temperatures, 300°–350°, to give benzene and hydrogen in a molecular process which is not inhibited by additives. This process was recognized,[2] and confirmed by the deuterium-labeling studies of Fleming and Wildsmith,[3] as the concerted symmetry-allowed 1,4-elimination of hydrogen [Eq. (4.2)]. Frey and Walsh[4] have reviewed similar 1,4-eliminations in cyclopentene, 2,5-dihydrofuran, and derivatives of 1,4-cyclohexadiene, and there is more recent work on the reversible hydrogenation of cyclopentadiene[5] and dehydrogenation of labeled cyclopentene.[6]

The difference between 1,2-dihydro- and 1,4-dihydro systems persists in high-temperature flash pyrolyses, as shown by the results of Brown *et al.*[7] in the 1-methylnaphthalene series. The 5,8-dihydro compound was smoothly dehydrogenated to 1-methylnaphthalene [Eq. (4.3)], whereas the 1,2-dihydro compound gave a complex mixture of products containing in addition to the products shown in Eq. (4.4) 2-methylnaphthalene, indene, methylindenes, and benzofulvene. Similar radical processes are evident in the pyrolysis of 2-ethyl-5-methyl-1-pyrroline [Eq. (4.5)] studied by McDaniel and Benisch.[8]

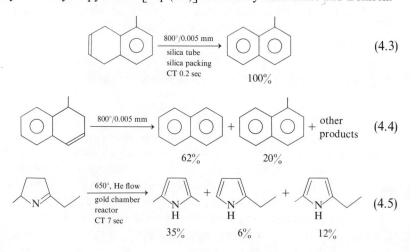

1,4-Elimination of hydrogen probably occurs in the conversion of intermediate methylcyanocyclopentadienes into benzonitrile, a major product of pyrolysis of the methylbenzotriazoles,[9] *o*-tolyl azide, and oxindole[10] at high temperatures [Eq. (4.6)]. Rearrangements similar to the postulated cyanofulvene → benzonitrile step are discussed in Chapter 9, Section IV,F. A formal 1,6-elimination of hydrogen is required for the pyrolytic generation of *p*-xylylene from *p*-xylene, although the key step in its formation is probably the disproportionation of *p*-methylbenzyl radicals[11] [Eq. (4.7)]. Errede and Lindrum[11] were able to prepare 0.12 M solutions of *p*-xylylene in toluene by pyrolysis of *p*-xylene at 1000° and 4 mm with a contact time of 0.008 sec.

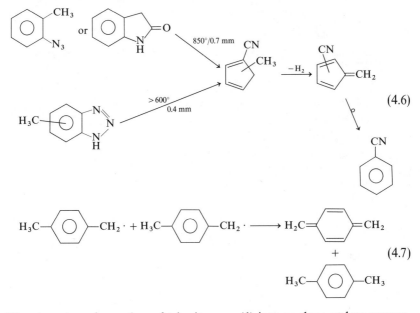

(4.6)

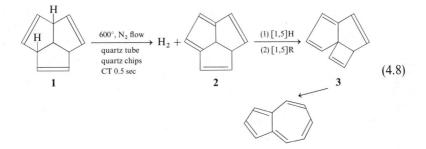

(4.7)

The clean transformation of triquinacene (**1**) into azulene at low conversion [Eq. (4.8)] requires both dehydrogenation and rearrangement. Scott

(4.8)

and Agopian[12] have suggested that dehydrogenation occurs by 1,4-elimination in triquinacene itself, and that sigmatropic rearrangements in the dehydrotriquinacene **2** followed by electrocyclic ring opening of the fused cyclobutene **3** then leads to azulene. At 700° a competing diradical process gives 1,2-dihydronaphthalene, naphthalene, and indene.

At very high temperatures, simple molecules may undergo 1,2-elimination of hydrogen. At temperatures approaching the softening point of quartz, methylamine can be dehydrogenated to methanimine[13] [Eq. (4.9)]. The microwave spectrum of methanimine generated in this way has been used for the identification of signals from methanimine in interstellar gas clouds.[14]

$$CH_3NH_2 \xrightarrow[\text{4 × 20 mm quartz tube}]{1000°-1500°/0.03 \text{ mm}} CH_2{=}NH + H_2 \qquad (4.9)$$

B. Dealkylations

The loss of methane by an apparent 1,2-elimination of H and CH_3 occurs readily in the high-temperature pyrolysis of highly branched compounds, and radical or radical chain processes are probably involved. In the presence of hydrogen donors the initial radical may abstract hydrogen to give a product formed by overall loss of CH_2. Processes of this kind are of some importance in steroid technology for the aromatization of ring A dienones, and a range of examples has been summarized by Djerassi and his students.[15] A favorable example,[16] the formation of Δ^6-dehydroestrone, is shown in Eq. (4.10).

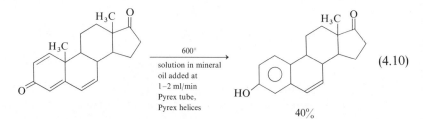

The conversion of *gem*-dimethyl compounds into exocyclic or terminal methylene compounds appears to be a standard pyrolytic process, although the exocyclic methylene compounds may undergo subsequent rearrangement. Equation (4.11) shows the composition of the pyrolyzate from 1,1-dimethylindene in which the major component, naphthalene, is probably formed by ring expansion of benzofulvene.[7] 3,3-Dimethyloxindole similarly gives 2-quinolone on pyrolysis at 850° [Eq. (4.12)].[17] King and Goddard[18] have studied the pyrolysis of isopropyl cyanide by the VLPP technique (see Chapter 2, Section II,D). At 978° with a collision number $Z = 19,550$ and a

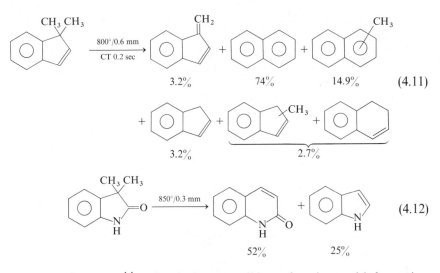

(4.11)

(4.12)

flow rate of 9.1×10^{14} molecules/sec overall loss of methane with formation of acrylonitrile predominated over loss of HCN [Eq. (4.13)]. The same paper gives a useful list of references to the pyrolysis of alkyl cyanides.

$$(CH_3)_2CHCN \xrightarrow{978°/10^{-4}-10^{-3}\ mm} \begin{array}{l} CH_2{=}CHCH_3 + HCN \\ 5.2\% \\[1em] CH_2{=}CHCN + CH_4 \\ 60.5\% \end{array}$$

(4.13)

Baron and DeCamp[19] have postulated formation of an arylcarbene by formal α-elimination of methane in the pyrolysis of 4-ethyltoluene [Eq. (4.14)]. The intermediacy of *p*-tolylcarbene was deduced from the isolation of styrene, a characteristic product of its rearrangement [see Chapter 5,

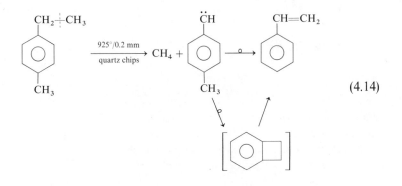

(4.14)

Section III,C, Eq. (5.42)], from the complex pyrolyzate. Both benzo-cyclobutene and styrene have been isolated from the pyrolysis of p-CH$_3$C$_6$H$_4$CH$_2$CO$_2$CH$_3$ at 810°–875°, a reaction which also must generate initial p-CH$_3$C$_6$H$_4$CH$_2$· and CH$_3$· radicals.[19] Tertiary alkyl radicals can be lost with ease from hindered benzylic positions, as in the dealkylation and rearrangement of the dihydrophosphaanthracene **4** to 10-phenyl-9-phosphaanthracene (**5**).[20]

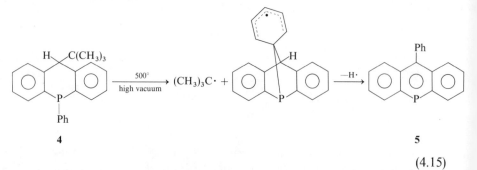

(4.15)

Methyl groups attached to heteroatoms can also be eliminated at high temperatures, as shown by the work of Hopkinson *et al.*[21] on the generation of phosphaethene and phosphaethyne for microwave spectroscopy by pyrolysis of dimethylphosphine [Eq. (4.16)]. In the nitrogen series the formation of methylaminoacetonitrile from tetramethylethylenediamine or from dimethylaminoacetonitrile [Eq. (4.17)] is closely related.[22]

$$(\text{CH}_3)_2\text{PH} \xrightarrow{\text{1000°/0.01–0.03 mm}} \text{CH}_4 + \text{CH}_2{=}\text{PH} \longrightarrow \text{H}_2 + \text{HC}{\equiv}\text{P} \quad (4.16)$$

$$(\text{CH}_3)_2\text{NCH}_2\text{CH}_2\text{N}(\text{CH}_3)_2 \xrightarrow{\text{820°/0.01 mm}} \text{CH}_3\text{NHCH}_2\text{C}{\equiv}\text{N} + \text{many other products} \quad (4.17)$$

$$(\text{CH}_3)_2\text{NCH}_2\text{CN} \xrightarrow{\text{820°/0.01 mm}}$$

III. α-ELIMINATIONS

A. α-Eliminations in Carboxylic Esters and Orthoesters

α-Elimination of a carboxylic acid from an ester in which normal β-elimination is not structurally possible may occur with rearrangement or through formation of an intermediate carbene, although temperatures somewhat higher than those used for typical β-eliminations are often required. Kwart and Hoster[23] found that gas-phase pyrolysis of neophyl acetate (**6**) at 600° gave a mixture of the rearranged alkenes **7** and **8** [Eq. (4.18)], and Kwart and Ling[24] later interpreted the formation of the conjugated alkene **10** from deuterated neophyl methylcarbonate as involving the five-membered transition state **9**, leading to an α-elimination concerted with rearrangement [Eq. (4.19)].

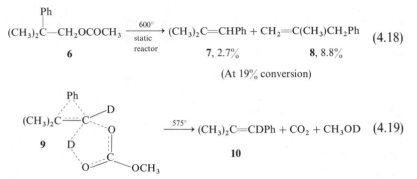

$$(CH_3)_2\overset{\overset{\displaystyle Ph}{|}}{C}\text{---}CH_2OCOCH_3 \xrightarrow[\text{reactor}]{\overset{600°}{\text{static}}} (CH_3)_2C\text{==}CHPh + CH_2\text{==}C(CH_3)CH_2Ph \quad (4.18)$$

$$\underset{\textbf{6}}{} \qquad \underset{\textbf{7}, 2.7\%}{} \qquad \underset{\textbf{8}, 8.8\%}{}$$

(At 19% conversion)

$$\xrightarrow{575°} (CH_3)_2C\text{==}CDPh + CO_2 + CH_3OD \quad (4.19)$$

9   **10**

Lehr and Wilson[25] in a study of the pyrolysis of the deuterated bis-(diarylmethyl) oxalate **11** found six products derived from major radical processes, and a seventh minor product, undeuterated 2-methylfluorene, which they showed must be formed by α-elimination and subsequent re-arrangement of the intermediate carbene **12** [Eq. (4.20)]. Such rearrange-ments of diarylcarbenes are very well documented [see Chapter 5, Section III,C, Eq. (5.51)]. α-Elimination in diarylmethyl carboxylates requires a high temperature; diphenylmethyl acetate does not eliminate acetic acid at 450°–460° and 10^{-4} mm.[26]

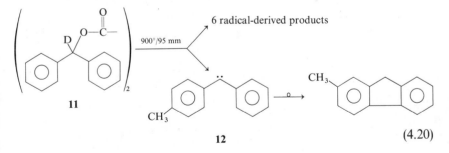

6 radical-derived products

900°/95 mm

11   **12**   (4.20)

Oele and Louw[27] have studied the gas-phase thermolysis of a number of derivatives of methyl acetate bearing alkoxy, phenoxy, or methylthio sub-stituents in the methoxyl group, using a microreactor GLC combination, and have found α-elimination of acetic acid to give alkoxycarbenes. A typical example is shown in Eq. (4.21). These findings are related to the sequence introduced by Eastwood and Crank[28] for the conversion of 1,2-diols to

$$\text{(4.21)}$$

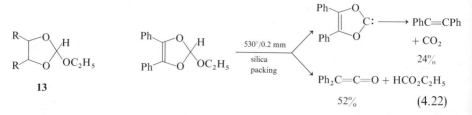

alkenes by heating the derived 2-ethoxy-1,3-dioxolanes **13** with a trace of a carboxylic acid. An example of a similar decomposition of a dioxolene in the gas phase[29] is shown in Eq. (4.22).

13

$$\text{(4.22)}$$

Oele and Louw[27] comment that their reactions are assisted by the special stabilization of the alkoxy- and dialkoxycarbenes first formed, and this is also true of the following preparatively more interesting examples. Trahanovsky and Park[30] discovered that furfuryl benzoate on pyrolysis at $640°–700°$ gives methylenecyclobutenone in 40% yield. A deuterium-labeling study[31] has shown that the product is *not* formed by direct α-elimination [Eq. (4.23)] but by α-elimination in 2-benzoyloxy-5-methylene-2,5-dihydrofuran **(14)** produced by successive 3,3 migrations of the benzoate group [Eq. (4.24)].

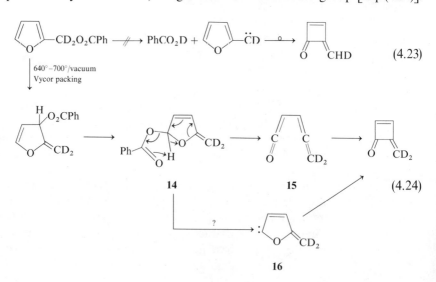

14   **15**   (4.24)

16

The authors[31] propose a concerted α-elimination and ring contraction of an intermediate ketene **15**, but formation and ring contraction of an intermediate carbene **16** must also be possible. 5-Methylfurfuryl benzoate gives 2,5-dimethylene-2,5-dihydrofuran by similar successive rearrangements followed by β-elimination of benzoic acid. In a further paper Trahanovsky and and Alexander[31a] have examined the pyrolysis of α-phenylfurfuryl acetate and have found both a direct α-elimination facilitated by the phenyl group, and the pathway involving initial migration of the acetate group.

The formation of cyclic 1,2-diketones by pyrolysis of benzoyloxylactones studied by Brown and co-workers[32] is clearly a closely related reaction. The benzoate **17** gives acenaphthenequinone in reasonable yield on flash pyrolysis at 560°, and in this case the final step is best formulated as ring contraction of an intermediate acyloxycarbene [Eq. (4.25)]. 3-Benzoyloxyphthalide similarly gives benzocyclobutenedione (33%).

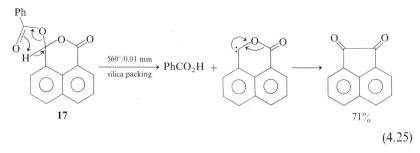

$$\text{(4.25)}$$

B. α-Eliminations in Alkyl Halides and Polyhalides and in Silanes

Reactions which might be interpreted as involving α-elimination of a hydrogen halide to give a carbene are occasionally observed in the pyrolysis of alkyl halides, but such reactions have usually been regarded as having polar or radical character. The formation of dimethylcyclopropane at low conversion in the pyrolysis of neopentyl chloride [Eq. (4.26)] is a case in point.[33] The useful synthesis of fumaronitrile and maleonitrile by high-temperature flow pyrolysis of bromo- or chloroacetonitrile, due to Hashimoto and co-workers,[34] is another ambiguous case [Eq. (4.27)]. There is evidence for the formation of cyanocarbene (**18**), based on the formation of benzonitrile in the presence of cyclopentadiene, but this does not really exclude

$$\underset{\underset{H_3C}{}{\overset{H_3C}{}}}{} \xrightarrow{444°} + \text{methylbutenes}$$

43% yield
at 5% conversion

$$\text{(4.26)}$$

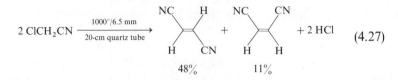

$$2 \ ClCH_2CN \xrightarrow[\text{20-cm quartz tube}]{1000°/6.5 \ mm} \quad \text{(structures)} \quad + 2 \ HCl \quad (4.27)$$

the possibility of a radical chain mechanism for the reaction of Eq. (4.27), with the coupling step as in Eq. (4.28).

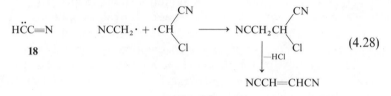

$$\overset{..}{H}C{\equiv}N \qquad NCCH_2{\cdot} + {\cdot}CH \overset{CN}{\underset{Cl}{}} \longrightarrow NCCH_2CH \overset{CN}{\underset{Cl}{}} \qquad (4.28)$$

18

$$NCCH{=}CHCN$$

The formation of carbenes by α-elimination is much better established in the case of the haloforms and related compounds, in which α-elimination of hydrogen halides occurs readily at 400°–600° and the dihalocarbenes thus generated can be employed in synthesis. The decomposition of chloroform in a stream of helium at 450°–525° gives mainly tetrachloroethylene and hydrogen chloride; Semeluk and Bernstein[35,36] favored a radical chain mechanism. Engelsma,[37] however, has summarized the state of kinetic and mechanistic work to 1965 and has concluded that the direct molecular decomposition [Eq. (4.29)] proposed by Shilov and Sabirova[38] best accounts for the kinetic data and for a large number of preparative results which he reports. The examples shown in Eqs. (4.30)–(4.35) involve interception of a dihalocarbene by addition to an alkene, diene, or heterodiene, or by insertion into the methylene group of fluorene. The final products are produced by further transformations of the primary adducts or insertion products. The origin of 2-chloropyridine [Eq. (4.33)] is uncertain; it may result from attack either at pyrrole nitrogen or at the α-carbon of pyrrole.

A very clear case of α-dehydrofluorination was discovered by Fuqua et al.[44] in the course of an attempt to generate tetrafluoro-p-xylyene by pyrolysis of

$$Cl_2C\overset{H}{\underset{Cl}{}} \ \rightleftharpoons \ Cl_2C{:} + HCl \qquad (4.29)$$

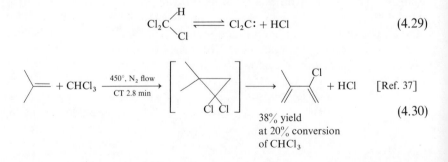

$$\text{(structure)} + CHCl_3 \xrightarrow[\text{CT 2.8 min}]{450°, \ N_2 \ flow} \left[\text{(intermediate)} \right] \longrightarrow \text{(product)} + HCl \qquad [\text{Ref. 37}]$$

38% yield
at 20% conversion
of $CHCl_3$

(4.30)

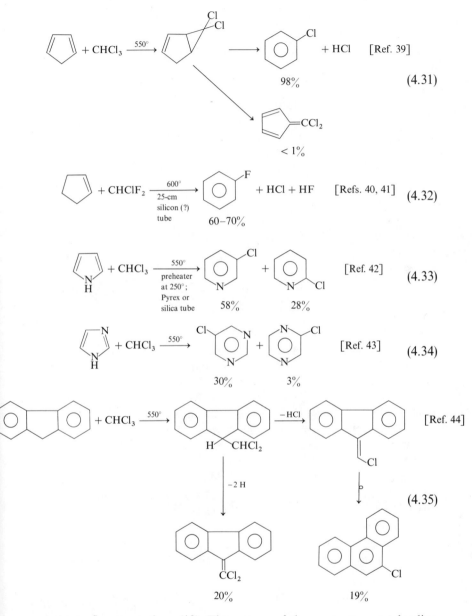

$$\alpha,\alpha,\alpha',\alpha'\text{-tetrafluoro-}p\text{-xylene (19)}.$$ The nature of the rearrangement leading to the major product, 2',2',4-trifluorostyrene (**22**) was obscure at the time, but it is now clear that this is a typical arylcarbene rearrangement [compare Chapter 5, Section III,C, Eq. (5.43)]. The minor product was formulated[42]

as 1,1,5-trifluorobenzocyclobutene, but the pathway shown in part in Eq. (4.36) suggests that it may be the 1,1,4-trifluoro compound **20** formed by carbene insertion into the difluoromethyl group of **21**.

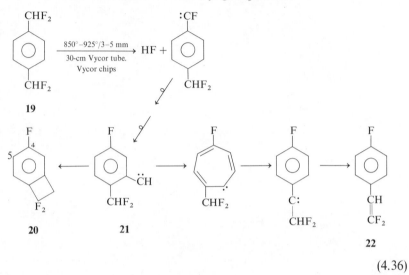

$$(4.36)$$

α-Elimination of a silane appears to be a facile process in the pyrolysis of certain derivatives of disilane studied by Barton and Juvet[45] and by Barton and Banasiak[46] [Eqs. (4.31) and (4.38)]. Dimethylsilylene was trapped by copyrolysis with cyclooctatetraene to give benzosilacyclopentadiene derivatives, and methylchlorosilylene was trapped with cyclopentadiene to give dihydrosilabenzene derivatives. Dichlorocarbene has similarly been generated in the gas phase from trichloromethyltrifluorosilane at temperatures above 120° [Eq. (4.39)].[47] Conlin and Gaspar[47a] have added dimethylsilylene, generated at 600° by the method of Eq. (4.37), to 2-butyne in the gas phase to produce tetramethylsilacyclopropene in 50% yield. This first example of a silacyclopropene was characterized by GC-MS, by nmr spectroscopy, and by its reactions with methanol.

$$(CH_3)_2Si\overset{OCH_3}{\underset{Si(CH_3)_2OCH_3}{<}} \quad \xrightarrow[N_2\ flow]{450°-550°} \quad (CH_3)_2Si: + (CH_3)_2Si(OCH_3)_2 \qquad (4.37)$$

$$\overset{H_3C}{\underset{Cl}{>}}Si\overset{Cl}{\underset{SiCl_2CH_3}{<}} \quad \xrightarrow[N_2\ flow]{600°} \quad \overset{H_3C}{\underset{Cl}{>}}Si: + CH_3SiCl_3 \qquad (4.38)$$

$$Cl_2C\overset{Cl}{\underset{SiF_3}{<}} \quad \xrightarrow{>120°} \quad Cl_2C: + SiClF_3 \qquad (4.39)$$

IV. β-ELIMINATIONS AND γ-ELIMINATIONS

A. β-Eliminations in Esters

As foreshadowed in the introductory section, β-elimination of carboxylic acids from esters to form alkenes is now covered very briefly because it has been so thoroughly reviewed by DePuy and King,[48] by Maccoll,[49] and by Smith and Kelly.[50]

The cyclic six-membered ring transition state **23** for ester pyrolysis in the gas phase first proposed by Hurd and Blunck[51] in 1938 has been generally accepted, and work in the succeeding 40 years has in the main added only detail to this model for the transition state. Controversy has mostly centred around the degree of ion pair character **24** which may be present in the transition state. The elimination of H and O_2CR is predominantly cis, and the direction of elimination in simple alkyl esters is controlled to a considerable degree by the number of β-hydrogen atoms available for each mode. Conjugating substituents attached to a β-carbon usually promote elimination to give a conjugated major product. Smith and Kelly[50] in summarizing theories of the mechanism of ester pyrolysis have stressed the difficulty of eliminating surface effects in kinetic work with esters, and it appears that discordant results and difficulties in the observation and interpretation of β-deuterium isotope effects may be caused by surface catalysis. The same point has recently been emphasized by Wertz and Allinger[52] who conclude that only in a few kinetic studies have precautions adequate to eliminate surface effects been taken. They consider that the experimental conditions employed in most preparative work must result in surface-catalyzed reaction, and that it is almost impossible to eliminate surface effects in glass apparatus. Under typical conditions (450–550°, with slow flow of ester and nitrogen carrier over Pyrex helices) ester pyrolysis may thus involve elimination through a surface ion pair, **25**, in which the β-hydrogen of an adsorbed carbonium ion can be removed either by the adsorbed carboxylate anion or by an active site on the surface itself. The arguments of Wertz and Allinger[52] are based on a detailed analysis of isomer ratios in the products of elimination from alkyl and cycloalkyl esters such as the 1,2-dimethylcyclopentyl acetates (Table 4.1, entries 4 and 5).

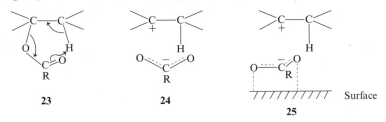

23   **24**   **25**   Surface

TABLE 4.1

Products of β-Elimination from Selected Esters

	Ester	Product(s), with product ratio or percent yield	Pyrolytic conditions	Reference
1.	[chain with OAc]	[alkene] : [alkene] : [alkene] 49 : 32 : 19 58 : 27 : 15	377°–437°; N₂ flow; glass microreactor; CT 15 sec	53
2.	CH₃—[cyclohexane]—OAc	[methylenecyclohexane] + [1-methylcyclohexene, CH₃] 74 : 26	510°; glass beads	54
3.	Ph—[chain]—OAc	Ph⌒ + Ph⌒ 75 : 25 (at 56% conversion)	450°; pyrex helices	55
4.	CH₃ / CH₃ / OAc [cyclopentane]	[CH₃–CH₃ cyclopentene] + [CH₃, methylenecyclopentane] 70–80 : 20–30	425°; Pyrex helices	56
5.	CH₃ / CH₃ / OAc [cyclopentane]	[CH₃–CH₃ cyclopentene] + [CH₃, methylenecyclopentane] 1–4 : 96–99	450°; N₂ flow; 55-cm Vycor tube; Pyrex helices	52, 57

#	Substrate	Product	Conditions	Ref.
6.	AcO⌇⌇OAc	⌇⌇⌇ 63–71%	575°; N_2 flow; Pyrex rings	58
7.	CH_2OAc / CH_2OAc	CH_2OAc 14% + CH_2OAc/CH_2OAc 30% + CH_2OAc/CH_2OAc 47%	525°–540°; 40-cm tube; carborundum chips	59
8.			500°; N_2 flow; 2 × 30 cm Pyrex tube; 1.5-mm Pyrex helices	60
9.	CD_2OAc	$=CD_2$ + $=CH_2$ D_2	480°; N_2 flow; 1-m Pyrex tube; Pyrex helices; CT 4 min	61
10.	O / OAc	O (indanone) 20%	550°; Pyrex helices	62
11.	N—O—C(=O)—OCH_3	$NH + CO_2 + CH_3OH$	550°/0.01 mm; Silica tube	63

(continued)

TABLE 4.1 *(Continued)*

Ester	Product(s), with product ratio or percent yield	Pyrolytic conditions	Reference
12.	$C_2H_4 +$ 71%	$525°$; glass helices	64
13.	$+$ $\longrightarrow$ 63%	$350°–360°$; N_2 flow; glass helices	65

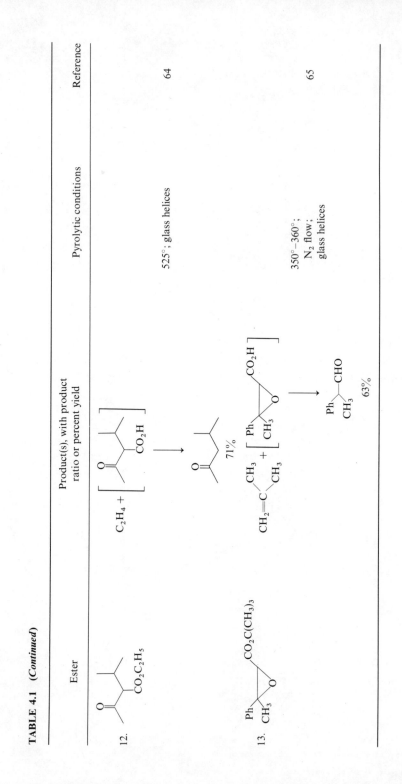

No attempt is made here to present comprehensive tables[48,50] of products of pyrolysis of esters. Table 4.1 merely shows a few examples relevant to the preceding brief discussion (entries 1–5), some synthetically interesting examples of the preparation of terminal or exocyclic methylene compounds (entries 6–9), two examples of the preparation of readily polymerized compounds (entries 10 and 11), and two examples of the pyrolytic fission of ethyl or *t*-butyl esters. 4,5-Dimethylenecyclohexene (entry 8) is thermo-dynamically unstable with respect to its isomer, *o*-xylene, but with attention to technique it was readily prepared free of *o*-xylene. The method is not so suitable in cases where the product can undergo secondary thermal rearrangement, as shown by the pyrolysis of the deuterated ester of entry 9, which gave a near-equilibrium mixture of exo- and endo-deuterated methylene-cyclobutanes. Entries 10 and 11 show the preparation of indenone and isoindole, two compounds which readily polymerize or decompose in the condensed state unless kept cold. The latter example was the first successful preparation and spectroscopic characterization of isoindole, although this method has now been superseded by the retro-Diels–Alder approach described in Chapter 8, Section III,D, and Eq. (8.62). Entries 12 and 13 illustrate the use of pyrolysis rather than hydrolysis for the generation and decarboxylation of labile carboxylic acids, which in the case of entry 12 gives a much higher yield of methyl isobutyl ketone than the conventional hydrolytic procedure. The pyrolysis and rearrangement of glycide *t*-butyl esters (see entry 13) has a special significance for the author, whose interest in pyrolysis was first kindled in 1960 by the experimental elegance and simplicity of this work in G. Büchi's laboratory. Comparison of entries 12 and 13 also shows the advantage of using *t*-butyl esters, which can undergo elimination at much lower temperatures than ethyl esters under otherwise similar conditions.

B. β-Eliminations Formally Related to Ester Eliminations

The preparatively important thermal decomposition of *S*-methylxanthate esters **26** to give alkenes[48,49] occurs so smoothly in the condensed phase near 200° that there has been little incentive for much study of the reaction in the gas phase. The kinetics of the decomposition of the related alkyl thionacetates **27** and thioacetates **28** in the gas phase were studied using flow apparatus by Bigley and Gabbott,[66,67] over the temperature ranges shown in parentheses. The behavior of dialkyl carbonates **29** is very similar to that of alkyl carboxylates, although somewhat lower temperatures suffice for elimination[68–70] The decomposition of alkyl *N,N*-dimethyl-carbamates such as **30** has been shown by Daly and co-workers[71] to involve

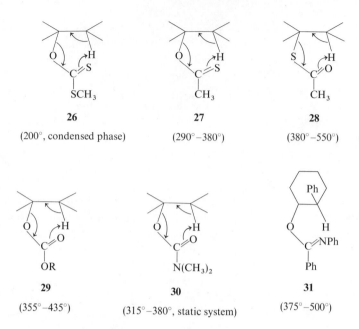

26

(200°, condensed phase)

27

(290°–380°)

28

(380°–550°)

29

(355°–435°)

30

(315°–380°, static system)

31

(375°–500°)

mainly attack of amide oxygen, rather than nitrogen, on the β-hydrogen. Marullo *et al.*[72] have studied the β-elimination of benzanilide from the *N*-phenylbenzimidate **31** in the vapor phase; the phenylcyclohexenes formed at 375° contained 85% of the conjugated isomer, 1-phenylcyclohexene.

Some less closely related pyrolytic β-eliminations are those of ketene-*N-t*-butylimines [Eq. (4.40)] reported by Ciganek,[73] and of acetoacetamides studied by Mukaiyama and co-workers.[74] The 6-center fission of aceto-acetamides provides a general pyrolytic route to isocyanates, although in the case of the *N*-cyclohexyl compound **32** β-elimination to give cyclohexene is a competing process [Eq. (4.41)]. Smith and Kelly[50] have summarized several other studies of the pyrolysis of amides and of diacylimides.

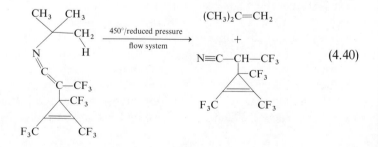

(4.40)

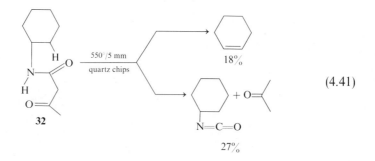

$$(4.41)$$

C. γ-Eliminations, and Eliminations with Rearrangement

Although ester pyrolysis is often regarded as a clean and efficient technique for effecting β-elimination, study of reviews[50,52] will show that this may be a gross oversimplification. 1,3-Elimination or the formation of rearranged products may predominate over β-elimination, and in many cases it is not clear whether the reaction is a true gas-phase reaction or whether it involves surface catalysis of the type proposed by Wertz and Allinger.[52] Three examples chosen from a large number are shown in Eqs. (4.42)–(4.44).

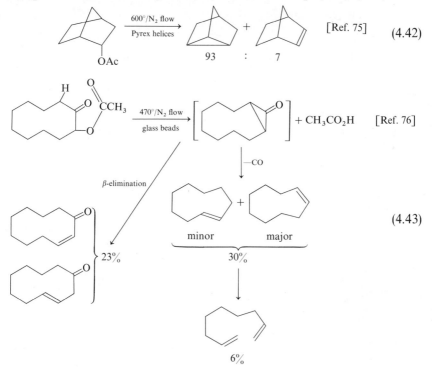

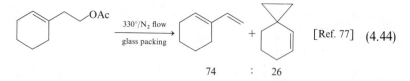

$$74 \quad : \quad 26$$

The classic case of the rearrangement of a tertiary ester is that of patchouli acetate (**33**) which on liquid-phase pyrolysis with distillation of acetic acid rearranged to give the α- and γ-patchoulenes **34** and **35** [Eq. (4.45)]. That the pyrolysis of patchouli acetate[78] and an apparently straightforward resynthesis of patchouli alcohol by Buchi and co-workers[79] from synthetic α-patchoulene involved similar rearrangements in opposite directions was not realized until proved by the crystallographic work of Dunitz's group[80] on the chromate ester of patchouli alcohol. Rearranged products were later found by Adams and Kovacic[81] on pyrolysis of 3-homoadamantyl acetate at 470°–600° over Vycor chips, and Kwart and Slutsky[82] have studied the same pyrolysis with a deuterated sample **36** and using a gold-coil microreactor[83] which is considered to minimize surface catalysis and to sustain pure gas-phase pyrolysis. Both six-membered and seven-membered cyclic transition states (TS) were proposed to explain the distribution of deuterium in the rearranged alkenes [Eq. (4.46)].

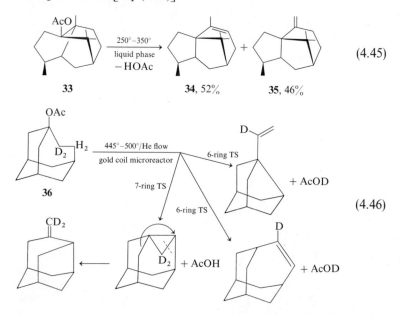

Similar 1,3-eliminations and eliminations with rearrangement have been encountered in the pyrolysis of certain methanesulfonic and arylsulfonic esters in which β-elimination is impossible or unfavorable. An important paper by Boyd and Overton[84] provides a concise review of rearrangements during ester pyrolysis, and describes the pyrolysis of 2-adamantyltoluene-*p*-sulfonate and methanesulfonate **37** [Eq. (4.47)] in the vapor phase over glass wool. The exit end of the hot tube and of the packing was coated with sodium carbonate to neutralize methanesulfonic acid and thus avoid secondary acid-catalyzed reactions, and both protoadamantene (**38**) and 2,4-dehydroadamantane (**39**) were obtained in 95% combined yield. 2-Adamantyl acetate was recovered after attempted pyrolysis at 560°, and the corresponding xanthate afforded only 4% of hydrocarbons formed by elimination on pyrolysis at 550°.

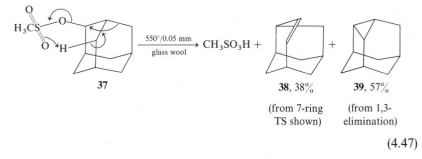

37

550°/0.05 mm
glass wool
→ CH₃SO₃H +

38, 38%

(from 7-ring
TS shown)

39, 57%

(from 1,3-
elimination)

(4.47)

Johnston and Overton[85] have used this method for rearrangement of the atisine derivative **40** into the enone **41** which has the alternative lycoctonine–aconitine skeleton. The toluene-*p*-sulfonate dissolved in acetone was injected into the silica pyrolysis tube through a septum, the needle being long enough to deliver the substrate directly into the hot zone. Three passes through the tube were necessary to achieve complete decomposition of the toluene-*p*-sulfonate, but the rearranged enone **41** was obtained in 77% yield [Eq. (4.48)].

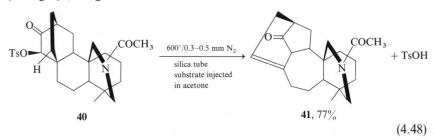

40

600°/0.3–0.5 mm N₂
silica tube
substrate injected
in acetone
→

41, 77%

+ TsOH

(4.48)

D. Elimination of Hydrogen Cyanide from Nitriles

In Section II,B, the loss of hydrogen cyanide from isopropyl cyanide on VLPP at 978° [Eq. (4.13)] was mentioned as being only a minor process which competes with the formation of acrylonitrile by loss of methane. This result illustrates the very great thermal stability of alkyl cyanides, and aryl cyanides are even more resistant to thermal decomposition; the latter are frequently found as terminal products of very high temperature pyrolysis. Tertiary alkyl cyanides are much less stable than primary or secondary compounds, and *t*-butyl cyanide decomposes readily in a flow system[86] at 600°–650° and 8–12 nm. Dastoor and Emovon[87] have studied the kinetics of the decomposition of EtCN,*i*-PrCN and *t*-BuCN in a seasoned silica flow system, and they find that *t*-butyl cyanide decomposes mainly by a molecular mechanism, probably involving a 4-center transition state, to give hydrogen cyanide and isobutene over the range 565°–654°.

Study of the patent literature shows that much effort has been devoted to the pyrolytic conversion of ethylene dicyanide, an unwanted by-product of acrylonitrile manufacture, into acrylonitrile. Some methods involve the use of supported ionic catalysts (e.g., charcoal impregnated with KCN/NaCN[88,88a]) but pyrolysis over inert material is also effective in eliminating hydrogen cyanide [Eq. (4.49)].[89,89a]

$$NCCH_2CH_2CN \xrightarrow[\text{inert packing}]{500°-650°} CH_2{=}CHCN + HCN \qquad (4.49)$$
$$94\%$$

Kuntz and Disselnkotter[90] have shown that secondary enamides (*N*-vinyl-amides) $R_2C{=}CR{-}NHCOR$ can readily be prepared in 40–80% yield by flow pyrolysis of 2-acylaminoalkyl cyanides with elimination of hydrogen cyanide. The preparation of *N*-vinylformamide shown in Eq. (4.50) gives 46% of the product in one pass, with 43% recovery of starting material. Pyrolysis through a steel tube over marble chips at 410°–420° gives *N*-vinylformamide in 82% yield.

$$CH_3\underset{\underset{CN}{|}}{C}HNHCHO \xrightarrow[\text{quartz tube, 3} \times \text{60 cm}]{600°/5\text{ mm}} CH_2{=}CHNHCHO + HCN \qquad (4.50)$$

E.   β-Elimination and Isomerization Reactions of Alkyl Halides

The pyrolysis of alkyl halides is another topic which is covered very briefly here because it has been so extensively reviewed elsewhere, and also because pyrolytic β-eliminations of hydrogen halides have so far proved of somewhat greater interest to kineticists than to preparative chemists. The subject is mechanistically very complicated, because reaction may occur by radical or radical chain pathways, by surface-catalyzed processes, or by unimolecular decomposition through transition states with some degree of heterolytic character. Early mechanistic work up to 1945 is summarized in the introductory section of the review by Maccoll[91] which presents a mass of quantitative data obtained in the period 1946–1967. The evidence is reviewed in terms of the concept of gas-phase heterolysis developed by Maccoll.[92] The work of this review is extended by that of Smith and Kelly[50] which again provides extensive tabulation of kinetic data, and a discussion of some topics not covered by Maccoll.

Alkyl chlorides, bromides, and iodides can all undergo *β*-elimination of hydrogen halide in the temperature range 200°–500°, tertiary halides at the lower end of this range and primary halides at the higher. The great strength of the carbon–fluorine bond makes it difficult to achieve *β*-elimination of hydrogen fluoride in this temperature range,[50] and in competitive situations other halides are preferentially eliminated [see Eq. (4.56)]. Academic interest in the pyrolysis of alkyl halides was kindled from 1949 by a long series of papers by Barton (Part I[93]) and by Barton and co-workers on elimination of hydrogen halides. This work arose from consideration of the industrial preparation of vinyl chloride by pyrolytic dehydrochlorination of 1,2-dichloroethane.

Unless unsaturated inhibitors are present, some alkyl chlorides and many alkyl bromides and iodides tend to decompose in the gas phase by radical or radical chain processes such as that proposed by Barton and Onyon[94] for the case of alkyl chlorides [Eq. (4.51)]. Benson[95] represents the key step in the radical decomposition of alkyl iodides essentially as in Eq. (4.52).

$$R \!\!\not|\!\! -Cl \longrightarrow R\cdot + Cl\cdot \text{ (initiation)}$$

$$Cl\cdot + H-\overset{|}{\underset{|}{C}}-\overset{|}{\underset{|}{C}}-Cl \longrightarrow ClH + \cdot\overset{|}{\underset{|}{C}}-\overset{|}{\underset{|}{C}}-Cl$$

$$\cdot\overset{|}{\underset{|}{C}}-\overset{|}{\underset{|}{C}}-Cl \longrightarrow {>}C{=}C{<} + Cl\cdot \qquad \text{(propagation)}$$

(4.51)

$$I\cdot + H-\overset{|}{\underset{|}{C}}-\overset{|}{\underset{|}{C}}-I \longrightarrow IH + {>}C{=}C{<} + I\cdot \qquad (4.52)$$

In clean glass or silica reactors rapid surface-catalyzed decomposition is usually observed initially; only in apparatus covered with a carbonaceous film can the true unimolecular elimination of hydrogen halide be studied. For example, Barton and Onyon[94] were able to obtain reproducible kinetics for the clean decomposition of *t*-butyl chloride to isobutene and hydrogen chloride over the temperature range 290°–341° only after conditioning of the static reactor by 20–40 previous pyrolytic runs. It will be evident from this that alkyl halides pyrolyzed in an unconditioned flow apparatus are likely to decompose by a number of competing mechanisms.

Unimolecular β-elimination of hydrogen halides in the gas phase was originally considered to be a 4-center cis-elimination (42) but Maccoll and co-workers[92,96] favor a highly polar transition state approaching a tight carbonium ion–halide ion pair (43). Smith and Kelly[50] give a critical discussion of the advantages and difficulties associated with such a model for the transition state, and discuss the polar semi-ion pair transition state of Benson and Bose.[97]

Equations (4.53–4.56) show four examples of β-eliminations of hydrogen halides under flow and static conditions.

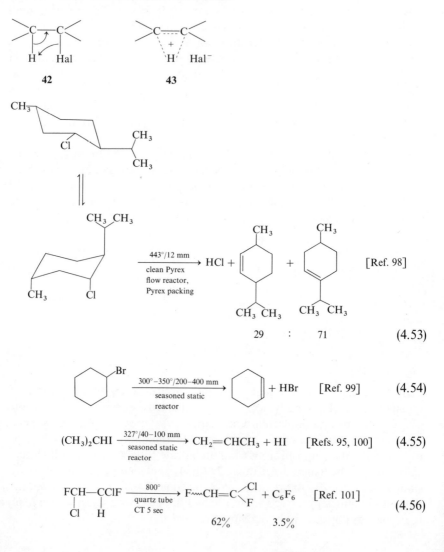

$$\text{(4.53)}$$

$$\text{(4.54)}$$

$$(CH_3)_2CHI \xrightarrow[\text{seasoned static reactor}]{327°/40-100\text{ mm}} CH_2=CHCH_3 + HI \qquad [\text{Refs. 95, 100}] \qquad (4.55)$$

$$\underset{\underset{Cl}{|}\ \underset{H}{|}}{FCH-CClF} \xrightarrow[\substack{\text{quartz tube}\\ \text{CT 5 sec}}]{800°} F\text{wwv}CH=C\underset{F}{\overset{Cl}{<}} + C_6F_6 \qquad [\text{Ref. 101}]$$
$$\qquad\qquad\qquad\qquad 62\% \qquad\quad 3.5\% \qquad\qquad\qquad\qquad (4.56)$$

Alkyl halides may also undergo thermal rearrangements in the gas phase very similar to those which occur in solution via ionic or ion pair intermediates. Fields *et al.*[102] have reported the isomerization of a number of chlorocyclopropanes to 3-chloropropenes at 500°–650° and have proposed concerted migration of chlorine [Eqs. (4.57) and (4.58)]. DePuy *et al.*[103] have found similar isomerizations of cyclopropyl acetates [e.g. Eq. (4.59)]. Similarly, Harding *et al.*[104] have shown that $(CH_3)_2C=CHCH_2Cl$ and $(CH_3)_2CClCH=CH_2$ are equilibrated in the gas phase in a seasoned Pyrex reactor at 245°–315°, as well as undergoing, respectively, 1,4- and 1,2-elimination of hydrogen chloride.

$$
\underset{}{\triangle}\!\!-\!\!Cl_2 \xrightarrow[\text{CT 0.1–0.2 sec}]{650°/2\ mm} Cl\text{-----}\triangle\!\!-\!\!Cl \longrightarrow ClCH_2CCl=CH_2 + SM \qquad (4.57)
$$

$$
\qquad\qquad\qquad\qquad\qquad\qquad\qquad\qquad 32\% \qquad\quad 57\%
$$

$$
Cl\!\!-\!\!\triangle\!\!-\!\!Cl_2 \xrightarrow[\text{CT 0.1–0.2 sec}]{500°/2\ mm} Cl\text{-----}\triangle\!\!-\!\!Cl_2 \longrightarrow Cl_2CHCH=CCl_2 \qquad (4.58)
$$

$$
\underset{}{\triangle}\!\!-\!\!OAc \xrightarrow[\text{Pyrex helices}]{480°,\ N_2\ flow} AcOCH_2CH=CH_2 + SM \qquad (4.59)
$$

$$
\qquad\qquad\qquad\qquad\qquad\qquad 44\% \qquad\qquad 55\%
$$

F. Miscellaneous β-Eliminations Involving Heteroatoms: O,S,N,P,F

Alcohols can undergo β-elimination of water to give alkenes by unimolecular 4-center elimination, by catalyzed dehydration with HBr or HCl in the gas phase, and by heterogeneous catalysis by passage over metal oxides such as alumina or thoria. The last process is of some value in the laboratory and is of very considerable industrial importance. The dehydration of alcohols under all three sets of conditions has been concisely reviewed by Knozinger.[105] The rate of uncatalyzed decomposition of *t*-butanol in the gas phase in a well-seasoned silica static reactor at 487°–620° was studied by Bernard[106] who observed a clean first-order elimination to give isobutene. Primary alcohols such as 1-butanol, however, underwent initial radical fission of the C_3H_7—CH_2OH bond which led to a complex mixture of products.[107] The direct dehydration of alcohols has been much less popular in the laboratory than the pyrolysis of their derived esters, probably because of a somewhat greater risk of rearrangement and of ether formation in the case of the alcohols.

The flash pyrolysis of thiols and of disulfides has recently attracted attention for the generation of some reactive sulfur compounds for photoionization and microwave spectroscopy. Bock and Mohmand[108] found that ethanethiol and diethyl sulfide underwent predominant elimination of hydrogen sulfide [Eq. (4.60)], whereas dimethyl disulfide decomposed at a lower temperature than the sulfide with formation of methylthio radicals which disproportionated to give methanethiol and methanethial [Eq. (4.61)], detected by photoelectron spectroscopy. Similarly, Tanimoto and Saito[109] generated ethenethiol (vinyl mercaptan) for microwave spectroscopy by elimination of hydrogen sulfide from ethane-1,2-dithiol at $600°-950°$ [Eq. (4.62)]; thiirane, ethanethial, and methanethial were also detected as the pyrolyzate flowed through a parallel plate microwave cell. Vinyl alcohol has similarly been observed in the pyrolyzate from ethane-1,2-diol.[110]

$$CH_3CH_2SH \xrightarrow[\text{50-cm quartz tube}]{547°/0.1 \text{ mm}} CH_2{=}CH_2 + H_2S \xleftarrow{817°} CH_3CH_2SCH_2CH_3 \qquad (4.60)$$

$$CH_3SSCH_3 \xrightarrow{617°} 2\,CH_3{\cdot} \longrightarrow CH_3SH + CH_2{=}S \qquad (4.61)$$

$$HSCH_2CH_2SH \xrightarrow[\text{quartz tube}]{600°-950°/0.02-0.04 \text{ mm}} H_2S + CH_2{=}CHSH + CH_3CH{=}S + CH_2{=}S$$

$$+ \overset{\triangle}{\underset{S}{}} \qquad (4.62)$$

The thermal decomposition of alkyl sulfoxides in solution at $100°-200°$ has frequently been used for the introduction of unsaturation by concerted cis β-elimination of a sulfenic acid RSOH. Sulfenic acids are highly reactive and usually unstable, and few such compounds have been isolated and characterized. Penn and Block's group[110a] has generated methanesulfenic acid by flash vacuum pyrolysis of t-butyl methyl sulfoxide at $400°$ and $0.1-0.2$ mm to give CH_3SOH and 2-methylpropene. The structure and geometry of methanesulfenic acid were determined by microwave spectrometry and the gas-phase dehydration of CH_3SOH produced at $400°$ to thioformaldehyde $CH_2{=}S$ at $750°$ was studied by FVP–microwave experiments using tandem pyrolysis tubes.

Davis and co-workers[111] have further shown that flash vacuum pyrolysis of aryl and alkyl t-butyl sulfoxides with collection of the pyrolyzate at $-196°$ can afford a wide range of sulfenic acids in moderate to good yields. The reactive sulfenic acids form thiolsulfinates RS(O)SR on thawing of the cold trap, but they can be trapped by cocondensation with propiolic ester [Eq. (4.63)]. An intermediate sulfenic acid must also be formed in the pyrolysis of ethyl α-phenylstyryl sulfoxide which is reported by Ando et al.[112] to give 3-phenylbenzothiophene accompanied by a little 1,1-diphenylethylene. It is

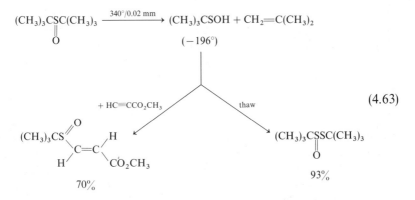

$$(4.63)$$

proposed that the benzothiophene is formed by cyclization of $Ph_2C{=}CS$. [Eq. (4.64)].

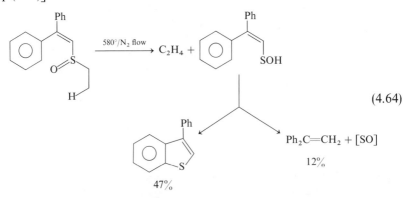

$$(4.64)$$

Alkylamines can undergo some alkene-forming eliminations on pyrolysis, but radical fissions occur readily and very complex mixtures of products tend to be produced. Patterson and co-workers[113] have examined the pyrolysis of β-phenylethylamine at 650° and 850° and have found a large number of products; only a few major components of the pyrolyzate are shown in Eq. (4.65). Hydrogen cyanide is a major product (50%) at 850°,

$$PhCH{=}CH_2 + PhCH_2CH_3$$

$$PhCH_2CH_2NH_2 \xrightarrow[\text{Berl saddles}]{650°/N_2 \text{ flow}} [PhCH_2{\cdot} + {\cdot}CH_2NH_2] \longrightarrow HCN \qquad (4.65)$$

$$PhCH_3 + PhCH{=}CHPh + PhCH_2CH_2Ph +$$

$$PhH + PhCH_2Ph + \text{anthracene} + \text{phenanthrene}$$

and Patterson has drawn attention to an early report[114] of the formation of hydrogen cyanide in 98% yield on pyrolysis of trimethylamine at 800° [Eq. (4.66)].

$$CH_3N\begin{matrix}CH_3\\CH_3\end{matrix} \xrightarrow{800°} 2\ CH_4 + HCN \qquad (4.66)$$

More recently Roberts and co-workers[115] have discovered that very ready elimination occurs in *N,N*-dichloroamines on gas chromatography at inlet temperatures as low as 210°–260°. Two types of process occur: alkene formation, with β-elimination of the elements of HCl and NCl as in Eq. (4.67) and dehydrochlorination leading to a chloroimine [Eq. (4.68)] or to a nitrile [Eq. (4.69)].

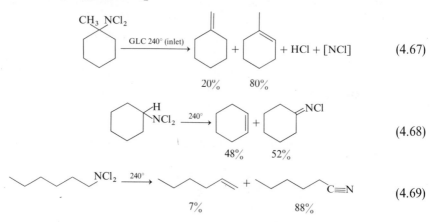

$$(4.67)$$

$$(4.68)$$

$$(4.69)$$

Finally, Eqs. (4.70)–(4.72) show the formation of carbon–sulfur, carbon–phosphorus, and carbon–nitrogen multiple bonds by elimination of hydrogen chloride and of carbonyl fluoride. Hydrogen chloride was removed from

$$CH_3SCl \xrightarrow[\text{40-cm quartz spiral}]{590°/0.1\ mm} CH_2{=}S + HCl \qquad [\text{Ref. 116}] \qquad (4.70)$$

$$CH_3PCl_2 \xrightarrow[\text{quartz tube}]{1000°/0.03\ mm} HCl + CH_2{=}PCl \qquad [\text{Ref. 21}]$$

$$\downarrow \qquad (4.71)$$

$$HCl + HC{\equiv}P$$

$$F_3C{\diagdown}N{-}C{\diagup}O \xrightarrow[\substack{\text{Ni tube,}\\\text{Ni packing}}]{575°} F_3CN{=}CF_2 + F_2C{=}O \qquad [\text{Ref. 117}] \qquad (4.72)$$

the pyrolyzate of Eq. (4.70) by injection of the stoichiometric amount of ammonia, and pure thioformaldehyde was obtained for photoelectron spectroscopic study. Pyrolysis of methanesulfinyl chloride $CH_3S(O)Cl$ at $600°$ similarly gives sulfine $CH_2=S=O$.[115a] The formation of phosphaethyne $HC\equiv P$ was detected by microwave spectroscopy.

V. FORMATION OF KETENES BY β-ELIMINATION

A. Ketenes from Acids, Acid Anhydrides, and Acid Chlorides

Ketene is produced industrially on a large scale by the direct pyrolysis of acetic acid in the temperature range $650°–750°$, often in the presence of phosphoric acid vapor produced by the cracking of added catalytic amounts of triethylphosphate. The rate of uncatalyzed recombination of ketene and water is sufficiently slow that it is possible to condense water and acetic acid from the gas stream by cooling it rapidly, and the substantially pure ketene stream can then be treated with acetic acid in a counterflow reactor to produce acetic anhydride. There is an extensive patent literature, and Eq. (4.73) shows a single example of a ketene process[118,118a] in which rapid cooling of the gas stream to $360°$ is used to slow recombination of the products. Bock and co-workers[119] have similarly generated thioketene for photoelectron spectroscopy by flash vacuum pyrolysis of dithioacetic acid [Eq. (4.74)]. The pyrolytic generation of ketenes from various starting materials has been reviewed by Hanford and Sauer,[120] by Borrmann,[121] and by Lacey.[122]

$$CH_3CO_2H \xrightarrow[\substack{(C_2H_5O)_3PO \\ \text{catalyst}}]{750°/300\ mm} CH_2=C=O + H_2O \qquad (4.73)$$

$$CH_3CS_2H \xrightarrow[\substack{1 \times 30\ cm \\ \text{quartz tube}}]{457°/0.08\ mm} CH_2=C=S + H_2S \qquad (4.74)$$

A very convenient method for the laboratory preparation of ketene is the pyrolysis of acetic anhydride at about $500°$ and at atmospheric pressure through a Vycor or silica tube [Eq. (4.75)]. Fisher and co-workers[123] describe an apparatus arranged for the rapid separation of the products acetic acid and ketene which can produce a continuous stream of ketene at the rate of $1–2$ moles/hr. In the author's laboratory the use of such hot-tube apparatus has been found much less troublesome than that of a ketene lamp

$$(CH_3CO)_2O \xrightarrow[\substack{54-cm \\ \text{Vycor tube}}]{500°–510°} CH_3CO_2H + CH_2=C=O \qquad (4.75)$$

in which acetone is cracked over a glowing Nichrome spiral. Methylketene can similarly be prepared by pyrolysis of propionic anhydride; Jenkins[124] found that pyrolysis at 550° and 5 mm with a contact time of 0.01 sec gave methylketene in about 90% yield. Baxter *et al.*[125] modified the method of Jenkins by pyrolyzing propionic anhydride through a packed silica tube at 550° and 0.005 mm using the apparatus of Chapter 2, Fig. 2.6, with removal of propionic acid by condensation at −78°. The methylketene after standing in ether at −78° gave the crystalline dimer **44** in 93% yield [Eq. (4.76)].

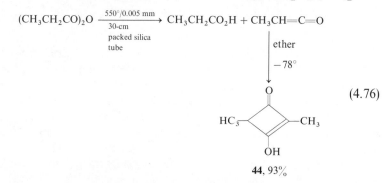

(4.76)

The formation of ketenes by elimination of carboxylic acids is not confined to saturated anhydrides. Brown and Ritchie[126] examined the pyrolysis of acrylic anhydride at 500° with a relatively long contact time (6 or 24 sec) and found that the pyrolyzate contained acrylic acid, carbon monoxide, acetylene, and a trace of acrolein. Blackman *et al.*[127] later showed that these conditions were too severe and that the initial process proposed by Brown and Ritchie, formation of acrylic acid and the transient species methyleneketene (**45**), occurs smoothly on flash vacuum pyrolysis. This has proved the simplest method of generation of methyleneketene for microwave spectroscopy [Eq. (4.77)]. The approach has been extended[128] to the generation of the next homologue, butatrienone (**47**), for microwave spectroscopy by pyrolysis of the crude mixed anhydride **46** [Eq. (4.78)].

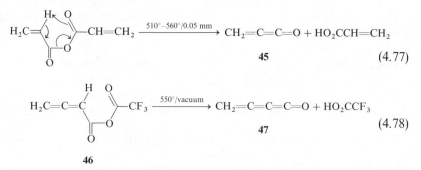

(4.77)

(4.78)

Acid chlorides can also undergo pyrolytic elimination of hydrogen chloride to give ketenes; Bock and co-workers[129] have shown that acetyl chloride gives ketene [Eq. (4.79)], but the conditions required for elimination may be too severe for survival of other ketenes [e.g. Eq. (4.80)]. The same is true of the pyrolysis of *N*-cinnamoyl-3,5-dimethylpyrazole which appears to generate benzylideneketene efficiently on flash vacuum pyrolysis, but only at a temperature at which decarbonylation and rearrangement to phenylacetylene occurs immediately [Eq. (4.81)].[130]

$$CH_3COCl \xrightarrow[\text{silica tube}]{547°/0.1 \text{ mm}} CH_2{=}C{=}O + HCl \qquad (4.79)$$

$$CH_2{=}CHCOCl \xrightarrow{637°/0.1 \text{ mm}} HC{\equiv}CH + CO + HCl \qquad (4.80)$$

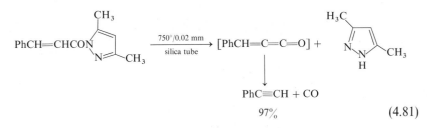

$$PhC{\equiv}CH + CO$$
$$97\% \qquad (4.81)$$

B. Ketenes from Esters

Hurd and Blunck[131] first found that phenyl acetate forms phenol and ketene (84%) on pyrolysis at 625°, and they proposed a radical chain pathway [Eq. (4.82)]. Barefoot and Carroll[132] have examined the pyrolysis of PhO_2CCD_3 through a Vycor tube at 635° in a stream of nitrogen, and have shown that the alternative concerted mechanism [Eq. (4.83)] which would lead to *o*-deuteration does not operate. Traces of *o*- and *p*-phenoxyphenol are also formed, which lends support to the radical chain mechanism, but presumably a direct 4-center elimination of phenol as in **48** is also possible.

$$PhO\cdot + CH_3CO_2Ph \longrightarrow PhOH + \cdot CH_2CO_2Ph$$
$$\cdot CH_2CO_2Ph \longrightarrow CH_2{=}C{=}O + PhO\cdot \qquad (4.82)$$

48

$$(4.83)$$

Acylketenes are very readily generated by pyrolytic elimination of methanol from methyl β-ketoesters, probably via the conformation of the enol shown in Eq. (4.84). Leyendecker[133] has used terminal unsaturation in such esters to intercept the acylketenes and further acylcarbenes by internal cycloaddition. Thus flash vacuum pyrolysis of methyl 3-oxohept-6-enoate gave cyclopentenone (80%), bicyclo[3.1.0]hexan-2-one (20%), and methyl acetate (78%) according to the scheme outlined in Eq. (4.85). Berkowitz and Ozorio[134] found that pyrolysis of ethyl cyclopropylacetate gave cyclopentenone by a vinylcyclopropane rearrangement of the intermediate cyclopropylketene. In this pyrolysis over glass wool elimination of ethanol occurs in preference to normal β-elimination of ethylene; the other major component of the pyrolyzate was starting material (42%).

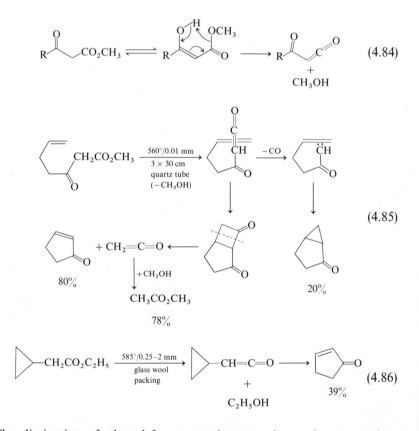

(4.84)

(4.85)

(4.86)

The elimination of ethanol from a urethane to give an isocyanate is a process isoelectronic with the formation of a ketene from an ester. Krantz and Hoppe[135] have reported an interesting sequence of isocyanate formation, $E \rightarrow Z$ isomerization, 1,5 hydrogen migration, and electrocyclic ring closure in the pyrolysis of the urethane **49**.

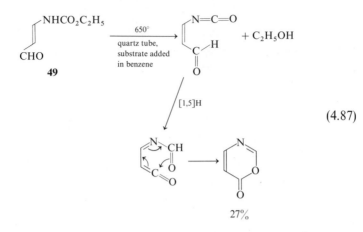

$$(4.87)$$

27%

VI. 1,4-ELIMINATIONS OF HX FROM SYSTEMS H—C—C=C—C—X

A. Alkyl Halides, Esters, and Acid Chlorides

There are now many examples of the pyrolytic elimination of HX from systems of the type **50**, but the first and simplest example of such an elimination in an alkenyl halide came from the work of Harding *et al.*[104] on γ,γ-dimethylallyl chloride. This elimination is complicated by the slow allylic isomerization of the γ,γ-dimethyl compound **51** to the α,α-dimethyl isomer **52**, but it was possible to show that the γ,γ-dimethyl compound undergoes elimination in the gas phase more rapidly than the α,α-isomer and that the kinetic parameters (notably a low frequency factor, log A = 12.03) were typical of a 6-centered cyclic process [Eq. (4.88)].

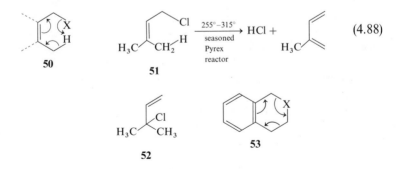

$$(4.88)$$

Several preparatively useful processes of similar type involve elimination of hydrogen halide from an *o*-alkylbenzyl halide. In the course of a study of retro-Diels–Alder reactions in the heterocycles **53** (X = O, NH, or S) and in their molecular ions Loudon *et al.*[136,137] developed a new synthesis of benzocyclobutene based on the formally related 1,4-elimination of hydrogen chloride from *o*-methylbenzyl chloride [Eq. (4.89)]. A spectacular example of the application of this method is the generation of [6]radialene (**54**) by flash vacuum pyrolysis of 2,4,6-tris(chloromethyl)mesitylene [Eq. (4.90)] reported by Schiess and Heitzmann[138] and by Boekelheide's group.[138a] Schiess and Heitzmann[138] collected [6]radialene from this pyrolysis as a white microcrystalline solid which could be purified on a small scale by sublimation in a high vacuum at − 10°. It was characterized by addition of bromine to give hexakis(bromomethyl)benzene, and by the formation of tris-adducts with dienophiles. At lower pyrolytic temperatures intermediate products with one or two fused four-membered rings were formed in addition to [6]radialene. Alternative approaches to **54** are the flash vacuum pyrolysis of the paracyclophane [2,2](3,6)benzo[1,2:4,5]dicyclobutene[138a] at 900°, of the appropriate tetracyclic trisulfone[138a] at 900° (compare Chapter 6, Section III,D) and of cyclododeca-1,5,9-triyne (see Chapter 9, Section V,F, Table 9.4).

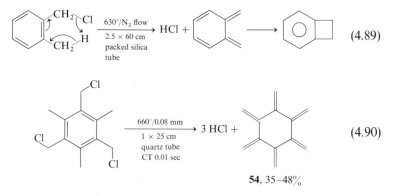

Related useful eliminations have been achieved in the case of *o*-alkyl-benzoyl chlorides. Schiess and Heitzmann[139] followed earlier work of Schiess and Radimerski[140] on 1,2-elimination of hydrogen chloride from *β,γ*-unsaturated acyl chlorides by examining the pyrolysis of *o*-toluoyl chloride. On pyrolysis at 630° it gave benzocyclobutenone in 28% yield [Eq. (4.91)] together with other decomposition products including benzyl chlo-

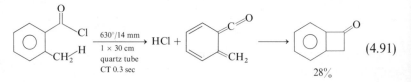

ride. Related compounds, however, gave the fused cyclobutenones in good yield [e.g., Eq. (4.92)], except for systems in which the intermediate *o*-quinonoid ketene could undergo a 1,5 hydrogen shift; *o*-alkenylbenzaldehydes were then obtained [Eq. (4.93)]. The formation of isocoumarin by pyrolysis of the aldehydo ester **55** probably involves 1,4-elimination of methanol although De Champlain *et al.*[141] consider also the possibility of cyclization of the enolized aldehyde instead.

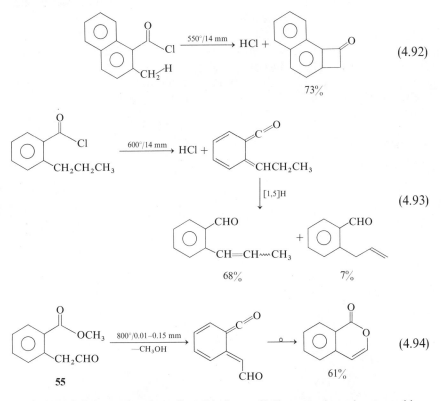

$$(4.92)$$

$$73\%$$

$$(4.93)$$

$$68\% \qquad 7\%$$

55 $\qquad (4.94)$

$$61\%$$

1,4-Eliminations of carboxylic acids from allylic esters have been used less commonly, and they are in any case likely to be ambiguous because of the possibility of 1,3 allylic shifts of the carboxylate group, with subsequent 1,2-elimination. The preparation of 2-acetoxyfuran[142] by pyrolysis of the 2,5-diacetoxy-2,5-dihydrofurans [Eq. (4.95)] is a possible example. Both stereoisomers afford 2-acetoxyfuran, though in somewhat different yield,

$$(4.95)$$

but the conditions of the pyrolysis appear such as to permit possible equi-libration of isomers and stereoisomers and the reaction may not in fact involve cis-1,4-elimination through a seven-membered transition state.

B. 1,4-Eliminations from o-Substituted Phenols and Aromatic Amines

The simplest example of this class of pyrolytic eliminations is the prepara-tion of o-quinone methide (**56**) from o-methoxymethylphenol due to Gardner and co-workers.[143] Pyrolysis at high dilution in nitrogen at 500°–650° gave a light yellow product [Eq. (4.96)] which after collection at $-196°$ liquefied at $-50°$, and could be reduced with ethereal $LiAlH_4$ to o-cresol in 15–20% yield. On warming to 0° monomeric **56** trimerized to **57**;[144] this trimer could be prepared in 70% yield by more rapid pyrolysis.

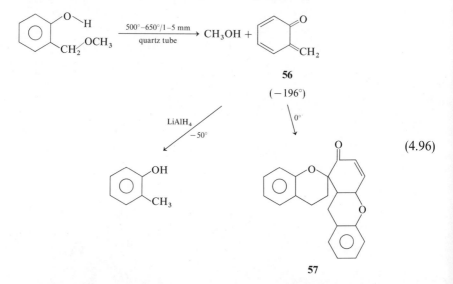

(4.96)

The pyrolysis of methyl esters of salicyclic and hydroxynaphthoic acids has been investigated by several groups[145,146,147] but in no case has the formation of the expected primary product of 1,4-elimination of methanol from methyl salicylate, **58**, been detected either by coupled pyrolysis–mass spectrometry or by trapping of **58** with ethanol. It appears that either the oxoketene **58** decarbonylates very rapidly, or else elimination of methanol is somehow concerted with decarbonylation and ring contraction so that the ketene **59** is formed directly from methyl salicylate [Eq. (4.97)]. Small amounts of phenol, indene, and naphthalene are also formed. The prepara-tive pyrolyses of Eqs. (4.98) and (4.99) are typical; only the esters formed by

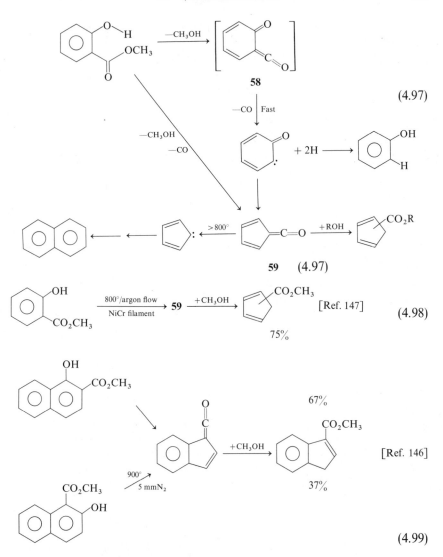

$$(4.97)$$

58

59　(4.97)

$$(4.98)$$

[Ref. 147]

75%

67%

[Ref. 146]

37%

$$(4.99)$$

trapping with excess of methanol are shown, although phenols, indene, and naphthalene are also produced.

By contrast, the formation of *o*-quinonoid imines in the pyrolysis of various *o*-substituted aromatic amines is very readily detected. Thus De Champlain *et al.*[141] showed that pyrolysis of methyl anthranilate at 840° with simultaneous introduction of ethanol between the furnace and the cold finger gave ethyl anthranilate, and thiophenol gave the corresponding phenylthioester. Alternatively, the pyrolyzate could be collected at $-196°$

and the infrared spectrum then indicated the presence of the iminoketene **60** (2080 and 1710 cm^{-1}), the corresponding β-lactam **61** (1825 and 1790 cm^{-1}), and cyanocyclopentadiene (2220 cm^{-1}). Equations (4.101) and (4.102) show two further examples from the same paper in which the intermediate *o*-quinonoid species undergo facile 1,5 hydrogen migration to give aromatic products.

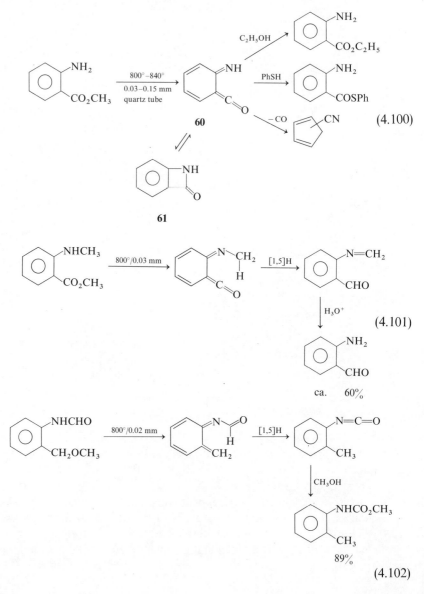

REFERENCES

1. Ellis, R. J., and Frey, H. M. (1966). *J. Chem. Soc. (A)*, p. 553.
2. Woodward, R. B., and Hoffmann, R. (1970). "The Conservation of Orbital Symmetry," pp. 143–144. Academic Press, New York.
3. Fleming, I., and Wildsmith, E. (1970). *Chem. Commun.*, p. 223.
4. Frey, H. M., and Walsh, R. (1969). *Chem. Rev. 69*, 103.
5. Anet, F. A. L., and Leyendecker, F. (1973). *J. Am. Chem. Soc.* **95**, 156.
6. Knecht, D. A. (1973). *J. Am. Chem. Soc.* **95**, 7933.
7. Brown, R. F. C., Gream, G. E., Peters, D. E., and Solly, R. K. (1968). *Aust. J. Chem.* **21**, 2223.
8. McDaniel, D. M., and Benisch, S. (1976). *J. Heterocycl. Chem.* **13**, 405.
9. Crow, W. D., and Wentrup, C. (1970). *Tetrahedron* **26**, 3965.
10. Brown, R. F. C., and Butcher, M. (1972). *Aust. J. Chem.* **25**, 149.
11. Errede, L. A., and Landrum, B. F. (1957). *J. Am. Chem. Soc.* **79**, 4952.
12. Scott, L. T., and Agopian, G. K. (1974). *J. Am. Chem. Soc.* **96**, 4325.
13. Johnson, D. R., and Lovas, F. L. (1972). *Chem. Phys. Lett.* **15**, 65.
14. Godfrey, P. D., Brown, R. D., Robinson, B. J., and Sinclair, M. W. (1973). *Astrophys. Lett.* **13**, 119.
15. Djerassi, C., ed. (1963). "Steroid Reactions: An Outline for Organic Chemists," p. 382. Holden-Day, San Francisco, California.
16. Kaufmann, S., Pataki, J., Rosenkrantz, G., Romo, J., and Djerassi, C. (1950). *J. Am. Chem. Soc.* **72**, 4531.
17. Brown, R. F. C., and Butcher, M. (1973). *Aust. J. Chem.* **26**, 369.
18. King, K. D., and Goddard, R. D. (1975). *J. Am. Chem. Soc.* **97**, 4504.
19. Baron, W. J., and DeCamp, M. R. (1973). *Tetrahedron Lett.*, p. 4225.
20. Jongsma, C., Lourens, R., and Bickelhaupt, F. (1976). *Tetrahedron* **32**, 121.
21. Hopkinson, M. J., Kroto, H. W., Nixon, J. F., and Simmons, N. P. C. (1976). *J. Chem. Soc. Chem. Commun.*, p. 513.
22. Brown, R. F. C., Hooley, N., and Irvine, F. N. (1974). *Aust. J. Chem.* **27**, 671.
23. Kwart, H., and Hoster, D. P. (1967). *Chem. Commun.*, p. 1155.
24. Kwart, H., and Ling, H. G. (1969). *Chem. Commun.*, p. 302.
25. Lehr, R. E., and Wilson, J. M. (1971). *J. Chem. Soc. Chem. Commun.*, p. 666.
26. Hoffmann, R. W., Schüttler, R., and Loof, I. H. (1977). *Chem. Ber.* **110**, 3410.
27. Oele, P. C., and Louw, R. (1972). *Tetrahedron Lett.*, p. 4941.
28. Eastwood, F. W., and Crank, G. (1964). *Aust. J. Chem.* **17**, 1392.
29. Moss, G. I., Crank, G., and Eastwood, F. W. (1970). *J. Chem. Soc. Chem. Commun.*, p. 206.
30. Trahanovsky, W. S., and Park, M.-G. (1973). *J. Am. Chem. Soc.* **95**, 5412.
31. Trahanovsky, W. S., and Park, M.-G. (1974). *J. Org. Chem.* **39**, 1448.
31a. Trahanovsky, W. S., and Alexander, D. L. (1979). *J. Am. Chem. Soc.* **101**, 142.
32. Brown, R. F. C., Eastwood, F. W., Lim, S. T., and McMullen, G. L. (1976). *Aust. J. Chem.* **29**, 1705.
33. Shapiro, J. S., and Swinbourne, E. S. (1967). *Chem. Commun.*, p. 465.
34. Hashimoto, N., Matsumura, K., and Morita, K. (1969). *J. Org. Chem.* **34**, 3410.
35. Semeluk, G. P., and Bernstein, R. D. (1954). *J. Am. Chem. Soc.* **76**, 3793.
36. Semeluk, G. P., and Bernstein, R. D. (1957). *J. Am. Chem. Soc.* **79**, 46.
37. Engelsma, J. W. (1965). *Rec. Trav. Chim. Pays-Bas* **84**, 187.
38. Shilov, A. E., and Sabirova, R. D. (1960). *Russ. J. Phys. Chem.* **34**, 408.
39. Busby, R. E., Iqbal, M., Langston, R. J., Parrick, J., and Shaw, C. J. G. (1971). *J. Chem. Soc. Chem. Commun.*, p. 1293.

40. Nefedov, O. M., and Ivashenko, H. A. (1968). *Isv. Akad. Nauk. S.S.S.R.*, *Ser. Khim.*, p. 446.
41. Platonov, V. E., and Yakobson, G. G. (1976). *Synthesis*, p. 374.
42. Baker, F. S., Busby, R. E., Iqbal, M., Parrick, J., and Shaw, C. J. G. (1969). *Chem. Ind.*, p. 1344.
43. Busby, R. E., Iqbal, M., Parrick, J., and Shaw, C. J. G. (1969). *J. Chem. Soc. Chem. Commun.*, p. 1344.
44. Fuqua, S. A., Parkhurst, R. M., and Silverstein, R. M. (1964). *Tetrahedron* **20**, 1625.
45. Barton, T. J., and Juvet, M. (1975). *Tetrahedron Lett.*, p. 3893.
46. Barton, T. J., and Banasiak, D. S. (1977). *J. Am. Chem. Soc.* **99**, 5199.
47. Birchall, J. M., Gilmore, G. N., and Hazeldine, R. N. (1974). *J. Chem. Soc. Perkin Trans. 1*, p. 2530.
47a. Conlin, R. T., and Gaspar, P. P. (1976) *J. Am. Chem. Soc.* **98**, 3715.
48. DePuy, C. H., and King, R. W. (1960). *Chem. Rev.* **60**, 431.
49. Maccoll, A. (1964). *In* "The Chemistry of Alkenes" (S. Patai, ed.), p. 203. Wiley (Interscience), New York.
50. Smith, G. G., and Kelly, F. W. (1971). *In* "Progress in Physical Organic Chemistry" (A Streitwieser and R. W. Taft, eds.), Vol. 8, p. 75. Wiley (Interscience), New York.
51. Hurd, C. D., and Blunck, F. H. (1938). *J. Am. Chem. Soc.* **60**, 2419.
52. Wertz, D. H., and Allinger, N. L. (1977). *J. Org. Chem.* **42**, 698.
53. Scheer, J. C., Kooyman, E. C., and Sixma, F. L. J. (1963). *Rec. Trav. Chim. Pays-Bas* **82**, 1123.
54. Haag, W. O., and Pines, H. (1959). *J. Org. Chem.* **24**, 877.
55. Bailey, W. J., and Hale, W. F. (1959). *J. Am. Chem. Soc.* **81**, 647.
56. Bailey, W. J., and King, C. (1956). *J. Org. Chem.* **21**, 858.
57. Froemsdorf, D. H., Collins, C. H., Hammond, G. S., and DePuy, C. H. (1959). *J. Am. Chem. Soc.* **81**, 643.
58. Benson, R. E., and McKusick, B. C. (1963). "Organic Syntheses" (N. Rabjohn, ed.), Vol. 4, p. 746. Wiley, New York.
59. Blomquist, A. T., Wolinsky, J., Meinwald, Y. C., and Longone, D. T. (1956). *J. Am. Chem. Soc.* **78**, 6057.
60. Bailey, W. J., and Rosenberg, J. (1955). *J. Am. Chem. Soc.* **77**, 73.
61. Doering, W. von E., and Gilbert, J. C. (1966). *Tetrahedron*, Supplement 7, p. 397.
62. Marvel, C. S., and Hinman, C. W. (1954). *J. Am. Chem. Soc.* **76**, 5435.
63. Bonnett, R., Brown, R. F. C., and Smith, R. G. (1973). *J. Chem. Soc. Perkin Trans. 1*, p. 1432.
64. Bailey, W. J., and Daly, J. J. (1957). *J. Org. Chem.* **22**, 1189.
65. Blanchard, E. P., and Büchi, G. (1963). *J. Am. Chem. Soc.* **85**, 955.
66. Bigley, D. B., and Gabbott, R. E. (1975). *J. Chem. Soc. Perkin Trans. 2*, p. 317.
67. Bigley, D. B., and Gabbott, R. E. (1973). *J. Chem. Soc. Perkin Trans. 2*, p. 1293.
68. Bigley, D. B., and Wren, C. M. (1972). *J. Chem. Soc. Perkin Trans. 2*, p. 926.
69. Bigley, D. B., and Wren, C. M. (1972). *J. Chem. Soc. Perkin Trans. 2*, p. 1744.
70. Bigley, D. B., and Wren, C. M. (1972). *J. Chem. Soc. Perkin Trans. 2*, p. 2359.
71. Daly, N. J., Heweston, G. M., and Ziolkowski, F. (1973). *Aust. J. Chem.* **26**, 1259.
72. Marullo, N. P., Smith, C. D., and Terapane, J. F. (1966). *Tetrahedron Lett.*, p. 6279.
73. Ciganek, E. (1969). *Tetrahedron Lett.*, p. 5179.
74. Mukaiyama, T., Tokizawa, M., Nohira, H., and Takei, H. (1961). *J. Org. Chem.* **26**, 4381.
75. Sonnenberg, F. M., and Stille, J. K. (1966). *J. Org. Chem.* **31**, 3441.
76. Carlson, R. G., and Bateman, J. H. (1967). *J. Org. Chem.* **32**, 1608.
77. Hanack, M., Schneider, H.-J., and Schneider-Bernlöhr, H. (1967). *Tetrahedron* **23**, 2195.

78. Büchi, G., Erickson, R. E., and Wakabayashi, N. (1961). *J. Am. Chem. Soc.* **83**, 927.
79. Büchi, G., MacLeod, W. D., and Padilla, O., J. (1964). *J. Am. Chem. Soc.* **86**, 4438.
80. Dobler, M., Dunitz, J. D., Gubler, B., Weber, H. P., Büchi, G., and Padilla, O., J. (1963). *Proc. Chem. Soc.*, p. 383.
81. Adams, B. L., and Kovacic, P. (1975). *J. Am. Chem. Soc.* *97*, 2829.
82. Kwart, H., and Slutsky, J. (1976). *J. Org. Chem.* **41**, 1429.
83. Kwart, H., Sarner, S. F., and Olson, J. H. (1969). *J. Phys. Chem.* **73**, 4056.
84. Boyd, J., and Overton, K. H. (1972). *J. Chem. Soc. Perkin Trans. 1*, p. 2533.
85. Johnston, J. P., and Overton, K. H. (1972). *J. Chem. Soc. Perkin Trans. 1*, p. 1490.
86. Hunt, M., Kerr, J. A., and Trotman-Dickenson, A. F. (1965). *J. Chem. Soc.*, p. 5074.
87. Dastoor, P. N., and Emovon, E. U. (1973). *Can. J. Chem.* **51**, 366.
88. Harris, C. R. (1946) U.S. Patent No. 2,410,820.
88a. Harris, C. R. (1947). *Chem. Abstr.* **41**, 1235.
89. Harris, C. R. (1947). U.S. Patent No. 2,429,459.
89a. Harris, C. R. (1948). *Chem. Abstr.* **42**, 1962b.
90. Kuntz, P., and Disselnkötter, H. (1972). *J. Liebigs. Ann. Chem.* **764**, 69.
91. Maccoll, A. (1969). *Chem. Rev.* **69**, 33.
92. Maccoll, A. (1965). *In* "Advances in Physical Organic Chemistry" (V. Gold, ed.), Vol. 3, p. 91. Academic Press, New York.
93. Barton, D. H. R. (1949). *J. Chem. Soc.*, p. 148.
94. Barton, D. H. R., and Onyon, P. F. (1949). *Trans. Faraday Soc.* **45**, 725.
95. Benson, S. W. (1963). *J. Chem. Phys.* **38**, 1945.
96. Maccoll, A., and Thomas, P. J. (1967). *Prog. React. Kinet.* **4**, 119.
97. Benson, S. W., and Bose, A. N. (1963). *J. Chem. Phys.* **39**, 3463.
98. Barton, D. H. R., Head, A. J., and Williams, R. J. (1952). *J. Chem. Soc.*, p. 453.
99. Green, J. H. S., and Maccoll, A. (1955). *J. Chem. Soc.*, p. 2449.
100. Jones, J. L., and Ogg, R. A. (1937). *J. Am. Chem. Soc.* **59**, 1939.
101. Christe, K. O., and Pavlath, A. E. (1964). *Chem. Ber.* **97**, 2092.
102. Fields, R., Hazeldine, R. N., and Peter, D. (1967). *Chem. Commun.*, p. 1081.
103. DePuy, C. H., Zabel, D. E., and Wiedeman, W. (1968). *J. Org. Chem.* **33**, 2198.
104. Harding, C. J., Maccoll, A., and Ross, R. A. (1969). *J. Chem. Soc. (B)*, p. 634.
105. Knözinger, H. (1971). *In* "The Chemistry of the Hydroxyl Group" (S. Patai, ed.), Part 2, p. 641. Wiley (Interscience), New York.
106. Barnard, J. A. (1959). *Trans. Faraday Soc.* **55**, 947.
107. Barnard, J. A. (1957). *Trans. Faraday Soc.* **53**, 1423.
108. Bock, H., and Mohmand, S. (1977). *Angew. Chem. Int. Ed. Engl. 16*, 104.
109. Tanimoto, M., and Saito, S. (1977). *Chem. Lett.*, p. 637.
110. Saito, S. (1976). *Chem. Phys. Lett. 42*, 399.
110a. Penn, R. E., Block, E., and Revelle, L. K. (1978). *J. Am. Chem. Soc.* **100**, 3622.
111. Davis, F. A., Yocklovitch, S. G., and Baker, G. S. (1978). *Tetrahedron Lett.*, p. 97.
112. Ando, W., Oikawa, T., Kishi, K., Saiki, T., and Migita, T. (1975). *J. Chem. Soc. Chem. Commun.*, p. 704.
113. Patterson, J. M., Haidar, N. F., Papadopoulos, E. P., and Smith, W. T. (1973). *J. Org. Chem. 38*, 663.
114. Voerkelius, G. A. (1909). *Chem.-Ztg.* **23**, 1078–1090.
114a. Voerkelius, G. A., (1910). *Chem. Abstr.* **4**, 1653.
115. Roberts, J. T., Kovacic, P., Tonnis, J. A., and Scalzi, F. V (1977). *J. Chem. Soc. Chem. Commun.*, p. 418.
115a. Block, E., Penn, R. E., Olsen, R. J., and Sherwin, P. F. (1976). *J. Am. Chem. Soc.* **98**, 1264.
116. Solouki, B., Rosmus, P., and Bock, H. (1976). *J. Am. Chem. Soc.* **98**, 6054.

117. Young, J. A., Simmons, T. C., and Hoffmann, F. W. (1956). *J. Am. Chem. Soc.* **78**, 5637.
118. Van Bogaert, G. E. (1968). U.S. Patent No. 3,378,583.
118a. Van Bogaert, G. E. (1968). *Chem. Abstr.* **69**, 51651v.
119. Bock, H., Solouki, B., Bert, G., and Rosmus, P. (1977). *J. Am. Chem. Soc.* **99**, 1663.
120. Hanford, W. E., and Sauer, J. C. (1946). *In* "Organic Reactions" (R. Adams, ed.), Vol. 3, p. 108. Wiley, New York.
121. Borrmann, D. (1968). *In* "Methoden der Organischen Chemie (Houben-Weyl)" (E. Müller, ed.), Vol. 7, Part 4, p. 52. Thieme, Stuttgart.
122. Lacey, R. N. (1964). *In* "The Chemistry of Alkenes" (S. Patai, ed.), p. 1161. Wiley (Interscience), New York.
123. Fisher, G. J., MacLean, A. F., and Schnizer, A. W. (1953). *J. Org. Chem.* **18**, 1055.
124. Jenkins, A. D. (1952). *J. Chem. Soc.*, p. 2563.
125. Baxter, G. J., Brown, R. F. C., Eastwood, F. W., and Harrington, K. J. (1977). *Aust. J. Chem.* **30**, 459.
126. Brown, A. L., and Ritchie, P. D. (1968). *J. Chem. Soc. C*, p. 2007.
127. Blackman, G. L., Brown, R. D., Brown, R. F. C., Eastwood, F. W., McMullen, G. L., and Robertson, M. L. (1978). *Aust. J. Chem.* **31**, 209.
128. Brown, R. D., Brown, R. F. C., Eastwood, F. W., Godfrey, P. D., and McNaughton, D. (1979). *J. Am. Chem. Soc.*, **101**, 4705.
129. Bock, H., Hirabayashi, Mohmand, S., and Solouki, B. (1977). *Angew. Chem. Int. Ed. Engl.* **16**, 105.
130. Brown, R. F. C., and Colmanet, S. J. (1977). Unpublished work, Monash University.
131. Hurd, C. D., and Blunck, F. H. (1938). *J. Am. Chem. Soc.* **60**, 2419.
132. Barefoot, A. C., and Carroll, F. A. (1974). *J. Chem. Soc. Chem. Commun.*, p. 357.
133. Leyendecker, F. (1976). *Tetrahedron* **32**, 349.
134. Berkowitz, W. F., and Ozorio, A. A. (1975). *J. Org. Chem.* **40**, 527.
135. Krantz, A., and Hoppe, B. (1975). *J. Am. Chem. Soc.* **97**, 6590.
136. Loudon, A. G., Maccoll, A., and Wong, S. K. (1969). *J. Am. Chem. Soc.* **91**, 7577.
137. Loudon, A. G., Maccoll, A., and Wong, S. K. (1970). *J. Chem. Soc. (B)*, p. 1733.
138. Schiess, P., and Heitzmann, M. (1978). *Helv. Chim. Acta* **61**, 844.
138a. Harruff, L. G., Brown, M., and Boekelheide, V. (1978). *J. Am. Chem. Soc.* **100**, 2893.
139. Schiess, P., and Heitzmann, M. (1977). *Angew. Chem. Int. Ed. Engl.* **16**, 469.
140. Schiess, P., and Radimerski, P. (1974). *Helv. Chim. Acta* **57**, 2583.
141. De Champlain, P. Luche, J. L., Marty, R. A., and de Mayo, P. (1976). *Can. J. Chem.* **54**, 3749.
142. Clauson-Kaas, N., and Elming, N. (1952). *Acta Chem. Scand.* **6**, 560.
143. Gardner, P. D., Sarrafizadeh R., H., and Brandon, R. L. (1959). *J. Am. Chem. Soc.* **81**, 5515.
144. Cavitt, S. B., Sarrafizadeh R., H., and Gardner, P. D. (1962). *J. Org. Chem.* **27**, 1211.
145. Grützmacher, H.-F., and Hübner, J. (1971). *J. Liebigs Ann. Chem.* **748**, 154.
146. Grützmacher, H.-F., and Hübner, J. (1973). *J. Liebigs Ann. Chem.*, p. 793.
147. Mamer, O. A., Rutherford, R. G., and Seidewand, R. J. (1974). *Can. J. Chem.* **52**, 1983.

Chapter 5

Generation of Carbenes and Nitrenes and Their Rearrangements in the Gas Phase

I. INTRODUCTION

The study of the chemistry of carbenes $R_2C:$ and nitrenes $R\ddot{N}:$ as transient reactive intermediates is a major theme of recent organic chemistry, and its importance is reflected by a number of monographs dealing with these species. Systematic accounts of carbene chemistry have been provided by Hine[1] and Kirmse[2] and developments in carbene chemistry are reviewed in a series edited by Jones and Moss.[3] Nitrenes have similarly been covered systematically by Lwowski.[4] The involvement of such reactive intermediates in pericyclic reactions is analyzed in detail in the series edited by Lehr and Marchand,[5] and a more general account of their chemistry has been given by Gilchrist and Rees.[6]

Most studies have been concerned with the generation of carbenes and nitrenes in solution through thermal or photochemical fragmentation reactions. The interpretation of reactions in solution may occasionally be ambiguous because of the involvement of organometallic or coordinated species (termed *carbenoids* by G. Köbrich[7]) which form products the same as those to be expected from free carbenes. The behavior of reactive intermediates may also be modified by interaction with donor solvents. Finally, carbenes and nitrenes in solution may be trapped so efficiently by reagents or solvents that a range of potential intramolecular reactions will not occur. By contrast, free carbenes and nitrenes generated by gas-phase pyrolysis at low pressures are less likely to undergo intermolecular reactions, whereas intramolecular rearrangements are commonplace.

The deep-seated rearrangements of aryl- and heteroarylcarbenes and nitrenes which occur almost exclusively in the gas phase at high temperatures have attracted most attention.[8] It is convenient, however, to begin by describing some less complex examples of pyrolytic reactions which appear to involve the formation and rearrangement of carbenes.

II.   FORMATION AND REARRANGEMENT OF CARBENES

A.   Pyrolysis of α-Diazocarbonyl Compounds

The   thermal   Wolff   rearrangement   of   alkoxycarbonylcarbenes $R—\overset{..}{C}—CO_2R'$ to ketenes $R'O—CR{=}C{=}O$ cannot be achieved in solution at temperatures below 150°, and products of addition to alkenes, or dimers, are formed instead.[9,10,11] This reluctance to rearrange can be overcome in flash pyrolytic experiments, although in the case of dimethyl diazomalonate Richardson et al.[12] have shown that the Wolff rearrangement is favored in a temperature range (420°–540°) higher than the minimum temperature (280°) required for initial loss of nitrogen. At the lower temperature the major constituent (92%) of the pyrolyzate is methyl acrylate, formed by an initial carbene insertion into a C—H bond followed by fission of the intermediate β-lactone [Eq. (5.1)]; such a sequence has not been observed in solution. Methyl acetate (3%) and methyl vinyl ether (5%) are also present in the pyrolyzate.

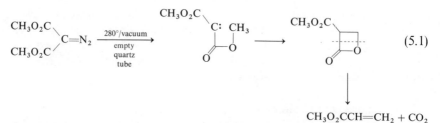

$$CH_3O_2CCH{=}CH_2 + CO_2$$

(5.1)

At higher temperatures methyl acrylate is still formed but products most readily explained by initial Wolff rearrangement and decarbonylation appear in the pyrolyzate [Eq. (5.2)]. The major product is methyl vinyl ether, formed by subsequent carbene insertion and β-lactone fission, but methyl pyruvate and methyl acetate are also present as products of a sequence of decarbonylations and rearrangements.

A related sequence has been invoked by Kammula et al.[13] to explain the spectacular   fragmentation   of   5-diazo-2,2-dimethyl-1,3-dioxane-4,6-dione

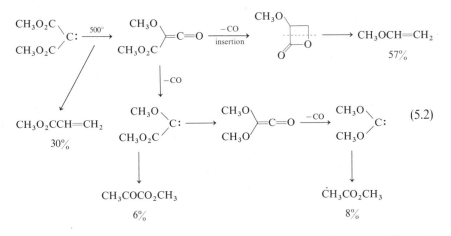

(5.2)

(diazo Meldrum's acid) which gives mainly nitrogen, carbon monoxide, and acetone [Eq. (5.3)]. The formation of the intermediate ketenes **1** and **2** has been proposed, and the former has been generated and trapped by photolysis of diazo Meldrum's acid in methanol.

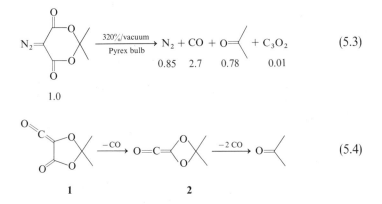

(5.3)

(5.4)

The Wolff rearrangement of aromatic cyclic α-diazocarbonyl compounds has been achieved by flash pyrolysis over a Nichrome spiral and the intermediate ketenes have been trapped in the gas phase with methanol and other alcohols.[14] The pyrolysis of 2-diazo-1-indanone was inefficient because it sublimed with difficulty but the product was the ring-contracted ester **3** with a little 1-indanone [Eq. (5.5)]. 2-Diazoindan-1,3-dione similarly underwent Wolff ring contraction and reaction with methanol to yield mainly the keto ester **4** [Eq. (5.6)]. This keto ester could not be obtained by photolysis of 2-diazoindan-1,3-dione in methanol because this led to dimethyl homophthalate (a product of further photochemical methanolysis) and other

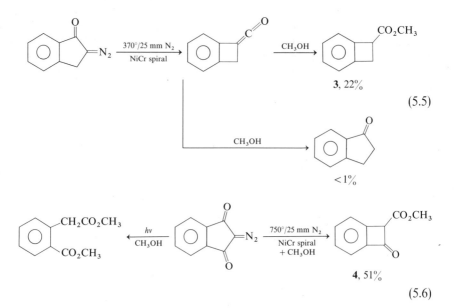

(5.5)

(5.6)

compounds. Spangler and co-workers[14] suggest that gas-phase pyrolysis of
this type may be useful for the synthesis of other strained and photochemically unstable systems.

B. Decarbonylation of Ketenes from Derivatives of Meldrum's Acid, and Related Reactions

The generation of dimethylketene by pyrolysis of the dimethyl derivative
of Meldrum's acid was first noted by Ott[15] at a time when such compounds
were believed to possess the β-lactone structure **5** rather than the dioxane-
dione structure **6** established by Davidson and Bernhard[16] in 1948. Flash
vacuum pyrolysis of these compounds over the temperature range 400°–
600° is a useful general method for the generation of a variety of substituted
ketenes, some of which undergo decarbonylation with great ease so that only
products derived from the corresponding carbenes are detected.

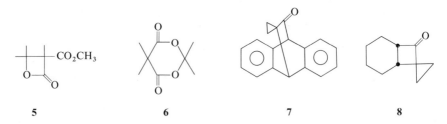

5   6   7   8

The strained ketene carbonylcyclopropane (dimethyleneketene, **10**) cannot be prepared by elimination of hydrogen chloride from cyclopropane-carbonyl chloride with tertiary amines, apparently because of increasing I-strain in forming the product,[17] although it is formed in solution by photolysis of the bridged cyclopropyl ketone **7**.[18] Flash vacuum pyrolysis of ketone **7**[18] at 750° or of the tricyclic ketone **8**[19] at 700° gives vinylketene (identified by trapping with methanol), propyne, and allene; diradical intermediates have been suggested in each case. However, pyrolysis of the spiro-dione **9** readily gives carbonylcyclopropane[20] which can be detected by infrared spectroscopy after deposition on sodium chloride at $-196°$, trapped with aniline, or isolated as the dimer, dispiro[2,1,2,1]octane-4,8-dione.

The presence of allene in this 500° pyrolyzate [Eq. (5.7)] can also be detected by infrared spectroscopy, and at 600° decarbonylation of the ketene and ring opening of the carbene to allene is complete [Eq. (5.8)].

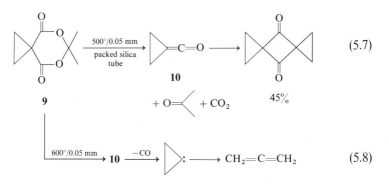

$$(5.7)$$

$$(5.8)$$

In the case of the pyrolysis of the corresponding spiroxirane **11** at 510° the formation of the intermediate oxiraneketene has not been detected, and ketene itself is obtained[21] as the final product of ring opening of the carbenaoxirane [Eq. (5.9)]. The same carbenaoxirane is formed in the chelotropic elimination [Eq. (5.10)] studied by Hoffmann and Schüttler,[22] who showed by ^{14}C-labeling that this carbene opens directly to ketene without scrambling of the carbon atoms through formation of oxirene.

The 5-methylthio derivative of 5-methyl Meldrum's acid fragments at 410° and 0.2 mm to give acetone, carbon dioxide and the ketene

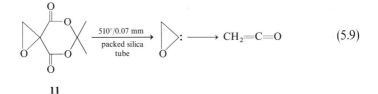

$$(5.9)$$

11

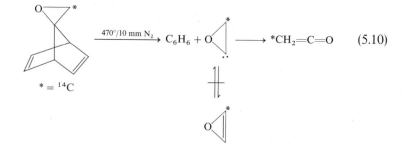

CH_3S—$C(CH_3)$=C=O, whereas at 590° decarbonylation to CH_3S—$\ddot{C}$—CH_3 and rearrangement to CH_3S—CH=CH_2 is complete.[21] The corresponding 5-phenylthio derivative gives 2-methylbenzo[*b*]thiophen-3[2*H*]-one and phenyl vinyl sulfide in the ratio 2:1 over the temperature range 450°–600° [Eq. (5.11)].

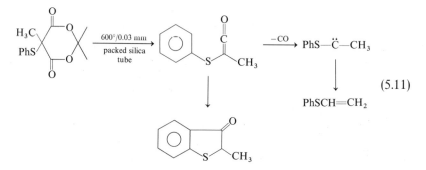

(5.11)

The pyrolytic behavior of the related 5-acyloxy derivatives investigated by Brown and co-workers[23] is strikingly different from that of the thioethers above. Fragmentation with decarbonylation occurred even at 450°–460°, and the intermediate ketenes, if formed at all, must decarbonylate very readily to acyloxycarbenes.

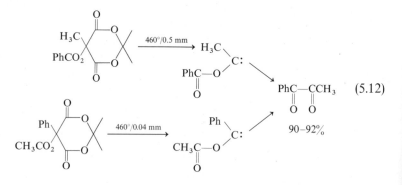

(5.12)

No vinyl esters are produced from the acyloxycarbenes RCO_2—$\overset{..}{C}$—CH_3, but instead 1,2-diketones are formed in high yield [Eq. (5.12)] by insertion of the carbene center into the O—CO bond, equivalent to overall acyl migration. The absence of vinyl esters in the pyrolyzate suggests a modification of the true carbene structure **12** by the adjacent oxygen, and the dipolar structure **13** may be important. Acyloxycarbenes RCO_2—$\overset{..}{C}$—H have been generated in solution by treatment of esters RCO_2CH_2Cl with the hindered base lithium 2,2,6,6-tetramethylpiperidide; Olofson and co-workers[24] found that these carbenes added normally to alkenes to give cyclopropyl esters, and ^{18}O-labeling experiments showed no sign of the equilibration of the acyloxycarbene with the cyclic structure **14**.

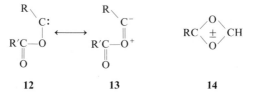

      12   **13**   **14**

Cyclic acyloxycarbenes have been generated in the gas phase by pyrolytic α-elimination of benzoic acid from benzoyloxy lactones [Eq. (5.13)]. The ring contraction leading to benzocyclobutenedione in this example[23] is the reverse of the well-established photochemical formation of the acyloxycarbene from this dione in solution.[25]

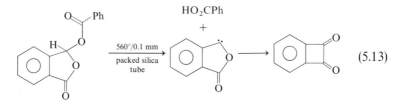

                                                         (5.13)

C. Formation of Methylenecarbenes

Methyleneketenes R_2C=C=C=O can be formed in the gas phase by pyrolysis of methylene or substituted methylene derivatives **15** of Meldrum's

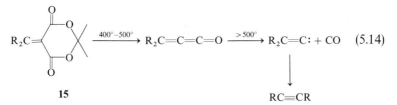

$$\text{15}$$
$$RC \equiv CR$$

acid,[21,26] usually in the temperature range $400°-500°$. At somewhat higher temperatures ($500°-650°$) substituted methyleneketenes[27] and the parent $CH_2=C=C=O$[21] decarbonylate to form methylenecarbenes $R_2C=C:$. In the absence of alternative intramolecular pathways such methylenecarbenes rearrange in the gas phase to give acetylenes $RC≡CR$.

The sequence of condensation of an aromatic aldehyde with Meldrum's acid, and flash vacuum pyrolysis of the resulting arylmethylene derivative at $550°-600°$ provides a convenient synthesis of substituted phenylacetylenes $RC_6H_4C≡CH$ (R = m- or p-H, CH_3, CH_3O, Cl, CN, or o-CH_3O) in 64–98% yield.[27] However the presence of an o-alkyl group may lead to rearrangement of the methyleneketene to a 2-naphthol (see Chapter 9, Section II,C). A ^{13}C-labeling experiment[27] has shown that hydrogen migration is favored over phenyl migration in the rearrangement of $PhCH=C:$, although the ratio of products obtained [Eq. (5.15)] probably does not accurately reflect the relative migratory aptitudes of H and Ph because under the conditions used the reaction may to a small extent be reversible (see Section II,D).

$$Ph^{13}CH=\underset{O}{\overset{O}{\diamondsquare}} \xrightarrow[\substack{\text{packed} \\ \text{silica tube}}]{560°/0.1 \text{ mm}} Ph^{13}C≡CH + PhC≡^{13}CH \qquad (5.15)$$
$$75 \quad : \quad 25$$

A further example of this sequence[28] is the condensation of methyl phenylglyoxylate $PhCOCO_2CH_3$ with Meldrum's acid and pyrolysis of the intermediate product to give $PhC≡CCO_2CH_3$ (73% yield at $600°$ and 1 mm).

The chemistry of unsaturated carbenes has recently been reviewed by Stang[29] who has also introduced the efficient method shown in Eq. (5.16) for the generation of methylenecarbenes in solution with trapping by an alkene. No intermolecular trapping of a thermally generated methylenecarbene with an alkene in the gas phase has been reported; presumably it

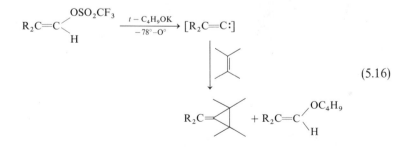

$$R_2C=C\overset{OSO_2CF_3}{\underset{H}{\diagdown}} \xrightarrow[-78°-0°]{t-C_4H_9OK} [R_2C=C:]$$

$$(5.16)$$

$$R_2C=\!\!\!\prec\!\!\!\diamondsuit + R_2C=C\overset{OC_4H_9}{\underset{H}{\diagup}}$$

would be difficult under flash pyrolytic conditions for intermolecular reactions to compete with rearrangement to an acetylene. One example of intramolecular reaction with a phenyl group[27] has been found in the 2-biphenylyl system and is discussed in Section II,D.

Cycloalkylidenecarbenes may be generated by pyrolysis of cycloalkylidene derivatives of Meldrum's acid at 480°–640°.[30] In this series decarbonylation occurs readily and products derived from intermediate cycloalkylideneketenes are not usually found. The behavior of the cycloheptylidene system [Eq. (5.17)] illustrates two major intramolecular reactions of gaseous cycloalkylidenecarbenes. Cycloheptylidenecarbene undergoes ring expansion to cyclooctyne, which has been prepared in 52% yield in this way.[30] This reaction is reversible at 500° so that direct generation of the carbene or repyrolysis of cyclooctyne leads also to *cis*-bicyclo[3.3.0]oct-2-ene and 1,3-cyclooctadiene via carbene insertion leading to the bicyclic cyclopropene **16**, ring opening to a new carbene **17**, and insertion into transannular or adjacent C—H bonds. These reactions are similar to those observed for cycloalkylidenecarbenes generated in solution by base catalyzed elimination,[31] except that secondary products of base catalysis are not found. In the gas phase cyclohexylidene- and cyclopentylidenecarbene are converted in good yield into 1,3-cycloheptadiene and 1,3-cyclohexadiene, respectively, but cyclobutylidenecarbene generated at 640° and 0.05 mm is converted into pent-1-en-3-yne (64%) [Eq. (5.18)]. The mechanism of this last reaction is uncertain, but electrocyclic ring opening of a cyclobutene derivative may be involved.[30]

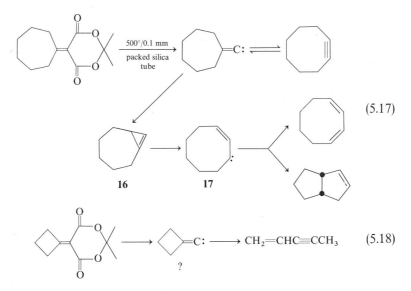

(5.17)

16   **17**

$$CH_2{=}CHC{\equiv}CCH_3 \qquad (5.18)$$

A synthetically versatile alternative approach to the generation of methylenecarbenes involves the pyrolysis of derivatives of 4-methyleneisoxazol-5-one. Wentrup and Reichen[32] have found that the 4-arylmethylene-3-methyl derivatives fragment smoothly on flash pyrolysis to give carbon dioxide, acetonitrile, and the arylmethylenecarbene which then rearranges to an arylacetylene [Eq. (5.19)]. In this reaction the methylenecarbene is formed directly rather than by decarbonylation of an intermediate ketene. The arylmethyleneisoxazolones are readily obtained by condensation of an aromatic aldehyde with acetoacetic ester and hydroxylamine,[33] and isolated yields of phenylacetylenes bearing Cl, OH, MeO, Me_2N, and methylenedioxy substituents range from 45 to 95%. Heteroarylacetylenes of the pyrrole, furan, thiophene, and pyridine series have been prepared similarly.[34]

$$\underset{\substack{\text{quartz tube}}}{\overset{\substack{520°-650°/ \\ 0.01-0.05\ mm}}{\longrightarrow}}\ CH_3CN + CO_2 + ArCH{=}C: \qquad (5.19)$$

$$\downarrow$$

$$ArC{\equiv}CH$$

The pyrolysis of isoxazolone derivatives **18** provides a general method for the generation of carbenes $Y{-}X{=}C:$ which has been vigorously exploited by Wentrup's group.[34] For example, 4-arylhydrazonoisoxazolones (**18**; $Y{-}X = ArNH{-}N$) fragment almost quantitatively between 400° and 500° at 10^{-3}–10^{-2} mm to give N-isocyanoamines, $ArNH{-}N{=}C: \leftrightarrow ArNH{-}^+N{\equiv}C^-$, which rearrange further to cyanamides and indazoles[35] [Eq. (5.21)]. The general method offers a route to ketenimines, to some aromatic isonitriles,[35a] and, in principle,[34] to fulminates, $RO{-}^+N{\equiv}C^-$.

$$\longrightarrow CH_3CN + CO_2 + Y{-}X{=}C: \qquad (5.20)$$

18

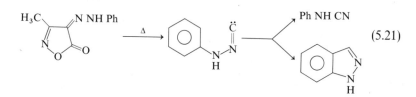

$$(5.21)$$

D. Thermal Equilibration of Acetylenes
    with Methylenecarbenes

Interest in the synthesis of arylacetylenes by generation and rearrangement of arylmethylenecarbenes led to the discovery[36] that this reaction is reversible at high temperatures (550°–720°), and it now appears probable that many acetylenes $RC{\equiv}CR$ are in equilibrium with methylenecarbenes $R_2C{=}C:$ at similar high temperatures. The resulting degenerate rearrangements have been detected by ^{13}C-labeling, and the involvement of carbenes is confirmed by the formation of products of intramolecular insertion reactions, but in some cases study of the degenerate rearrangement is difficult because other complex decompositions intervene.[37]

The rearrangement was first established[27] in the case of labeled phenyl-acetylene, $Ph^{13}C{\equiv}CH$, which on pyrolysis at 550° and 0.05 mm through a packed silica tube gave phenylacetylene containing 10% of $PhC{\equiv}^{13}CH$, detected by proton magnetic resonance (pmr) spectroscopy. On pyrolysis at 700° and 0.02 mm a pyrolyzate which contained $PhC{\equiv}^{13}CH$ and $Ph^{13}C{\equiv}CH$ in equal amounts was obtained, and this can most readily be explained in terms of the equilibrium of Eq. (5.22). An earlier attempt[38] to detect phenyl migration in $Ph^{13}C{\equiv}CH$ or its anion in solution below 200° was unsuccessful.

$$Ph^{13}C{\equiv}CH \underset{}{\overset{700°}{\rightleftarrows}} Ph^{13}CH{=}C: \rightleftarrows H^{13}C{\equiv}CPh \qquad (5.22)$$
$$\overset{13}{:}C{=}CHPh$$

Similar behavior has been established[37] for labeled 1-adamantylacetylene, $1\text{-AdC}{\equiv}^{14}CH$, at 780° and 0.1 mm (25% of rearranged product), but no rearrangement was found for $(CH_3)_3C^{13}C{\equiv}CH$ in one pass at 790° or for $Ph(CH_3)_2C^{13}C{\equiv}CH$ at 680°. It appears that rearrangement of alkyl-acetylenes requires a higher temperature than that of arylacetylenes, and in accord with this trend the rearrangement of doubly labeled acetylene, $HC{\equiv}^{13}CD$, led to about 40% of $DC{\equiv}^{13}CH$ in one pass at 850° and 0.05 mm.[39] The doubly labeled acetylene was generated in the flow system by a retro-Diels–Alder reaction of the bridged anthracene derivative **19** [Eq. (5.23)].

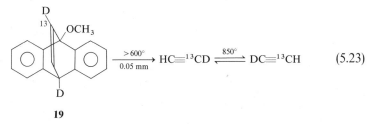

19

That free methylenecarbenes are indeed involved in these rearrangements is suggested by the formation of both phenanthrene and 1,2-benzazulene by pyrolysis of biphenyl-2-ylacetylene[27] [Eq. (5.24)]. Various pathways are possible for the formation of phenanthrene, and at 700° some may be formed

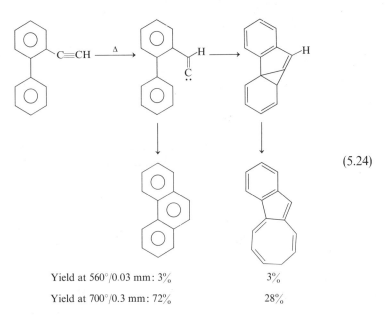

(5.24)

Yield at 560°/0.03 mm: 3%   3%

Yield at 700°/0.3 mm: 72%   28%

indirectly by isomerization of 1,2-benzazulene, but the formation of 1,2-benzazulene itself must surely involve trapping of the methylenecarbene by the adjacent phenyl substituent as shown. Carbene insertions into benzylic C—H bonds occur in the rearrangement of o-tolylacetylene to indene[37] (75% yield at 740° and 0.02 mm) and of 1-ethinyl-8-methylnaphthalene to phenalene.[37] The formation of acenaphthylene in 80% yield on pyrolysis of 1-naphthylacetylene (750° and 0.5 mm) suggests the occurrence of insertion into an aromatic C—H bond.[37]

The equilibrium between cyclic acetylenes and the corresponding cycloalkylidenecarbenes should be strongly influenced by ring size.[30] Cyclononyne is partly converted into products derived from cyclooctylidenecarbene on pyrolysis at 600° and 0.05 mm, and cyclooctyne behaves similarly at 480° and 0.1 mm.[30] In smaller ring systems the balance between ring strain in the cycloalkyne and the higher enthalpy of a methylenecarbene is not easily investigated directly. The pyrolysis of the tetrahydrofuranylidene compound **20** [Eq. (5.25)] leading to butatriene and formaldehyde as major products can be interpreted[30] as involving retro-Diels–Alder cleavage of the strained alkyne **21**, but alternative pathways from the initial methylenecarbene are also feasible.

Strausz et al. in an ab initio molecular orbital study of the $HC\equiv CH/CH_2=C$: and $FC\equiv CF/CF_2=C$: systems[40] estimate ΔH for isomerization to the carbene at $+37.9$ kcal/mole for C_2H_2 and $+36.4$ kcal/mole for C_2F_2. The activation energy for the isomerization of $HC\equiv CH$ via a cyclic transition

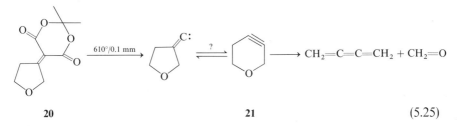

| **20** | **21** | (5.25) |

state **22** is estimated at 114.7 kcal/mole from this study, although the true activation energy cannot be so much greater than ΔH because, as these authors comment,[40] all experimental results suggest a low activation energy for the very rapid isomerization of $CH_2=C:$ to $HC\equiv CH$. The calculated energy of the corresponding cyclic transition state for the C_2F_2 system is about 59 kcal/mole higher than that for C_2H_2, in accord with the observed relative reluctance of photochemically generated $CF_2=C:$ to rearrange to $FC\equiv CF$ and its consequent reactivity in addition and insertion reactions in the gas phase.[41] More recently Dykstra and Schaefer[42] have calculated a barrier of 8.6 kcal/mole to the isomerization of $CH_2=C:$, and this leads to a much lower activation energy of about 47 kcal/mole for conversion of $HC\equiv CH$ to $CH_2=C:$.

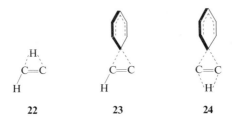

| **22** | **23** | **24** |

Cyclic transition states of type **22** seem reasonable as leading to the degenerate rearrangements and carbene intermediates of the high-temperature reactions reviewed above, but participation of unsaturated groups such as phenyl is also likely, as in structure **23**. The experimental observation of rearrangement alone would also be consistent with more symmetrical transition states such as the doubly bridged **24**.

E. Isonitrile–Nitrile Rearrangements

The thermal rearrangement of an isonitrile $R—^+N\equiv C^-$ to a nitrile $R—C\equiv N$ is formally closely related to the methylenecarbene–acetylene rearrangement. The nitrile is thermodynamically strongly favored[43] [Eq. (5.26)] and the thermal equilibration of an alkyl or aryl nitrile with the isonitrile has not as yet been detected.

$$CH_3—^+N{\equiv}C^- \longrightarrow CH_3—C{\equiv}N \qquad (5.26)$$

$$\Delta H = -15 \text{ kcal/mole}$$

The kinetics and mechanism of rearrangement of isonitriles in solution and in the gas phase have been reviewed by Maloney and Rabinovitch.[44] With the exception of isonitriles of special structure such as **25**, which appears to undergo rearrangement with racemization by a radical mechanism,[45] isonitrile rearrangements are concerted reactions involving transition states well represented by **26**, with little separation of charge. Dewar and Kohn[46] prefer to depict this as the π-complex **27** for CH_3NC and they calculate by the MINDO/2 method an activation energy of 34.3 kcal/mole, which compares reasonably well with the experimental value 38.4 kcal/mole.[47]

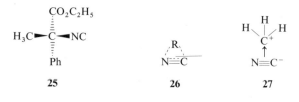

25   26   27

A consequence of these thermochemical considerations is that isonitriles rearrange rapidly in the gas phase at moderate temperatures (half-life for CH_3NC ca. 90 sec at 260°[44]) and isonitriles which may be obligatory intermediates in high-temperature processes usually appear in the pyrolyzates as nitriles. Thus the products of pyrolysis of the 4-arylhydrazonoisoxazolones (see Section II,C) are cyanamides $ArNHC{\equiv}N$ rather than isocyanoamines,[35] and similar examples [Eqs. (5.27),[48] (5.28),[49] and (5.29)[50]] have been encountered in the course of pyrolytic work in the author's laboratory.

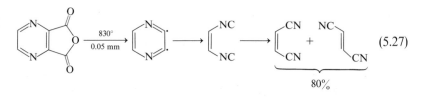

$$(5.27)$$

$$80\%$$

$$ArNHCOCH_2Cl \xrightarrow[0.1-0.4 \text{ mm}]{780°-850°} ArCN + CH_2O + HCl \qquad (5.28)$$

$$ArNHCH{=}S \xrightarrow[0.01-0.1 \text{ mm}]{700°-720°} ArCN + H_2S \qquad (5.29)$$

A specific example of the *N*-arylchloracetamide decomposition[49] will show the difficulty of establishing the involvement of free isonitriles. A possible mechanism for this reaction involves formation of an intermediate iminolactone which could fragment to an isonitrile and formaldehyde [Eq. (5.30)]. The 5-methyl-2-biphenylylchloroacetamide gave on pyrolysis the 5-methyl-2-carbonitrile and 2-methylphenanthridine [Eq. (5.31)], products consistent with rearrangement and insertion reactions of a 2-biphenylyl isonitrile. The 5-methyl group served only to label the phenanthridine ring; the unmethylated compound gave phenanthridine in the same yield. However, 2-biphenylyl isonitrile itself on pyrolysis under similar conditions gave the rearranged 2-carbonitrile with only 0.6% of phenanthridine. This result does not disprove the formation of isonitriles in *N*-arylchloroacetamide pyrolyses, but it does require consideration[49] of other more electrophilic intermediates for the formation of phenanthridine.

$$\text{ArN=} \quad\longrightarrow\quad \text{Ar}^+\text{N}{\equiv}\text{C}^- + CH_2O \qquad\qquad (5.30)$$

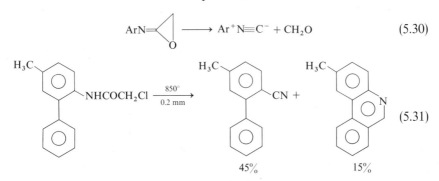

$$(5.31)$$

The thermal equilibration of nitrile and isonitrile structures has been achieved in the case of hydrogen cyanide itself by passing a stream of the gas through a series of nozzles heated to about 730°.[51] The small proportion of HNC formed was detected as the stream passed between the plates of a parallel plate microwave spectrometer. The characteristic $J = 1 \leftarrow 0$ microwave absorption of HNC at 90663.59 MHz was identified with a previously unassigned line from interstellar molecules.[51] Hydrogen cyanide is estimated to be 14.5 kcal/mole more stable than hydrogen isocyanide by *ab initio* calculation,[52] and at least 10.8 kcal/mole more stable from consideration of the limits of microwave detection.[53]

III. REARRANGEMENTS INVOLVING AROMATIC CARBENES AND NITRENES

A. Introduction

The rearrangements of arylcarbenes, arylnitrenes, and related species in the gas phase are of a complexity which rivals that of carbocations in solution or of gaseous ions in the mass spectrometer. The subject has been reviewed critically and in considerable detail by Wentrup,[8] and the present brief account must inevitably cover much the same ground. The aim of the following sections is to summarize the key experimental facts, to draw attention to leading papers, and to outline such firm conclusions as have emerged. The treatment is thus oversimplified and much detail is omitted. In particular, individual pyrolytic reactions are emphasized at the expense of complex high temperature equilibria involving many isomeric species. Nothing can properly replace extended study of the original publications. The reader should be aware, however, that research activity in this area is intense, that mechanistic theories have tended to be modified or demolished soon after their proposal, and that this situation is likely to continue for some years to come. Thus some parts of Wentrup's 1976 review must be read with caution in the light of further work[54] summarized in his lecture published[34] in 1977.

B. Methylenecyclohexadienylidene: Ring Contraction to Fulvenallene

Methylenecyclohexadienylidene (**29**) can be generated by pyrolysis of various cyclic precursors of the form **28** in the temperature range $500°-1000°$, with extrusion of stable fragment molecules X and Y such as CO, CO_2, or N_2 [Eq. (5.32)]. Under such conditions the major product is usually fulvenallene (**30**), the primary product of a Wolff-type ring contraction, though this is accompanied by ethinylcyclopentadiene (**31**) which has been shown[55,56] to be a product of secondary isomerization. Fulvenallene in the condensed phase is an unstable and highly reactive hydrocarbon which is often characterized by addition of a secondary amine to give a stable fulvene derivative **32**.

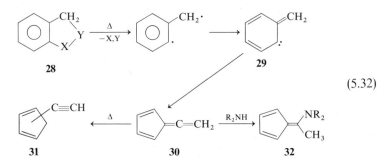

(5.32)

Under true flash vacuum pyrolytic conditions with a short contact time and at low conversion benzocyclopropene can be formed by cyclization of methylenecyclohexadienylidene as in the pyrolysis of benzocyclobutenone reported by Arnold *et al.*[57] [Eq. (5.33)]. Benzocyclopropene itself on pyrolysis at 515°–800° and 0.01–0.1 mm gives fulvenallene and ethinylcyclopentadiene, the latter in increased yield at the highest temperature.[55]

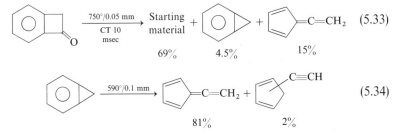

$$\text{(5.33)}$$

69%   4.5%   15%

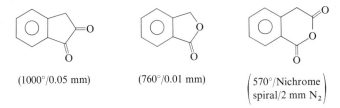

$$\text{(5.34)}$$

81%   2%

Conditions used to produce 70–72% yields of fulvenallene by pyrolysis of indan-1,2-dione,[56] of phthalide,[55] and of homophthalic anhydride[58,59] are shown below their structures. Wiersum and Nieuwenhuis[60] have described a recycling furnace which permits the pyrolysis of phthalide at 700°–750° to give a liquid pyrolyzate containing fulvenallene (60–70%) and benzene at the rate of 3 gm in 6–8 hr. These authors suggest that the benzene may be formed by an alternative cleavage of phthalide which leads initially to benzyne, carbon monoxide, and formaldehyde.

(1000°/0.05 mm)   (760°/0.01 mm) $\left(\begin{array}{l}570°/\text{Nichrome}\\ \text{spiral}/2 \text{ mm N}_2\end{array}\right)$

Wentrup, Wentrup-Byrne, and Müller[54] have shown by pyrolysis of [1-^{12}C]benzocyclopropene (depleted of ^{13}C) that at 800° the major pathway of ring contraction is indeed the direct Wolff-type process leading to terminally labeled fulvenallene (83%) although this is accompanied by fulvenallene in which the label has been scrambled by automerization [Eq. (5.35)].

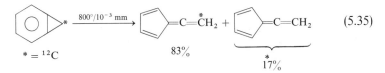

$$\text{(5.35)}$$

* = ^{12}C   83%   *17%

Labeled fulvenallene formed at 1000° is extensively scrambled (80%), and scrambling is complete after repyrolysis at 1000°–1050°, which suggests that scrambling occurs after initial ring contraction. Automerization at high temperatures is believed to involve a pool of interconverting C_7H_6 intermediates[34] including phenylcarbene, C_6H_5—$\overset{\cdot\cdot}{C}H$.

The rearrangement of a fulvenallene to a phenylcarbene has been experimentally established by a study of the pyrolysis of 6-methylphthalide,[54] which gave 2-methylfulvenallene at 700°. At 1000° 2-methylfulvenallene rearranged [Eq. (5.36)] to give a mixture of benzocyclobutene and styrene, characteristic products of the multistep gas-phase rearrangement of 4-tolylcarbene which is discussed in Section III,C.

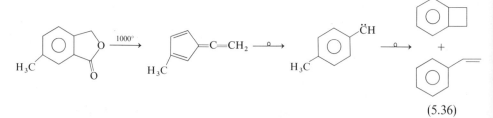

$$(5.36)$$

Crow, Lea, and Paddon-Row[61] found that pyrolysis of indazole at 700–800° and 0.01–0.05 mm also gives fulvenallene and ethinylcyclopentadiene, but in this case the mechanism of their formation is ambiguous because tautomerism of indazole may lead at high temperatures to either 3*H*- or 7a*H* tautomers. Loss of nitrogen from these tautomers should give methylenecyclohexadienylidene or phenylcarbene, respectively, [Eq. (5.37)] and each of these carbenes can form fulvenallene.

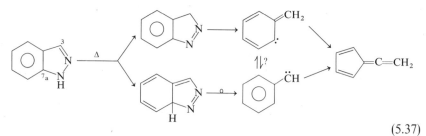

$$(5.37)$$

Crow and Paddon-Row[62] have summarized the results of extended investigations of the pyrolysis of indazoles using N-deuteration and 3-methylation,[63] and ^{13}C-labeling[62]; they conclude that both pathways are involved. This is consistent with the behavior of the corresponding azaindazole, pyrazolo[3,4-*b*]pyridine, which on pyrolysis at 700°–760° and 0.03 mm[61] gave a fraction containing an ethinylpyrrole (3*H*-pathway) and cyanocyclopentadiene. The latter is formed by a pathway leading from the 7a*H*-tautomer through 3-pyridylcarbene and phenylnitrene (see Section III,E).

This interpretation [Eq. (5.38)] depends on the reality of tautomerism to a 7a*H* tautomer, and alternative explanations which do not involve this

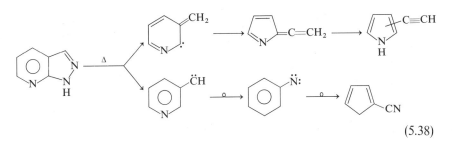

(5.38)

tautomer may be possible. Thus extension of the results of Wentrup and co-workers[54] on scrambling in labeled fulvenallene [Eq. (5.49)] to these indazoles and azaindazoles could lead to a different view of their reactions.

C. Arylcarbenes: Ring Expansion and Ring Contraction

Although the chemistry of phenylcarbene in solution had been studied extensively in the 1950s and 1960s,[1,2] it was not until 1970 that Hedaya's group at the Union Carbide Research Institute reported the varied capacity of phenylcarbene for thermal rearrangement in the gas phase.[64] In a mass spectrometric and chemical investigation they observed the formation of fulvenallene by pyrolytic α-elimination of hydrogen fluoride from benzyl fluoride at 1050° and 0.1 mm (contact time 1 msec); benzyl radicals were formed in a competing process. Comparison with the flash vacuum pyrolysis of phenyldiazomethane at 600°–1000° and 0.05 mm confirmed that phenylcarbene was a primary reactive intermediate, and it was found that the products from phenyldiazomethane collected at liquid nitrogen temperature varied markedly with the pyrolytic temperature. Lower temperatures favored the formation of a ring expanded product, heptafulvalene, whereas higher temperatures led mainly to ring-contracted products [Eq. (5.39)]. Ring

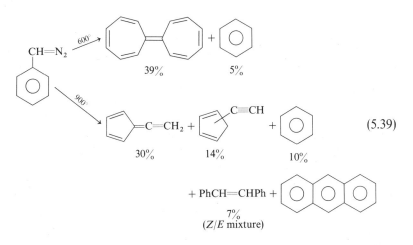

(5.39)

expansion of phenylcarbene to cycloheptatrienylidene had been discovered a little earlier by Joines, Turner, and Jones,[65] who pyrolyzed the sodium salts of the tosylhydrazones of benzaldehyde and of tropone in a stream of nitrogen at 250° and 40 mm and obtained heptafulvalene from each [Eq. (5.40)].

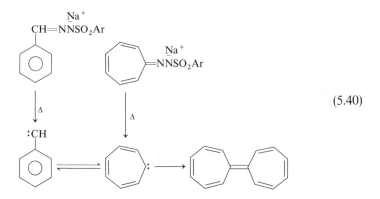

$$(5.40)$$

The chemistry of the interconversion of phenylcarbene and cycloheptatrienylidene was similarly explored by Wentrup and Wilczek.[66] Most work on arylcarbene rearrangements has involved the pyrolytic generation of transient aryldiazomethane either from a salt of a tosylhydrazone, or from a 5-aryltetrazole[67,68,69] [Eq. (5.41)]. Pyrolysis of 5-aryltetrazoles always gives some nitrile by cycloelimination of hydrazoic acid.

$$(5.41)$$

Theoretical investigations of the energetics and geometry of ring expansion of phenylcarbene are described by Wentrup and co-workers[68] (CNDO/2 and extended Hückel methods) and by Dewar and Landman[70] (MINDO/2 method). The isomerization of phenylcarbene to cycloheptatrienylidene is shown as reversible in Eq. (5.40); reversibility at high temperatures is required to explain the isomerization of *p*-, *m*-, or *o*-tolylcarbenes to the products benzocyclobutene and styrene through the intermediates *o*-tolylcarbene and methylphenylcarbene [Eq. (5.42)].

Hedaya and Kent[71] used ^{13}C-labeling to explore the rearrangement of *p*-tolylcarbene, and the result [Eq. (5.43)] is consistent with a mechanism proposed by Baron, Jones, and Gaspar[72] and later described by Crow and Paddon-Row[62] as the carbene isomerization cycle [Eqs. (5.44) and (5.45)].

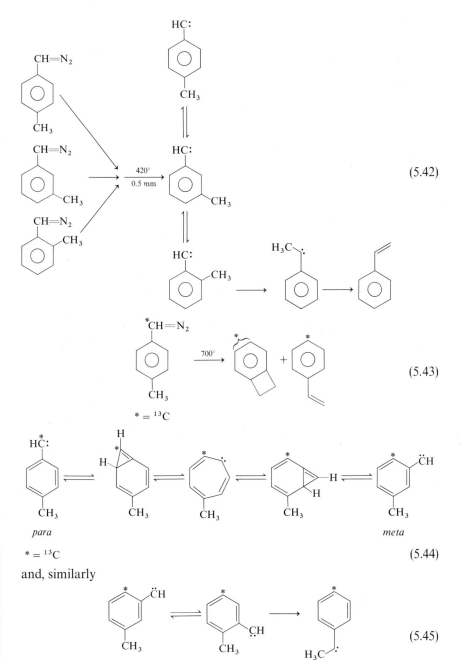

(5.42)

(5.43)

$* = {}^{13}C$

para   *meta*

$* = {}^{13}C$ (5.44)

and, similarly

meta

$* = {}^{13}C$

(5.45)

In the tolylcarbenes the methyl group acts as an irreversible trap for carbene centers. In phenylcarbene itself the ultimate products are fulvenallene and ethinylcyclopentadiene and the ^{13}C-labeling study of Crow and Paddon-Row[62] showed that at 770° the fulvenallene produced was randomly labeled [Eq. (5.46)] within the limits of accuracy of the analysis of the dimethylamine adduct by Fourier transform ^{13}C nmr spectrometry.

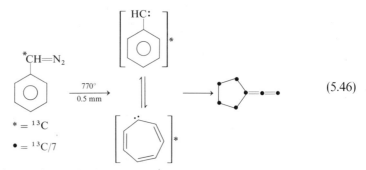

$$(5.46)$$

This result requires migration of hydrogen atoms in cycloheptatrienyli-dene or an equivalent bicyclic intermediate because otherwise C-1 of the starting material must always be bonded to carbene-CH of phenylcarbene.[62] The exact nature of the intermediate in which hydrogen migration occurs is likely to remain uncertain.

Crow and Paddon-Row,[62] in discussion of the pool of C_7H_6 isomers, accept ring contraction of phenylcarbene via a prefulvenoid diradical[64] which can collapse directly to an ethinylcyclopentadiene.

$$(5.47)$$

Wentrup[8] has criticized this view and has summarized evidence suggesting that fulvenallene is the primary product of ring contraction of phenyl-carbene, as it is of direct contraction of methylenecyclohexadienylidene. The further isomerization of fulvenallene to ethinylcyclopentadiene Wentrup attributes to chemical activation of fulvenallene in its exothermic formation from phenylcarbene; this can be partly suppressed by collisional deactiva-tion with nitrogen at 5–7 mm.

The discussion of Crow and Paddon-Row[62] is based on the assumption that the carbon skeleton of fulvenallene is not subject to further automeriza-tion under the pyrolytic conditions of their experiment (770° and 0.5 mm). However in 1977 Wentrup, Wentrup-Byrne, and Muller[54] showed that ter-minally labeled fulvenallene did undergo complete scrambling of the label on pyrolysis at 1050° and 10^{-3} mm, and they then reexamined the ring contraction of phenylcarbene using a ^{13}C-label at C-1 of the ring. Under mild conditions (590°, with 5–7 mm of N_2 for collisional deactivation of intermediates) the fulvenallene product was obtained with C-5 "almost three times more labeled than any other carbon atom," whereas pyrolysis at 700° and 10^{-3} mm gave extensively scrambled product and scrambling was complete in that formed at 900° and 10^{-3} mm [Eq. (5.48)].

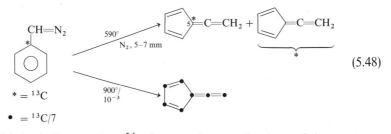

$$(5.48)$$

On this basis these authors[54] reject previous mechanisms of ring contraction, and propose that both ring contraction and hydrogen migration involve intermediate formation of bicyclo[3.2.0]hepta-1,3,6-triene (**33**). This proposal is partially outlined for the case of [1-^{13}C] phenylcarbene in Eq. (5.49), in which every process is shown as fully reversible, leading to scrambling of the label at high temperatures. The same mechanism had previously been considered and rejected (on grounds which may not now be adequate) by Crow and Paddon-Row.[62]

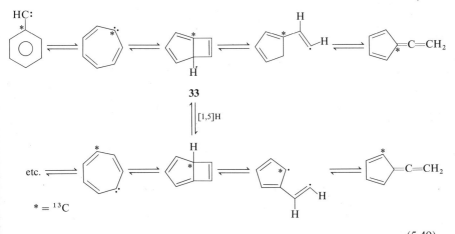

$$(5.49)$$

Pyrolysis of the isomeric hydrocarbon bicyclo[3.2.0]hepta-1,4,6-triene (**34**) at 900° and 0.1 mm[73] also leads to fulvenallene and ethinylcyclopentadiene, and the dideuterated compound **35** gives mainly (77%) the 7-deuterated fulvenallene[74] shown in Eq. (5.50). The ease of 1,5-sigmatropic

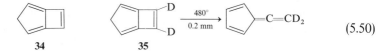

$$(5.50)$$

hydrogen shifts in cyclopentadiene rings is such that these experiments might well involve intermediate formation and cleavage of the 1,3,6-triene (**33**) of Eq. (5.49).

In previous examples, we have seen the effect of methyl groups in trapping carbene centers. An adjacent aryl group behaves similarly, so that diphenyl-carbene in the gas phase forms some fluorene by the operation of the carbene isomerization cycle. The early observation of Staudinger and Endle[75] that pyrolysis of diphenylketene at 600°–700° at water pump pressure effects decarbonylation leading to fluorene (43%) is probably the first example of this reaction, but recent work has used diaryldiazoalkanes as precursors.

Jones et al.[76] pyrolyzed diphenyldiazomethane at 330° and 3 mm in a stream of nitrogen and obtained fluorene (29%), dimeric products derived from combinations of diphenylcarbene and phenylcycloheptatrienylidene (tetraphenylethylene, 10%; triphenylheptafulvene, 13%; diphenylhepta-fulvalene, 4%), and benzophenone azine (18%). The operation of a multiple carbene–carbene rearrangement was shown by the formation of 2-substituted fluorenes alone from 4-methyl- or 4-methoxy-labeled diphenyldiazomethane [Eq. (5.51)], whereas direct cyclization of the carbene with hydrogen migra-tion would have given 3-substituted fluorenes. Wentrup and Wilczek[66] proved the same point using methyl groups at both positions 4 and 4′.

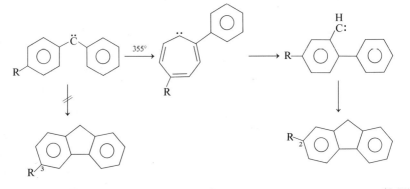

R = H, CH$_3$, or OCH$_3$ (5.51)

This tactic of interception of a carbene (or nitrene) center by an adjacent aryl group has been used repeatedly in this area.[8] Mayor and Wentrup[77] have examined the direction of rearrangement and cyclization of many arylheteroarylcarbenes; they propose that initial attack leading to ring ex-pansion is controlled by a synergic pair of interactions in which the carbene

σ electrons interact with the lowest unoccupied molecular orbital (σ–LUMO interaction) of the most electrophilic ring, and the vacant carbene p-orbital effects electrophilic attack on the o-position of the same ring (p–HOMO interaction). As a single example we may consider the pyrolysis of phenyl-4-pyridyldiazomethane at 500° and 10^{-3} mm which gives pure 2-azafluorene (70%). Labeling of the carbene center with ^{14}C and degradation of the product[77] revealed the existence of competing pathways [Eq. (5.52)].

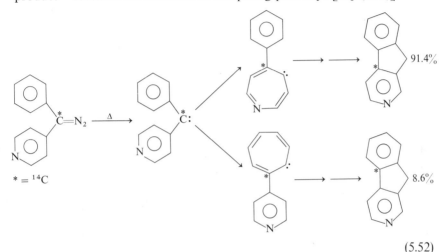

(5.52)

The major pathway requires initial expansion of the pyridine ring, with carbene σ electrons interacting effectively with the LUMO (large coefficient at C-4) and the vacant carbene p-orbital attacking at C-3 (large HOMO coefficient). This selectivity is finely balanced, however, and introduction of a 4-phenyl substituent into the phenyl ring changes it so that labeled (4-biphenylyl)(4-pyridyl)carbene undergoes rearrangement preferably into the substituted phenyl ring,[78] in accord with the availability of a low lying LUMO with a large coefficient at C-4 of this ring, adjacent to the carbene.

The work of Mayor and Wentrup[77] provides useful synthetic routes to many difficulty accessible or previously unknown ring systems. For example, pyrolysis of 5,7-dimethyl-v-triazolo[1,5-a]pyrimidine (a ring tautomer of a phenylpyrimidinyldiazomethane) [Eq. (5.53)] gave by rearrangement into the phenyl ring 2,4-dimethylpyrimido[2,1-a]isoindole (36), a highly reactive and unstable derivative of a previously unknown system. The minor product of this reaction, 2,4-dimethyl-5H-pyrido[3,2-b]indole (37), is formed by expansion of the pyrimidine ring and a carbene–nitrene rearrangement (see Section III,E).

In annelated systems carbene–carbene rearrangements can take place at much lower temperatures. The rearrangements of naphthylcarbenes and

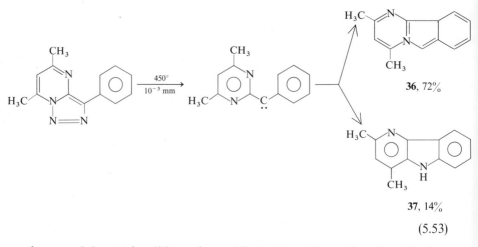

36, 72%

37, 14%

(5.53)

benzocycloheptatrienylidenes formed from the sodium salts of tosylhydrazones have been studied[76] both in the gas phase and in solution [Eq. (5.54)].

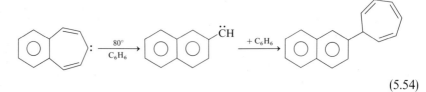

(5.54)

Rearrangement of naphthylcarbenes occurs toward the bond of highest order (1,2 bond), as expected for the involvement of discrete naphthocyclopropene intermediates [Eq. (5.55)]. Coburn and Jones[79] have detected

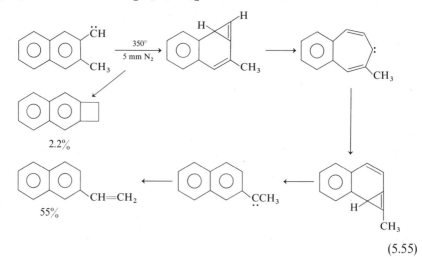

2.2%

55%

(5.55)

cyclopropene intermediates from annelated systems by trapping experiments in solution, and the structural requirements for the occurrence of carbene–carbene rearrangements in solution have been reviewed by Jones.[80]

D. Iminocarbenes: Ring Contraction of Iminocyclohexadienylidene to Cyanocyclopentadiene

In the course of a systematic study[81,82,83] of the pyrolysis of $1H$-1,2,3-triazoles to give iminocarbenes, R—C̈—CR=NR, Gilchrist *et al.*[81] found that pyrolysis of 1,5-diphenyl-*v*-triazole gave 1-phenylindole (1.5%), 2-phenylindole (25%), 3-phenylindole (13%), and the ketenimine PhCH=C=NPh which was isolated after hydration as phenylacetanilide (29%). Investigation of this reaction with the [5-^{13}C]triazole **38** showed[82] that the distribution of ^{13}C among the products required the formation of a $1H$-azirine as an intermediate in the interconversion of isomeric iminocarbenes [Eq. (5.56)].

This experiment was complicated by further interconversion of 2-phenyl- and 3-phenylindoles by 1,5 phenyl and hydrogen shifts, but ^{13}C nmr analysis

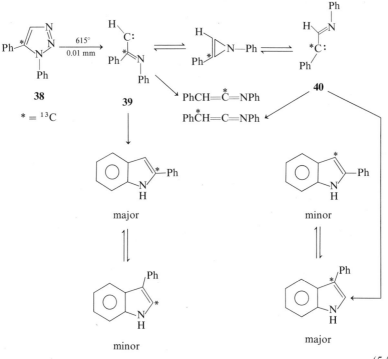

(5.56)

of the labeled products clearly showed that the major indoles were those formed directly by insertion reactions of the iminocarbenes **39** and **40**. The major ketenimine was PhCH=C̈=NPh, the product of Wolff rearrangement of the initial iminocarbene **39**. Formation of a 1*H*-azirine cannot be assumed in all such cases, however, because the iminocarbene Ph—C̈—CH=NPh formed directly by pyrolysis of [4-¹³C] 1,4-diphenyl-*v*-triazole showed little or no scrambling of the label in either the phenylindole or ketenimine products. The authors suggest[82] that formation of the 1*H*-azirine from the iminocarbene H—C̈—CPh=NPh is favored by relief of steric interaction between the two adjacent phenyl groups. In the isomeric iminocarbene Ph—C̈—CH=NPh Wolff rearrangement to PhCH=C=NPh is probably very favorable because it involves migration of hydrogen rather than phenyl.

Similar involvement of transient 1*H*-azirines is invoked to explain the products of pyrolysis of many 1-unsubstituted, 1-alkyl-, and 1-aryl-1*H*-1,2,3-triazoles.[81] No 1*H*-azirine has ever been isolated, and the attempted pyrolytic generation of a 1-phthalimido derivative of this antiaromatic 4π system gave only a rearranged 2*H*-azirine[83] [Eq. (5.57)].

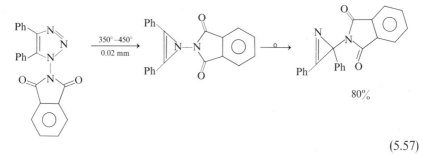

$$\text{(5.57)}$$

The aromatic equivalent of these systems is the diradical **41** which can react as iminocyclohexadienylidene (**42**) and can undergo both Wolff-type ring contraction and equilibration with 1*H*-benzazirine. The formation of

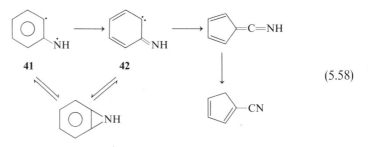

$$\text{(5.58)}$$

cyanocyclopentadiene from the precursors benzotriazole and isatin via this ring contraction was first reported by Crow and Wentrup[84] in 1968.

The pyrolysis of benzotriazole itself has been examined under a variety of conditions and with various inert carrier gases;[85] yields of cyanocyclopentadiene range from 86% to 99%, with traces of aniline also being formed. Aniline is believed to be a product of abstraction of hydrogen from other molecules, probably by the triplet state of the diradical **41**.

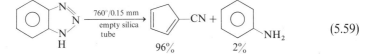

$$(5.59)$$

Benzotriazoles substituted with Cl, F, CF_3, CN, and CH_3 groups at C-4 or C-5 give substituted cyanocyclopentadienes and anilines[85] on pyrolysis at 550°–700°, as in Eq. (5.60). The tautomerism of the methylbenzotriazole must lead to the formation of both methyl-substituted diradicals, but both diradicals contract initially to the same 1-cyano-4-methylcyclopentadiene. The formation of the minor product, 1-cyano-2-methylcyclopentadiene is due to secondary sigmatropic migration of the substituents; under mild conditions (2% conversion at 500° and 0.02 mm) only the 4-methyl compound is obtained. The tautomerism of the 5(6)-methylbenzotriazole would lead to the formation of both *p*- and *m*-toluidine, as observed, even in the absence of 1*H*-benzazirine formation. At higher temperatures (700°–780°) yields of cyanomethylcyclopentadienes decline, and benzonitrile, a product of further dehydrogenation and rearrangement, is obtained in 70–75% yield.

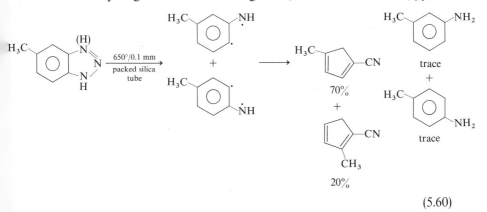

$$(5.60)$$

Annelated triazoles and triazolohetarenes behave similarly,[85] and two examples are shown below. In both of these systems [Eqs. (5.61) and (5.62)] the primary products are interconverted at 800°. Pyrolyzates formed from

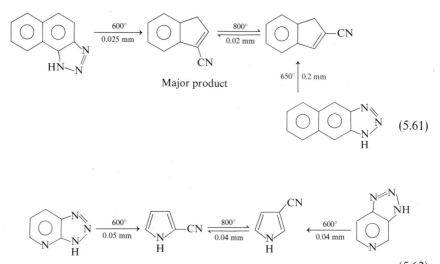

(5.61)

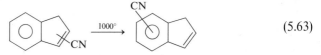

(5.62)

the naphthotriazoles at very high temperatures (800°–1000°) also contain products formed by further migration of cyano groups. The 2- and 3-cyanoindenes can be shown to isomerize to give 4(7)- and 5(6)-cyanoindenes [Eq. (5.63)], probably by a radical mechanism; similar migrations occur in phenylacetonitrile and related compounds.

(5.63)

Isatins are not subject to the same ambiguities arising from tautomerism which complicate description of the pyrolysis of benzotriazoles, but their

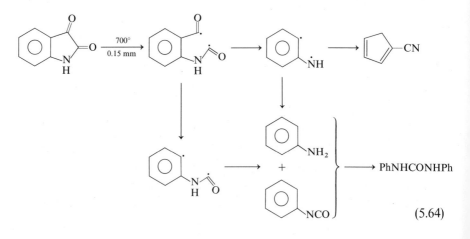

(5.64)

complete decomposition usually requires somewhat higher temperatures (ca. 700°). Isatin itself gives cyanocyclopentadiene in only 40% yield, together with aniline (10%) and diphenylurea (1.6%).[86] The last product is thought to arise from hydrogen migration in a partially decarbonylated diradical leading to phenylisocyanate, and thus to the urea [Eq. (5.64)].

The pyrolysis of methylisatins at 600°–900° and 0.005–0.1 mm through a packed silica tube gives cyanomethylcyclopentadienes, benzonitrile, and toluidines.[86] The toluidine formed from 5-methylisatin at 600° was the unrearranged *p*-isomer, and only 6% of the *m*-isomer was present in the toluidine mixture formed at 900° [Eq. (5.65)]. From this experiment Wentrup[86] estimated that 10–12% of the toluidine fraction was formed via the methyl-1*H*-benzazirine, but it gave no information as to the involvement of the 1*H*-benzazirine in the formation of the readily isomerized cyanomethylcyclopentadienes.

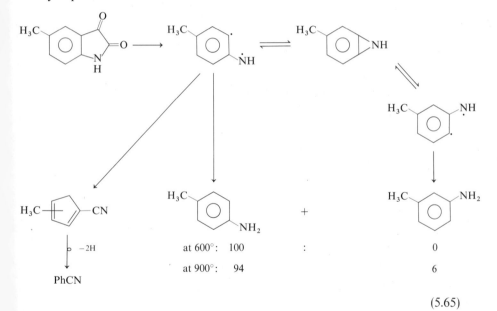

$$\begin{array}{ccc} \text{at } 600°: & 100 & : & 0 \\ \text{at } 900°: & 94 & & 6 \end{array}$$

(5.65)

Pyrolysis of [13C-7a]isatin[87] at 715° and 0.15 mm revealed that under these conditions equilibration of diradicals (or iminocyclohexadienylidenes) with 1*H*-benzazirine is complete for the species which undergo ring contraction to cyanocyclopentadiene. The 13C-label in the product was equally distributed between the cyano group and the ring carbons (ratio 13CN to 13C at each ring position = 5:1). Isotopic labels within the cyclopentadiene ring in this and other labeling experiments are distributed over all ring positions by final sigmatropic shifts.

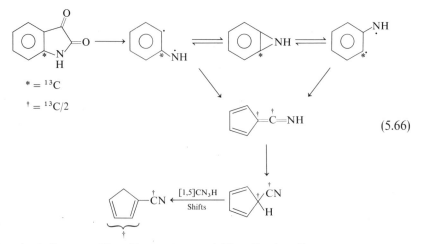

$* = {}^{13}C$

$\dagger = {}^{13}C/2$

(5.66)

E. Arylnitrenes: Ring Expansion and Ring Contraction

The whole field of the chemistry of arylcarbene and arylnitrene rearrangements in the gas phase has grown from the initial experiments of Crow and Wentrup[88] and of Hedaya and co-workers[89] on the generation of phenylnitrene by flash pyrolysis of phenyl azide at 330°–750°, which gave products including cyanocyclopentadiene, aniline, azobenzene, pyridine, benzene, biphenyl, benzonitrile, diphenylamine, and carbazole. Both groups noted that the formation of cyanocyclopentadiene is favored by severe conditions (high temperatures; rapid rates of sample introduction; explosive reaction with uncontrolled increase in pressure) whereas mild conditions favor azobenzene and aniline formation.

Aryl azides are the only useful direct precursors of arylnitrenes. N-Sulfinylaniline, PhN=S=O, also gives cyanocyclopentadiene (4%) on pyrolysis[90] at 1000°, but complex reactions other than those involving phenylnitrene also occur [Eq. (5.67)]. Aryl isocyanates ArN=C=O do not undergo pyrolytic decarbonylation.[88]

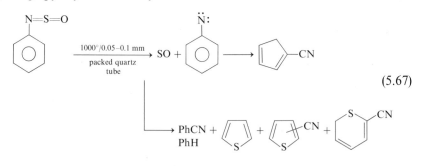

(5.67)

We shall be concerned first with the direct generation of phenylnitrene from phenyl azide [Eq. (5.68)] and later with its indirect generation by rearrangement of pyridylcarbenes, as in Eq. (5.69).

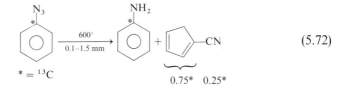

$$\text{Direct:} \quad \text{(ring–}N_3) \longrightarrow N_2 + \text{(ring–}\ddot{N}\text{:)} \tag{5.68}$$

$$\text{Indirect:} \quad \longleftrightarrow \quad \longleftrightarrow \quad \tag{5.69}$$

Nitrenes are formed in the pyrolysis of simple alkyl and alkenyl azides such as methyl azide[91] and vinyl azide,[92] but vinylnitrenes have been studied mostly in solution.[93]

$$CH_3N_3 \xrightarrow[\text{0.1 mm}]{900^\circ} N_2 + CH_3\ddot{N}\text{:} \longrightarrow CH_2{=}NH,\ HCN,\ H_2,\ NH_3 \tag{5.70}$$

$$CH_2{=}CHN_3 \xrightarrow[\text{0.01–0.1 mm}]{400^\circ} CH_2{=}CH{-}\ddot{N}\text{:} \longrightarrow CH_2{-}CH \backslash N \tag{5.71}$$

Much of the work on phenylnitrene has been aimed at explaining the effect of conditions of generation on its energy and reactivity, and at understanding the mechanism of ring contraction to cyanocyclopentadiene. The pyrolysis of phenyl azide is not a preparatively useful reaction; a 50% yield of cyanocyclopentadiene at 700° and 0.07 mm has been reported with one apparatus,[89] but yields are usually much lower.[8]

The formation of azobenzene and aniline in the pyrolysis of phenyl azide is considered[88,89] to result from dimerization and hydrogen abstraction reactions of phenylnitrene with minimal excess vibrational energy and a lifetime which permits the initial singlet state to cross over to the ground-state triplet.[94] Crow and Paddon-Row[95] showed that no significant amount of scrambling by cycloperambulation of the nitrogen had occurred in aniline formed by the pyrolysis of [1-^{13}C]phenyl azide, whereas both [13]C- and ^{14}C-labeling[96] studies showed extensive scrambling in the cyanocyclopentadiene [Eq. (5.72)]. The cyano group carried 25–32% of the label; complete randomization of carbon atoms would require 16.7%.

$$\underset{* = {}^{13}C}{\overset{N_3}{\text{(ring)}}} \xrightarrow[\text{0.1–1.5 mm}]{600^\circ} \underset{}{\overset{NH_2}{\text{(ring)}}} + \text{(cyclopentadiene–CN)} \tag{5.72}$$

0.75*   0.25*

This result rules out direct ring contraction of phenylnitrene to cyano-cyclopentadiene exclusively via a 2,6-bonded intermediate such as **43**, and requires consideration of bicyclic and ring expanded species **44** and **45** and species related to these by hydrogen migration, and cycloperambulation of the nitrogen around the ring.[96] No very firm conclusions as to the mechanism of ring contraction have emerged from these studies with labeled phenyl azide,[95,96] and attention has been concentrated on the reactions of phenylnitrene generated indirectly but in a more controllable manner.

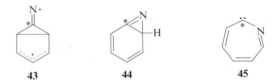

    43   **44**   **45**

The possibility of ring expansion of a phenylnitrene was shown by Wentrup[97] by the vigorous pyrolysis of 2,6-dimethylphenyl azide, which gave 6-methyl-2-vinylpyridine in 8% yield [Eq. (5.73)]. Ring expansion of the initial nitrene leads to a 2-azepinylidene which can undergo alternative ring contraction to a methyl(2-pyridyl)carbene; irreversible trapping of this carbene center by the methyl group forms the 2-vinyl group of the product.

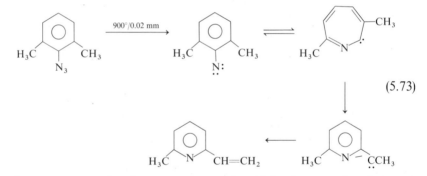

$$(5.73)$$

Rearrangements in the reverse sense (pyridylcarbene → phenylnitrene) proceed smoothly under mild conditions and are common; both experimental experience and thermochemical arguments[98] suggest that the equilibrium favors the phenylnitrene. The indirect route to phenylnitrene requires initial generation of a pyridylcarbene by pyrolysis of a *v*-triazolo[1,5-*a*]-pyridine, the sodium salt of a pyridine 3- or 4-aldehyde tosylhydrazone, or a 5-(2-, 3-, or 4-pyridyl)tetrazole. Crow and Wentrup[99] found that pyrolysis of *v*-triazolo[1,5-*a*]pyridine, the ring tautomer of 2-pyridyldiazomethane, leads after rearrangement to phenylnitrene carrying minimal excess energy which forms azobenzene and aniline [Eq. (5.74)], but no cyanocyclopentadiene over the temperature range 450°–850°.[67]

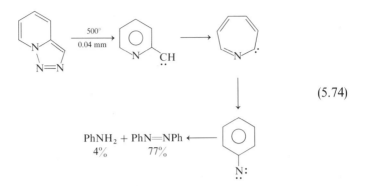

$$ (5.74) $$

The 6-methyl derivative **46** behaves similarly at 500° and 0.04 mm, giving 3,3′-dimethylazobenzene and *m*-toluidine, but at 800° and 0.1 mm it affords a complex mixture of hydrocarbons, methyl- and vinylpyridines, and nitriles qualitatively similar to that formed from *m*-tolyl azide under the same conditions. The 3-methyl derivative **47** fails to show carbene–nitrene rearrangement; the initial carbene rearranges instead to give 2-vinylpyridine in quantitative yield. However the 6-methyl-3-phenyl derivative undergoes carbene–nitrene rearrangement in preference to rearrangement into the phenyl substituent [compare the discussion of phenyl-4-pyridylcarbene in Section III,C and Eq. (5.52)] and 2-methylcarbazole is formed in high yield[99] [Eq. (5.75)].

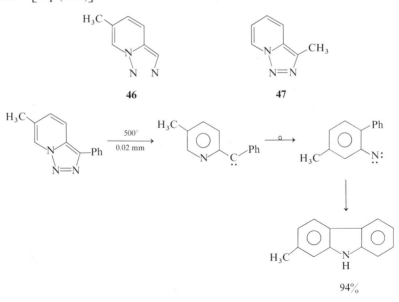

$$ (5.75) $$

Crow and co-workers[100] have used 5-pyridyltetrazoles and sodium salts of the aldehyde tosylhydrazones as precursors of pyridylcarbenes. Because of the occurrence of preliminary carbene–carbene rearrangements both 4-pyridyl- and 3-pyridylcarbene also rearrange through 2-pyridylcarbene to give, ultimately, products derived from phenylnitrene.[100] Cyanocyclopentadiene (33–58%) and aniline (6–19%) are the major products [Eq. (5.76)], but benzene, benzonitrile, and pyridines are also formed.

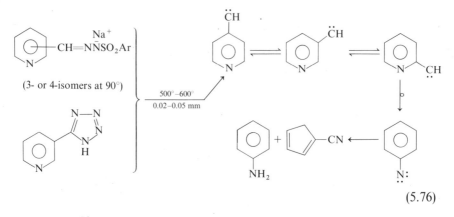

$$(5.76)$$

Wentrup[98] has discussed the strikingly different behavior of the phenylnitrene generated from v-triazolo[1,5-a]pyridine and that from the pyridyltetrazoles in terms of their thermochemistry; he concludes that at low pressures (true flash vacuum pyrolytic conditions) phenylnitrene formed by the pyridyltetrazole route will possess about 20 kcal/mole of excess energy with respect to that from the triazolopyridine route. At higher pressures (≥ 0.1 mm) this chemically activated phenylnitrene can be collisionally deactivated so that cyanocyclopentadiene formation is suppressed and azobenzene formation occurs.

Competition between carbene–methyl insertion, carbene–carbene rearrangement, and carbene–nitrene rearrangement in the isomeric picolylcarbenes has been investigated by Crow and co-workers.[101] The distribution of ^{13}C in aniline formed by rearrangement of the three pyridyl[^{13}C]carbenes has been analyzed,[100] and the results are mainly consistent with the sequence of isomerizations shown in Eq. (5.76). 2-Pyridyl[^{13}C]carbene formed from the pyridyltetrazole at 600° and 0.02–0.05 mm gives aniline with 86% of the label at the o-positions [100% expected from Eq. (5.76)], 13% at the m- and 1% at the p-position. This small degree of scrambling is ascribed to some cycloperambulation of the ^{13}CH group in the 2-pyridylcarbene before rearrangement to phenylnitrene.[100]

Both iminocyclohexadienylidene (Section III,D) and phenylnitrene with excess energy undergo ring contraction to cyanocyclopentadiene. Thétaz and Wentrup[87,98] have proposed that phenylnitrene undergoes ring contraction indirectly by first isomerizing to 1*H*-benzazirine and then to iminocyclohexadienylidene and cyanocyclopentadiene. This unifying theory is based on the results of ^{13}C-labeling experiments under carefully controlled conditions, with analysis of the products by ^{13}C nmr spectrometry. Pyrolysis of 5-(2-pyridyl)[^{13}C]tetrazole at 400° with nitrogen (1 mm) present for collisional deactivation of phenylnitrene led to azobenzene labeled exclusively in the *o*-positions [Eq. (5.77)]. At 610° cyanocyclopentadiene, azobenzene, and aniline were formed, and the last two products were still labeled only in the *o*-positions, showing that no scrambling occurred in phenylnitrene itself. Finally, pyrolysis at 670° and 0.001 mm (no deactivating carrier gas) gave cyanocyclopentadiene in which the ^{13}C label was distributed between the cyano group and each ring carbon in the ratio $^{13}CN/^{13}C(ring) = 1.7$. The mechanism of Eq. (5.79) would require this ratio to be $25:15 = 1.67$. This evidence for ring contraction via iminocarbenes is supported by similar results from the indirect generation of 2-naphthylnitrene from 1-isoquinolyl[^{13}C]carbene.

The problem for the student of arylnitrene rearrangements is to decide to what extent the theory base on these controlled experiments can be applied to other reactions such as the pyrolysis of phenylazide itself. Thétaz and Wentrup[87] claim that "there are just two intramolecular reactions of phenylnitrene" [ring expansion, Eq. (5.73), and 1*H*-benzazirine/iminocyclohexadienylidene formation, Eq. (5.79)]. If phenyl azide underwent ring contraction only via 1*H*-benzazirine then the cyanocyclopentadiene formed from [1-^{13}C]phenyl azide at 600° in the experiment of Crow and Paddon-Row[95] should have shown a ratio $^{13}CN/^{13}C(ring)$ of 5, whereas the experimental result in different runs ranged from 1.67 [Eq. (5.72)] to 2.35. Papers

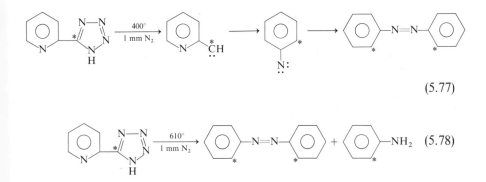

$$(5.77)$$

$$(5.78)$$

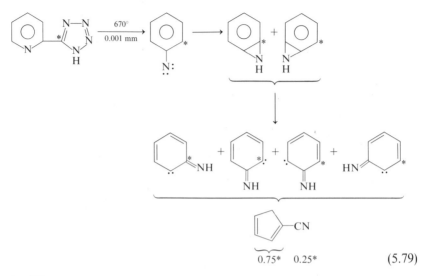

$$\text{(5.79)}$$

$* = {}^{13}C$

from Crow's group have emphasized[95,100,101] that excess energy carried by phenylnitrene may lead to changes in reaction pathway, and a reaction equivalent in its effect to the cycloperambulation scheme of Crow and Paddon-Row[95] [Eq. (5.80)] is needed to explain the considerable scrambling arising from highly energetic phenylnitrene.

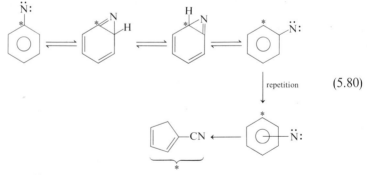

$$\text{(5.80)}$$

$* = {}^{13}C$

F.   Heteroarylnitrenes: Ring Expansion, Ring Contraction, and Ring Fission

2-Pyridylnitrene can be generated by pyrolysis of tetrazolo[1,5-*a*]pyridine (**48**)[102] or of 2*H*-1,2,4-oxadiazolo[2,3-*a*]pyridin-2-one (**49**).[103] In each case

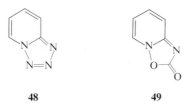

48   49

the major products isolated are cyanopyrroles, but there is some variation in the minor products.

Tetrazolo[1,5-*a*]pyridine on pyrolysis under conditions as mild as 380° and 0.05 mm gives 2-cyanopyrrole, 3-cyanopyrrole, 2-aminopyridine, and glutacononitrile, $NCCH_2CH\!=\!CHCN$[102,104] 2-Cyanopyrrole and 2-amino-pyridine are the major products at low temperatures, and 3-cyanopyrrole and glutacononitrile are increasingly formed above 500°. 2-Cyano- and 3-cyanopyrrole interconvert readily at high temperatures.[85] The formation of unexpected isomers from the pyrolysis of methyl derivatives of **48** under conditions where interconversion did not occur suggested the involvement of a symmetrical ring-expanded intermediate, and this was proved for the parent compound[8,104] by ^{15}N-labeling experiments [Eq. (5.81)].

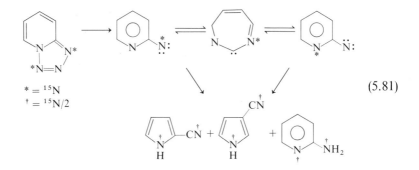

$* = {}^{15}N$
$† = {}^{15}N/2$

(5.81)

Ring contraction of ^{15}N-labeled 5-methyl-2-pyridylnitrene has been shown not to involve *N*-cyanopyrrole formation;[105] each of the two isomeric nitrenes which are formed via ring expansion contracts directly to a 2-cyano-pyrrole with complete isotopic specificity, whereas scrambling would have resulted from intermediate formation of an *N*-cyanopyrrole [Eq. (5.82)].

Pyrimidylnitrenes and pyrazinylnitrenes, however, do undergo contraction to *N*-cyano compounds.[106,107] Labeling studies[108] have shown that the pyrazinylnitrene undergoes direct ring contraction to *N*-cyanoimidazole [Eq. (5.84)], whereas 2,6-dimethyl-4-pyrimidylnitrene [Eq. (5.85)]ring contracts only after ring expansion and isomerization to a pyrazinylnitrene.

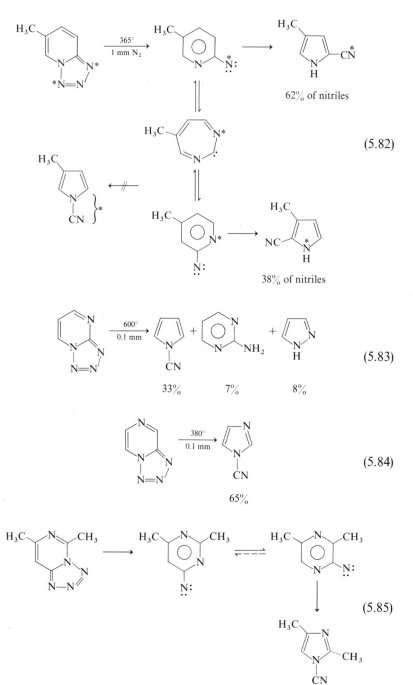

(5.82)

62% of nitriles

38% of nitriles

(5.83)

33%   7%   8%

(5.84)

65%

(5.85)

It thus appears that 2-pyridylnitrenes can ring contract directly to 2-cyanopyrroles without passing through *N*-cyanopyrroles, and probably without the involvement of azirine and iminocarbene intermediates.[105] An iminocarbene intermediate is definitely excluded in the case of pyrolysis of labeled 4-cyanotetrazolo[1,5-*a*]pyridine to give as major product 2,3-dicyanopyrrole, because there is no *o*-hydrogen in the nitrene first formed.

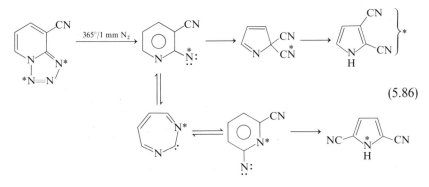

$$(5.86)$$

$$* = {}^{15}N$$

It is not clear whether this direct contraction to 2-cyanopyrroles should be represented as a direct shift of the N—C-2 bond to C-3 (**50**) or as involving a 1,3-bonded intermediate **51**; distortion from the planarity of the pyridyl-nitrene is implied in either case. Wentrup[98] has pointed out that the ring contraction process must be strongly exothermic, and that at low pressures the 2-cyanopyrrole products isomerize by cyano and even methyl shifts induced by chemical activation.

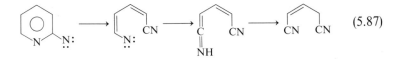

50   **51**

A minor product of the pyrolysis of tetrazolo[1,5-*a*]pyridine is glutacono-nitrile, which is formed by ring fission of 2-pyridylnitrene, isomerization of the new nitrene to an imine, and hydrogen migration [Eq. (5.87)]. This process is much more prominent in the pyrolysis of the oxadiazolone **49** and indeed Wentrup[8] has questioned the formation of free pyridylnitrene in this

$$(5.87)$$

case, on the basic of a thermochemical discussion of the relatively high yields of 3-cyanopyrrole and glutacononitrile and the absence of 2-amino-pyridine [Eq. (5.88)].

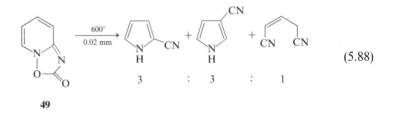

$$ \text{(5.88)} $$

$$ 3 \quad : \quad 3 \quad : \quad 1 $$

49

The gas-phase chemistry of 2-quinolylnitrene, whether generated from a tetrazolo compound[109] at 510° and 0.01 mm or from an oxadiazolone[103] at 600° and 0.1 mm, involves ring expansion, ring contraction to 1-iso-quinolylnitrene, and quantitative ring fission to a vinylnitrene [Eq. (5.89)].

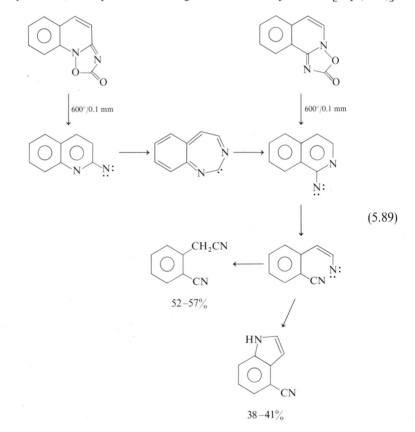

$$ \text{(5.89)} $$

52–57%

38–41%

The final products, 4-cyanoindole and homophthalonitrile, are formed by subsequent insertion and hydrogen migration reactions of the vinylnitrene. This interpretation has been confirmed by ^{14}C- and ^{15}N-labeling experiments,[103] and by the trapping of the isoquinolylnitrene with an adjacent phenyl group[108,110] [Eq. (5.90)]. Direct generation of 1-isoquinolylnitrene leads to the same products, 4-cyanoindole and homophthalonitrile, in similar yields.

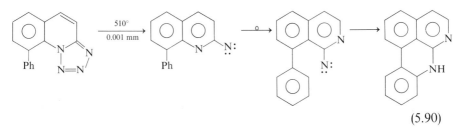

$$(5.90)$$

3-Isoquinolynitrene, however, ring contracts smoothly to 1-cyanoisoindole[103] [Eq. (5.91)], and the behavior of 2-quinolylnitrene is modified by annelation to benzo[f]quinolynitrene[103] or to 6-phenanthridylnitrene[109] [Eqs. (5.92) and (5.93)]. The factors controlling expansion, contraction, and fission in heteroarylnitrenes are discussed by Wentrup.[8,108]

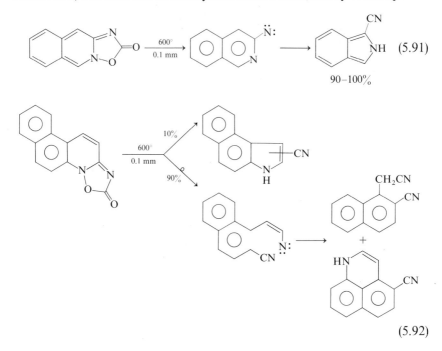

$$(5.92)$$

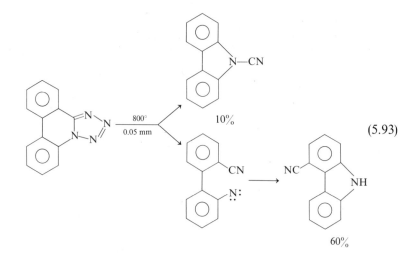

$$(5.93)$$

Finally, it is convenient, if somewhat out of context, to mention the ring fission of nitrenes derived from polymethylenetetrazoles, investigated by Wentrup.[111] The nitrene from pentamethylenetetrazole, for example, does not contract directly to *N*-cyanopiperidine. Instead it undergoes radical fission and hydrogen transfer which forms *N*-cyano-1-aminopent-4-ene as the primary product; this, however, readily cyclizes under the conditions of the reaction to give *N*-cyano-2-methylpyrrolidine [Eq. (5.94)].

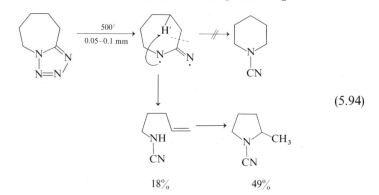

$$(5.94)$$

IV.   SULFONYLNITRENES AND CARBAMOYLNITRENES

A.   Sulfonylnitrenes

The chemistry of sulfonylnitrenes in solution has been extensively studied, and their reactions have been reviewed by Abramovitch and Sutherland.[112] Abramovitch and Holcomb[113] have reported a remarkable reaction in which dihydropyridine **54** is formed by flash pyrolysis of β-phenylethyl-sulfonyl azide. Generation of β-phenylethylsulfonylnitrene in diglyme gives the sultam **52** and the sulfonamide **53** by insertion and by hydrogen abstraction from the solvent [Eq. (5.95)] and the sultam is also obtained, with many other products, on flash pyrolysis of the sulfonyl azide at 300° and 3 mm. Pyrolysis at 650° and 3 mm, however, gives dihydropyridine **54** as the major product, with smaller amounts of styrene, indoline, and indole [Eq. (5.96)]. The last two compounds are probably secondary products formed from the sultam **52**, as suggested by the pyrolysis of **52** itself.

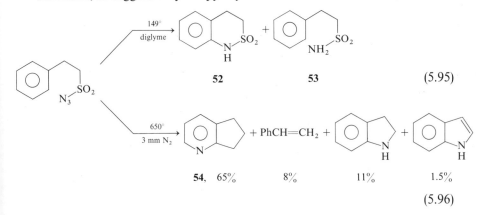

(5.95)

(5.96)

Abramovitch and Holcomb[113] have proposed the mechanism shown in Eq. (5.97), based on the results of the pyrolysis of substituted compounds.

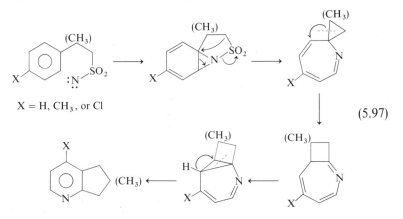

$X = H, CH_3,$ or Cl

(5.97)

B.   Carbamoylnitrenes

Reichen[114] has effected the Curtius rearrangement of dialkylamino-carbonyl azides by flash vacuum pyrolysis to give good yields of dipolar

1,2,4-triazolidine-3,5-diones formed by dimerization of the intermediate isocyanates.

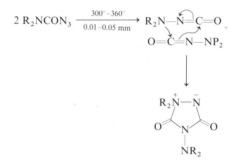

$$(5.98)$$

REFERENCES

1. Hine, J. S. (1964). "Divalent Carbon." Ronald Press, New York.
2. Kirmse, W. (1964). "Carbene Chemistry." Academic Press, New York.
3. Jones, Jr., M., and Moss, R. A., eds. (1973). "Carbenes," Vol. 1. Wiley (Interscience), New York.
4. Lwowski, W., ed. (1970). "Nitrenes." Wiley (Interscience), New York.
5. Lehr, R. E. and Marchand, A. P., eds. (1977). "Pericyclic Reactions," Vol. 1. Academic Press, New York.
6. Gilchrist, T. L., and Rees, C. W. (1969). "Carbenes Nitrenes and Arynes." Nelson, London.
7. Köbrich, G. (1967). *Angew. Chem. Int. Ed. Engl.* **6**, 41.
8. Wentrup, C. (1976). *Top. Current Chem.* **62**, 173.
9. Do Minh, T., and Strausz, O. P. (1970). *J. Am. Chem. Soc.* **92**, 1766.
10. Thornton, D. E., Gosari, R. K., and Strausz, O. P. (1970). *J. Am. Chem. Soc.* **92**, 1768.
11. Gelhaus, J., and Hoffmann, R. W. (1970). *Tetrahedron* **26**, 5901.
12. Richardson, D. C., Hendrick, M. E., and Jones, Jr, M. (1971). *J. Am. Chem. Soc.* **93**, 3790.
13. Kammula, S. L., Tracer, H. L., Shevlin, P. B., and Jones, Jr, M. (1977). *J. Org. Chem.* **42**, 2931.
14. Spangler, R. J., Kim, J. H., and Cava, M. P. (1977). *J. Org. Chem.* **42**, 1697.
15. Ott, E. (1913). *Justus Liebigs Ann. Chem.* **401**, 159.
16. Davidson, D., and Bernhard, S. A. (1948). *J. Am. Chem. Soc.* **70**, 3426.
17. Walborsky, H. M. (1952). *J. Am. Chem. Soc.* **74**, 4962.
18. Ripoll, J. L. (1977). *Tetrahedron* **33**, 389.
19. Rousseau, G., Bloch, R., Le Perchec, P., and Conia, J. M. (1973). *J. Chem. Soc. Chem. Commun.*, p. 795.
20. Baxter, G. J., Brown, R. F. C., Eastwood, F. W., and Harrington, K. J. (1975). *Tetrahedron Lett.*, p. 4283.
21. Brown, R. F. C., Eastwood, F. W., and McMullen, G. L. (1977). *Aust. J. Chem.* **30**, 179.
22. Hoffmann, R. W., and Schüttler, R. (1975). *Chem. Ber.* **108**, 844.
23. Brown, R. F. C., Eastwood, F. W., Lim, S. T., and McMullen, G. L. (1976). *Aust. J. Chem.* **29**, 1705.
24. Olofson, R. A., Lotts, K. D., and Barber, G. N. (1976). *Tetrahedron Lett.*, p. 3381.

25. Staab, H. A., and Ipaktschi, J. (1968). *Chem. Ber.* **101**, 1457.
26. Brown, R. F. C., Eastwood, F. W., and Harrington, K. J. (1974). *Aust. J. Chem.* **27**, 2373.
27. Brown, R. F. C., Eastwood, F. W., Harrington, K. J., and McMullen, G. L. (1974). *Aust. J. Chem.* **27**, 2393.
28. Baxter, G. J., and Brown, R. F. C. (1975). *Aust. J. Chem.* **28**, 1551.
29. Stang, P. J. (1978). *Chem. Rev.* **78**, 383.
30. Baxter, G. J., and Brown, R. F. C. (1978). *Aust. J. Chem.* **31**, 327.
31. Erickson, K. L., and Wolinsky, J. (1965). *J. Am. Chem. Soc.* **87**, 1142.
32. Wentrup, C., and Reichen, W. (1976). *Helv. Chim. Acta* **59**, 2615.
33. Donleavy, J. J., and Gilbert, E. E. (1937). *J. Am. Chem. Soc.* **59**, 1072.
34. Wentrup, C. (1977). *Chimia* **31**, 258.
35. Reichen, W., and Wentrup, C. (1976). *Helv. Chim. Acta* **59**, 2618.
35a. Wentrup, C., Stutz, U., and Wollweber, H.-J. (1978). *Angew. Chem. Int. Ed. Engl.* **17**, 688.
36. Brown, R. F. C., Harrington, K. J., and McMullen, G. L. (1974). *J. Chem. Soc. Chem. Commun.*, p. 123.
37. Brown, R. F. C., Eastwood, F. W., and Jackman, G. P. (1977). *Aust. J. Chem.* **30**, 1757.
38. Casanova, J., Geisel, M., and Morris, R. N. (1969). *J. Am. Chem. Soc.* **91**, 2156.
39. Brown, R. F. C., Eastwood, F. W., and Jackman, G. P. (1978). *Aust. J. Chem.* **31**, 579.
40. Strausz, O. P., Norstrom, R. J., Hopkinson, A. C., Schoenborn, M., and Csizmadia, I. G. (1973). *Theor. Chim. Acta* **29**, 183.
41. Norstrom, R. J., Gunning, H. E., and Strausz, O. P. (1976). *J. Am. Chem. Soc.* **98**, 1454.
42. Dykstra, C. E., and Schaefer, H. F. (1978). *J. Am. Chem. Soc.* **100**, 1378.
43. Benson, S. W. (1965). *J. Chem. Ed.* **42**, 502.
44. Maloney, K. M., and Rabinovitch, B. S. (1970). *In* "Isonitrile Chemistry" (I. Ugi, ed.), pp. 41–64. Academic Press, New York.
45. Yamada, S., Takashima, K., Sato, T., and Terashima, S. (1969). *Chem. Commun.*, p. 811.
46. Dewar, M. J. S., and Kohn, M. C. (1972). *J. Am. Chem. Soc.* **94**, 2704.
47. Schneider, F. W., and Rabinovitch, B. S. (1962). *J. Am. Chem. Soc.* **84**, 4215.
48. Brown, R. F. C., Crow, W. D., and Solly, R. K. (1966). *Chem. Ind.* (*London*), p. 343.
49. Brown, R. F. C., Butcher, M., and Fergie, R. A. (1973). *Aust. J. Chem.* **26**, 1319.
50. Brown, R. F. C., Coddington, J. M., Rae, I. D., and Wright, G. J. (1976). *Aust. J. Chem.* **29**, 931.
51. Blackman, G. L., Brown, R. D., Godfrey, P. D., and Gunn, H. I. (1976). *Nature* (*London*) **261**, 395.
52. Pearson, P. K., Blackman, G. L., Schaefer, H. F., Roos, B., and Wahlgren, U. (1973). *Astrophys. J.* **184**, L19.
53. Blackman, G. L., Brown, R. D., Godfrey, P. D., and Gunn, H. I. (1975). *Chem. Phys. Lett.* **34**, 241.
54. Wentrup, C., Wentrup-Byrne, E., and Müller, P. (1977). *J. Chem. Soc. Chem. Commun.*, p. 210.
55. Wentrup, C., and Müller, P. (1973). *Tetrahedron Lett.*, p. 2915.
56. Hedaya, E., and Kent, M. E. (1970). *J. Am. Chem. Soc.* **92**, 2149.
57. Arnold, D. R., Hedaya, E., Merritt, V. Y., Karnischky, L. A., and Kent, M. E. (1972). *Tetrahedron Lett.*, p. 3917.
58. Spangler, R. J., and Kim, J. H. (1972). *Tetrahedron Lett.*, p. 1249.
59. Spangler, R. J., Beckmann, B. G., and Kim, J. H. (1977). *J. Org. Chem.* **42**, 2989.
60. Wiersum, U. E., and Nieuwenhuis, T. (1973). *Tetrahedron Lett.*, p. 2581.
61. Crow, W. D., Lea, A. R., and Paddon-Row, M. N. (1972). *Tetrahedron Lett.*, p. 2235.
62. Crow, W. D., and Paddon-Row, M. N. (1973). *Aust. J. Chem.* **26**, 1705.

63. Crow, W. D., and Paddon-Row, M. N. (1972). *Tetrahedron Lett.*, p. 3207; p. 2217 (1973).
64. Schissel, P., Kent, M. E., McAdoo, D. J., and Hedaya, E. (1970). *J. Am. Chem. Soc.* **92**, 2147.
65. Joines, R. C., Turner, A. B., and Jones, W. M. (1969). *J. Am. Chem. Soc.* **91**, 7754.
66. Wentrup, C., and Wilczek, K. (1970). *Helv. Chim. Acta* **53**, 1459.
67. Crow, W. D., Paddon-Row, M. N., and Sutherland, D. S. (1972). *Tetrahedron Lett.*, p. 2239.
68. Wentrup, C., Mayor, C., and Gleiter, R. (1972). *Helv. Chim. Acta* **55**, 2628 and 3066.
69. Gleiter, R., Rettig, W., and Wentrup, C. (1974). *Helv. Chim. Acta* **57**, 2111.
70. Dewar, M. J. S., and Landman, D. (1977). *J. Am. Chem. Soc.* **99**, 6179.
71. Hedaya, E., and Kent, M. E. (1971). *J. Am. Chem. Soc.* **93**, 3283.
72. Baron, W. J., Jones, M., and Gaspar, P. P. (1970). *J. Am. Chem. Soc.* **92**, 4739.
73. D'Amore, M. B., Bergman, R. G., Kent, M. E., and Hedaya, E. (1972). *J. Chem. Soc. Chem. Commun.*, p. 49.
74. Henry, T. J., and Bergman, R. G. (1972). *J. Am. Chem. Soc.* **94**, 5103.
75. Staudinger, H., and Endle, R. (1913). *Ber. Deut. Chem. Ges.* **46**, 1437.
76. Jones, W. M., Joines, R. C., Myers, J. A., Mitsuhashi, T., Krajka, K. E., Waali, E. E., Davis, T. L., and Turner, A. B. (1973). *J. Am. Chem. Soc.* **95**, 826.
77. Mayor, C., and Wentrup, C. (1975). *J. Am. Chem. Soc.* **97**, 7467.
78. Lân, N. M., and Wentrup, C. (1976). *Helv. Chim. Acta* **59**, 2068.
79. Coburn, T. T., and Jones, W. M. (1974). *J. Am. Chem. Soc.* **96**, 5218.
80. Jones, W. M. (1977). *Acc. Chem. Res.* **10**, 352.
81. Gilchrist, T. L., Gymer, G. E., and Rees, C. W. (1975). *J. Chem. Soc. Perkin Trans. 1*, p. 1.
82. Gilchrist, T. L., Rees, C. W., and Thomas, C. (1975). *J. Chem. Soc. Perkin Trans. 1*, p. 8.
83. Gilchrist, T. L., Gymer, G. E., and Rees, C. W. (1973). *J. Chem. Soc. Perkin Trans. 1*, p. 555.
84. Crow, W. D., and Wentrup, C. (1968). *Chem. Commun.*, p. 1026.
85. Wentrup, C., and Crow, W. D. (1970). *Tetrahedron* **26**, 3965.
86. Wentrup, C. (1972). *Helv. Chim. Acta* **55**, 1613.
87. Thétaz, C., and Wentrup, C. (1976). *J. Am. Chem. Soc.* **98**, 1258.
88. Crow, W. D., and Wentrup, C. (1967). *Tetrahedron Lett.*, p. 4379.
89. Hedaya, E., Kent, M. E., McNeil, D. W., Lossing, F. P., and McAllister, T. (1968). *Tetrahedron Lett.*, p. 3415.
90. Wentrup, C. (1971). *Tetrahedron* **27**, 1027.
91. Rice, F. O., and Grelecki, C. J. (1957). *J. Phys. Chem.* **61**, 830.
92. Ford, R. G. (1977). *J. Am. Chem. Soc.* **99**, 2389.
93. Smolinsky, G., and Pryde, C. A. (1968). *J. Org. Chem.* **33**, 2411.
94. Smolinsky, G., Wasserman, E., and Yager, W. A. (1962). *J. Am. Chem. Soc.* **84**, 3220.
95. Crow, W. D., and Paddon-Row, M. N. (1975). *Aust. J. Chem.* **28**, 1755.
96. Crow, W. D., and Paddon-Row, M. N. (1972). *Tetrahedron Lett.*, p. 2231.
97. Wentrup, C. (1969). *Chem. Commun.*, p. 1386.
98. Wentrup, C. (1974). *Tetrahedron* **30**, 1301.
99. Crow, W. D., and Wentrup, C. (1968). *Tetrahedron Lett.*, p. 6149.
100. Crow, W. D., Khan, A. N., Paddon-Row, M. N., and Sutherland, D. S. (1975). *Aust. J. Chem.* **28**, 1763.
101. Crow, W. D., Khan, A. N., and Paddon-Row, M. N. (1975). *Aust. J. Chem.* **28**, 1741.
102. Crow, W. D., and Wentrup, C. (1968). *Chem. Commun.*, p. 1082.
103. Brown, R. F. C., and Smith, R. J. (1972). *Aust. J. Chem.* **25**, 607.
104. Crow, W. D., and Wentrup, C. (1969). *Chem. Commun.*, p. 1387.
105. Harder, R., and Wentrup, C. (1976). *J. Am. Chem. Soc.* **98**, 1259.
106. Wentrup, C., and Crow, W. D. (1970). *Tetrahedron* **26**, 4915.

107. Wentrup, C. (1972). *Helv. Chim. Acta* **55**, 565.
108. Wentrup, C., Thétaz, C., and Gleiter, R. (1972). *Helv. Chim. Acta* **55**, 2633.
109. Wentrup, C. (1971). *Tetrahedron* **27**, 367.
110. Brown, R. F. C., Irvine, F., and Smith, R. J. (1973). *Aust. J. Chem.* **26**, 2213.
111. Wentrup, C. (1971). *Tetrahedron* **27**, 1281.
112. Abramovitch, R. A., and Sutherland, R. G. (1970). *Top. Current Chem.* **16**, 1.
113. Abramovitch, R. A., and Holcomb, W. D. (1975). *J. Am. Chem. Soc.* **97**, 676.
114. Reichen, W. (1976). *Helv. Chim. Acta* **59**, 2601.

Chapter 6

Fragmentation of Cyclic Structures with Elimination of Small Molecules

I. INTRODUCTION

This chapter deals with the pyrolytic cleavage of thermodynamically stable small molecules such as CO, CO_2, COS, N_2, RCN, and SO_n from a somewhat heterogeneous collection of cyclic and heterocyclic compounds. The arrangement is largely by structural type of the starting material rather than by mechanism, and concerted and stepwise diradical or dipolar mechanisms will be found side by side. Many of the compounds considered contain an aromatic ring which survives pyrolysis while a second fused, often heterocyclic, ring undergoes fission; the primary product is then an aromatic system with two ortho substituents which may have radical, carbene, nitrene, or dipolar character. Such compounds have been at the center of the revival of interest in the pyrolysis of organic compounds, and curiosity about the relationship between their mass spectral, photochemical, and thermal fragmentations has been an important stimulus to research in this area.

II. PYROLYSIS OF CARBONYL COMPOUNDS AND SOME RELATED THIONYL AND SULFONYL COMPOUNDS

A. Ketones

The pyrolysis of cyclobutanones may involve competition between α-cleavage with decarbonylation and 2 + 2 cleavage to alkenes and ketenes, and is discussed in Chapter 8, Section II,C. The special case of the decarbonylation of benzocyclobutenone to benzocyclopropene and fulvenallene is mentioned in Chapter 5, Section III,B [see Eq. (5.33)]. Cyclopentanone in a static gas-phase system appears to undergo mainly α-cleavage at 532°–581° [Eq. (6.1)], although this kinetic study by Delles *et al.*[1] did not permit detailed mechanistic proposals. On the other hand the flash vacuum thermolysis of cyclohexanone at 1050° studied by de Mayo and Verdun[2] gave products including methyl vinyl ketone (25%) and ethylene which could be explained only by assuming major β-cleavage. Smaller amounts of ethane, acetylene, propene, allene, 1-butene, 1,3-butadiene, 1,5-hexadiene, cyclopentane, cyclopentadiene, acrolein, and benzene were also formed. The occurrence of at least 65% of direct β-cleavage was confirmed by the pyrolysis of cyclohexanone deuterated at the 2-, 3-, and 4-positions; the decomposition of cyclohexanone-2,2,6,6-d_4 is shown in Eq. (6.2). Methylenecyclohexane-2,2,6,6-d_4 similarly gave isoprene and ethylene, and the authors prefer to depict these reactions as involving direct β-cleavage rather than by initial enolization of cyclohexanone-d_4 followed by retro-Diels–Alder reaction of the enol **1**.

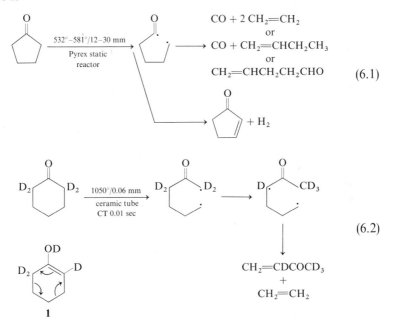

Similar β-cleavage, assisted by an additional C=C bond, occurs in the pyrolytic ring contraction of cyclooct-4-enone at 720° [see Chapter 9, Section III,B; Eq. (9.54)] and in the flash pyrolytic fragmentation of camphor studied by Sato and co-workers.[3,4] The flow pyrolysis of camphor-3,3-d_2[4] at 800° gave dihydrocarvone (**2**), the methyleneketone **3**, and methyl isopropenyl ketone with unchanged deuterium content, but the process shown

in Eq. (6.3) is evidently an oversimplification because the isoprene formed
contained 21% of d_1 species, whereas Eq. (6.3) would require that it contained no deuterium. Other products including benzene, toluene, m-xylene,

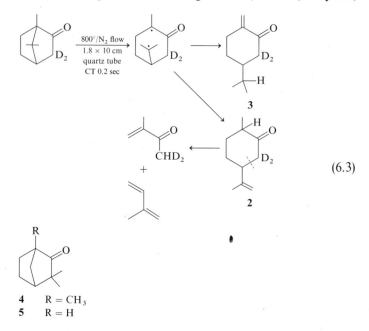

(6.3)

styrene, m-methylethylbenzene, and o-cresol were also formed. The pyrolysis
of fenchone (**4**) and camphenilone (**5**) under the same conditions was investigated;[4] these ketones gave almost identical complex mixtures of hydrocarbons probably formed by initial β-cleavage, but the mechanism of
fragmentation was not studied in detail. In passing it may be noted that
flow pyrolysis of camphor oxime at 500°[5] gave the major product **6** by

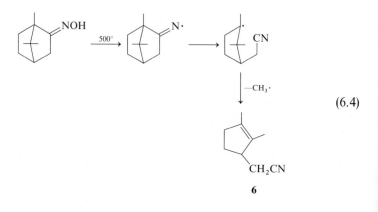

(6.4)

α-cleavage, although in this case rearrangement is probably initiated by breaking of the N—O bond [Eq. (6.4)].

Cookson and co-workers[6] showed that the cage ketone **7** decomposed only by initial cleavage of a cyclobutane ring (**7**, as marked) at 450° and the major product dihydroindene was then formed by chelotropic elimination of CO from the intermediate bridged cyclopentenone **8** [Eq. (6.5)]. By contrast the heavily phenylated and methylated cage ketone **9** underwent quantitative decarbonylation with cleavage to the diene **10** at 450°.[7] In this case α-cleavage of the carbonyl group is probably assisted by participation of the 2,3-bond between the phenylated centers. Similarly 1,1,3,3-tetramethylindan-2-one readily underwent α-cleavage at the 1,2 and 2,3 benzylic bonds at 600°–800°[8] to give naphthalene and many other rearranged products probably via the primary intermediate **11** [Eq. (6.7)].

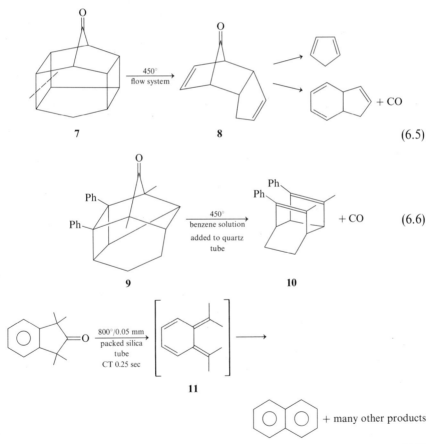

(6.5)

(6.6)

(6.7)

B.   1,2-Diones and Related Systems

Early interest in the relationship between thermal and mass spectral fragmentations is shown by the work of Brown and Solly[9] on the flash vacuum pyrolysis and mass spectrometry of indantrione. In each case stepwise decarbonylation occurred, and the products isolated from pyrolysis at 600° included benzocyclobutenedione, anthraquinone, phenanthrenequinone, fluorenone, biphenylene, triphenylene, and many other polycyclic hydrocarbons. In this case the reaction undoubtedly starts with the breaking of an OC—CO bond as shown in Eq. (6.9), whereas in some other cases shown in Table 6.1 it is not clear whether an OC—CO bond or an ArCR$_2$—CO bond breaks first. Several pyrolytic syntheses are now available for benzocyclobutenedione; Eq. (6.10) shows the pyrolysis of S,S-disubstituted-N-phthalimidosulfoximides studied by Anderson *et al.*[10] The dimethyl compound

Mass spectrum: m/e 160 $\xrightarrow{-CO}$ m/e 132 $\xrightarrow{-CO}$ m/e 104 $\xrightarrow{-CO}$ m/e 76 $\qquad$ (6.8)

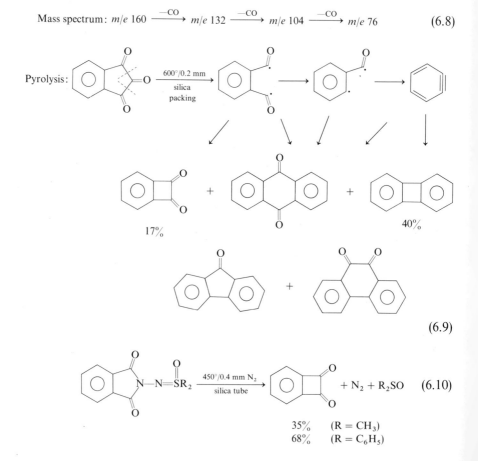

(6.9)

(6.10)

35%   (R = CH$_3$)
68%   (R = C$_6$H$_5$)

TABLE 6.1

Products of Pyrolysis of 1,2-Diones and Ketolactones

	1,2-Dione	Product(s) (% yield)	Pyrolytic conditions	Reference

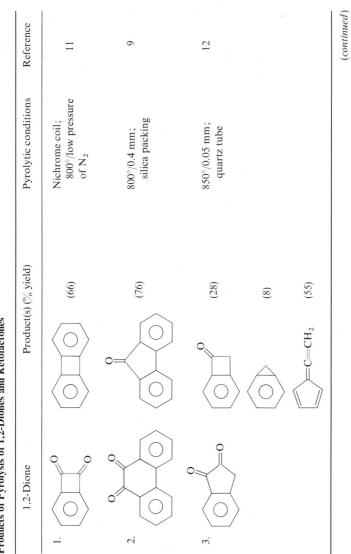

1. (66) Nichrome coil; 800°/low pressure of N_2 11

2. (76) 800°/0.4 mm; silica packing 9

3. (28) (8) (55) 850°/0.05 mm; quartz tube 12

(continued)

TABLE 6.1 (*Continued*)

1,2-Dione	Product(s) (% yield)	Pyrolytic conditions	Reference

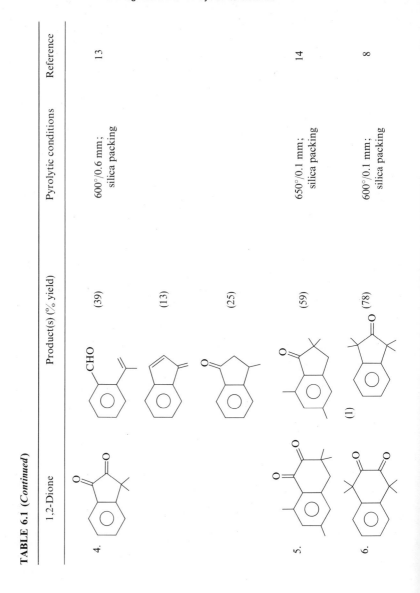

4.

(39)

(13)

(25)

600°/0.6 mm;
silica packing

13

5.

(59)

(1)

650°/0.1 mm;
silica packing

14

6.

(78)

600°/0.1 mm;
silica packing

8

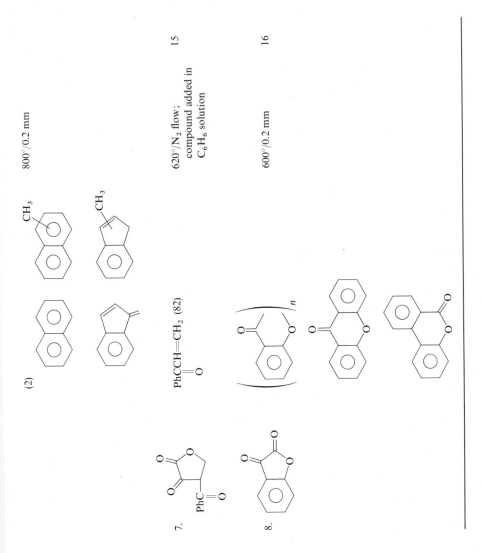

800°/0.2 mm

(2)

7.

8.

$PhCCH=CH_2$ (82)

620°/N$_2$ flow; compound added in C$_6$H$_6$ solution   15

600°/0.2 mm   16

n

gave a poor yield of the dione because of hydrogen transfer from the S-methyl groups which led to phthalimide, but the diphenyl compound gave 68% of benzocyclobutenedione. An alternative approach via retro-Diels–Alder generation of phthalazinedione is described in Chapter 8, Section III,B, Eq. (8.38).

Table 6.1 shows some examples of the pyrolysis of 1,2-diones and related α-ketolactones. Both of the indandiones (entries 3 and 4) give the corresponding o-quinonoid ketene as a primary pyrolytic intermediate. The parent intermediate **12** then either cyclizes, decarbonylates and cyclizes to benzocyclopropene, or decarbonylates and rearranges to fulvenallene [see Chapter 5, Eq. (5.33)]. The dimethyl intermediate **13** forms o-isopropenylbenzaldehyde through a 1,5 hydrogen shift, and the other products have been shown to be formed by secondary reactions of this aldehyde. The formation of 3-methylindanone (entry 4) is interesting but not immediately explicable, and the representation of its formation as **14** does not seem very satisfactory. The methylindenol **15** has been obtained from attempted gas chromatography of o-isopropenylbenzaldehyde, and is a proven pyrolytic precursor of benzofulvene.

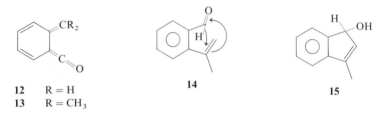

12   R = H
13   R = CH$_3$

14

15

The tetramethylindanone formed from the tetralindione (entry 6) at 600° undergoes further decarbonylation at 800° to give first the o-quinonoid intermediate **11** [see Eq. (6.7)] and then probably o-isopropenylisopropylbenzene **16** by a 1,5 hydrogen shift in **11**. Comparison of 800° pyrolyzates from various related hydrocarbons with that from the tetralindione of entry 6 suggests that cyclization of **16** to an indane (**17**) may occur on the pathway to naphthalene, with later elimination of methane and ring expansion.

16

17

The literature summarized by Haddon[17] suggests that 1,2-diones of the type **18** and its benzo and dibenzo derivatives are reluctant to decompose

18

by concerted loss of C_2O_2, whereas the corresponding monocarbonyl-bridged systems decarbonylate with great ease.

The pyrolysis of isatin (**19**), which probably involves initial breaking of an OC—CO bond, leads ultimately to iminocyclohexadienylidene and cyano-cyclopentadiene; this pyrolysis is discussed in Chapter 5, Section III,D.

19

C. Quinones and Polycyclic Aromatic Carbonyl Compounds

The possibility that cyclobutadienes might be produced by bisdecarbonyl-ation of benzoquinones has been a spur to flash pyrolytic work with qui-nones. Their pyrolysis is discussed in some detail in the lively review article of Hageman and Wiersum,[18] which contains some experimental results which do not otherwise seem to have been published.

Another elusive molecule is monomeric cyclopentadienone. DeJongh and co-workers[19] showed that pyrolysis of *o*-benzoquinone at 800° and 10 mm N_2 over a Nichrome coil gave the dimer **20** of cyclopentadienone, and the same dimer could also be obtained by pyrolysis of *o*-phenylene sulfite[19,20] or *o*-phenylene carbonate.[21] Chapman and McIntosh[22] used a simple low-temperature cell with a sodium chloride plate at −196° to record the infrared spectrum of cyclopentadienone formed by flash vacuum pyrolysis of *o*-benzoquinone [Eq. (6.11)]. Dimerization of this highly reactive dienone occurred even below −80°. DeJongh *et al.*[23] have studied the pyrolysis of tetrachloro-*o*-benzoquinone, which at 650° and 17 mm N_2 over a Nichrome

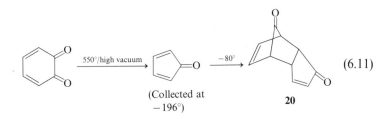

550°/high vacuum   −80°

(Collected at −196°)

20

(6.11)

coil gave the octachloro derivative of **20** in 47% yield. Pyrolysis at 710°–720° through a quartz tube, however, gave mainly chlorocarbons formed by bisdecarbonylation [Eq. (6.12)]. The major enyne **21** could possibly be formed from tetrachlorocyclobutadiene, but there is no firm evidence for this rather than for the diradical shown in Eq. (6.12). At 850° the diyne **22** and the enyne **21** were obtained in the ratio 2:1, showing that formation of **22** is a secondary process.

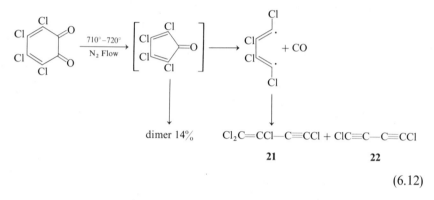

dimer 14% $Cl_2C\!=\!CCl\!-\!C\!\equiv\!CCl + ClC\!\equiv\!C\!-\!C\!\equiv\!CCl$

                        21   **22**

(6.12)

The pyrolysis of 9,10-phenanthrenequinone is noted in Table 6.1 of Section II,B, and that of 1,2-naphthoquinone is in Table 6.2, entry 1.

Hageman and Wiersum[24] found that the pyrolysis of *p*-benzoquinone at 850° gives mainly vinylacetylene (70%) with smaller amounts of benzene, styrene, phenylacetylene, and indene, and traces of toluene and naphthalene [Eq. (6.13)]. No cyclopentadienone or its dimer was detected in any experiment, and the authors suggest[18] that the bicyclic ketone **23** is the precursor of the C_4H_4 species which rearranges to vinylacetylene. There is no evidence to distinguish between cyclobutadiene, tetrahedrane, or diradical structures for the intermediate C_4H_4. Tetrachloro-*p*-benzoquinone on flash vacuum pyrolysis at 825°[18] similarly gives the chlorinated enyne **21** (90%) and the diyne **22** (10%).

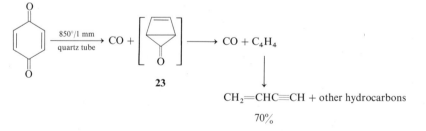

                        23

$CH_2\!=\!CHC\!\equiv\!CH$ + other hydrocarbons

70%

(6.13)

2-Phenylbenzoquinone on flash vacuum pyrolysis at 800° [25] gave starting material (40%), naphthalene (32%), 2-hydroxydibenzofuran (5%), and 2-phenyl-1,4-hydroquinone (5%). The same question arises as to the structure of the presumed PhC_4H_3 intermediate which leads to naphthalene; is it a cyclobutadiene, a tetrahedrane, or a diradical? The formation of vinyl-acetylene from the parent quinone, however, suggests that naphthalene may be formed by an electrocyclic reaction of an enyne such as **24**, a process analogous to the synthetically useful cyclization of 1-arylbutadienes (see Chapter 9, Section II,C).

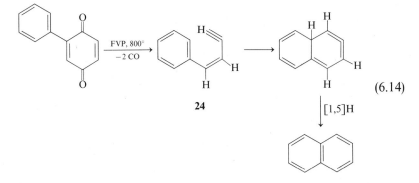

(6.14)

The products of pyrolysis of some further quinones, extended quinones, and other aromatic carbonyl compounds are shown in Table 6.2. The simple preparation of dibenzopentalene from the readily available chrysene-6,12-quinone (entry 5) is particularly noteworthy. Entries 6–8 show the decarbonylation of some tropone derivatives taken from the extensive work of Mukai *et al.*[26] on tropones and tropolones. The decarbonylation of such compounds may involve loss of carbon monoxide from a fused cyclopropanone such as **25** derived from tropone [Eq. (6.15)]. Similar decarbonylation of the vapors of many quinones and aromatic carbonyl compounds has been achieved by Suhr's group[27] by passing the vapors through glow discharge plasmas.

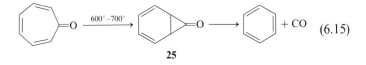

(6.15)

D. Cyclic Sulfites and Carbonates

The pyrolytic decomposition of cyclic sulfites and carbonates is conveniently considered here, following the account of the pyrolysis of quinones,

TABLE 6.2

Products of Pyrolysis of Quinones and Other Aromatic Carbonyl Compounds

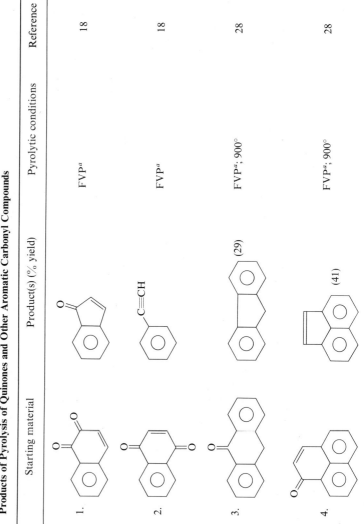

	Starting material	Product(s) (% yield)	Pyrolytic conditions	Reference
1.			FVP[a]	18
2.		C≡CH	FVP[a]	18
3.		(29)	FVP[a], 900°	28
4.		(41)	FVP[a], 900°	28

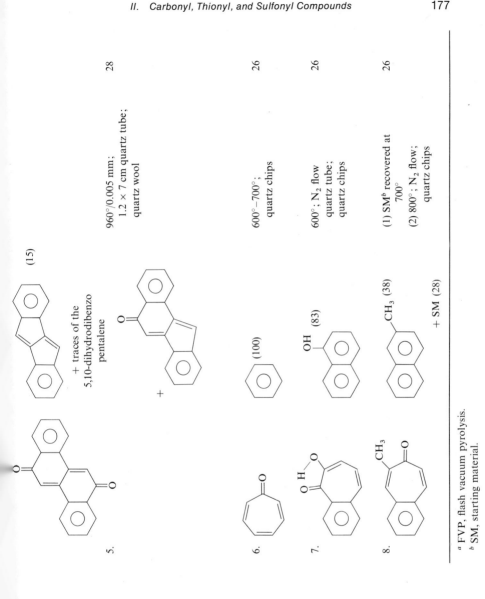

because some such aromatic esters form quinones as primary products of pyrolysis. Before discussing these aromatic esters, however, we briefly mention the cyclic sulfites formed from aliphatic compounds.

Price and Berti[29] and Coxon and co-workers[30,31] have investigated the decomposition of cyclic sulfites and some carbonates of 1,2-diols in the condensed phase at $250°-300°$ and have found products including carbonyl compounds, epoxides, and unsaturated alcohols. A synthetically useful application of sulfite pyrolysis developed by Matlack and Breslow[32] is the decomposition of the sulfite **26** from 2,2-bis(chloromethyl)-1,3-propanediol which gave the allylic dichloride **27** in high corrected yield (91%) though at low conversion (59% recovery). In all such decompositions the sulfur fragment is SO_2.

$$\text{26} \xrightarrow[\substack{2.5 \times 35 \text{ cm} \\ \text{Vycor tube} \\ \text{quartz chips}}]{\substack{500°/15 \text{ mm}}} \text{27, 37\%} + CH_2O + SO_2 \tag{6.16}$$

The pyrolysis of aromatic cyclic sulfites and carbonates has been the subject of extended study by DeJongh's group since 1967,[19] and these substances have provided an unusually interesting and complete set of data for the correlation of mass spectral and pyrolytic fragmentations. The problem of the direction of fragmentation is illustrated by the pyrolytic behavior of *o*-phenylene sulfite and carbonate [Eqs. (6.17) and (6.19)]. The sulfite decomposes by loss of SO to give *o*-benzoquinone, and ultimately cyclopentadienone isolated as the dimer.[20] This fragmentation closely parallels the 70 eV mass spectral fragmentation,[19,20] in which loss of SO is the major primary process [Eq. (6.18)] characterized by strong metastable peaks, and there is only very minor loss of SO_2. By contrast the carbonate shows

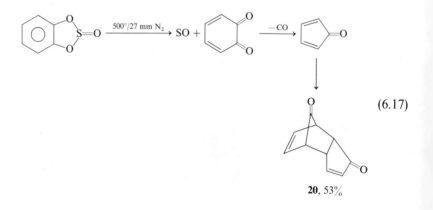

$$\tag{6.17}$$

20, 53%

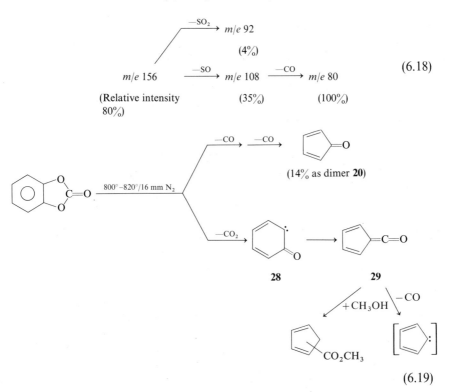

$$\begin{array}{c} \xrightarrow{-SO_2} \quad m/e\ 92 \\ (4\%) \end{array}$$

$$m/e\ 156 \xrightarrow{-SO} m/e\ 108 \xrightarrow{-CO} m/e\ 80$$

(6.18)

(Relative intensity 80%)   (35%)   (100%)

(14% as dimer **20**)

28   **29**

(6.19)

major pyrolytic loss of CO_2 to give the ketene **29** which can be trapped with methanol [Eq. (6.19)], and minor loss of CO; this also parallels the mass spectrum, which displays a strong metastable peak corresponding to loss of CO_2 [Eq. (6.20)]. In the absence of methanol the ketene **29** undergoes decarbonylation, dimerization, and rearrangement to give hydrocarbons including indene, naphthalene, and traces of azulene.

$$\begin{array}{c} \xrightarrow{-CO} m/e\ 108 \xrightarrow{-CO} m/e\ 80 \\ (0.3\%) \qquad (1.8\%) \end{array}$$

$$m/e\ 136$$

$$\begin{array}{c} \xrightarrow{-CO_2} m/e\ 92 \xrightarrow{-CO} m/e\ 64 \\ (58.5\%) \qquad (100\%) \end{array}$$

(80%)

(6.20)

These results are in accord with the hypothesis of Dougherty[33,34] that mass spectral processes which show strong metastable peaks usually occur from ions in low-lying electronic states and should thus be similar to thermal rather than to photochemical fragmentations.

DeJongh and Thomson[35] have made a semiempirical molecular orbital study of the decomposition of o-phenylene sulfite and carbonate and of their molecular ions. They used a CNDO/2 program to calculate Mulliken overlap populations and total energies of the various possible product species. Overlap populations, taken as an indication of total bond strengths, were found to be lowest for the Ar—OCO and ArO—SO bonds, in agreement with the observed major fragmentations. Similarly, the sum of the energies of the ketocarbene **28** + CO_2 was found to be 21.7 kcal lower than the sum of the energies of o-benzoquinone + CO for the neutral species; for the same pairs but with the carbocyclic systems as $+1$ charged species this difference favored loss of CO_2 by 65.6 kcal. Calculation of energies for the cases of loss of SO or SO_2 from o-phenylene sulfite led to a predicted preference for loss of SO of 47.9 or 4 kcal for the neutral and charged species, respectively.

Three further examples of this parallelism are shown in Table 6.3, although the correlation is by no means so close in every case. Loss of SO from the molecular ion of biphenylylene 2,2′-sulfite (entry 1) became relatively more important at lower ionizing voltages, so that the mass spectral process of lower energy showed a reasonable correlation with the more important pyrolytic process. The formation of 3,4-benzocoumarin (entry 3) is considered to be due to a rearrangement of the primary diradical produced by loss of SO_2; more direct processes lead to xanthone and dibenzofuran.

Wentrup[37] has estimated the heats of formation of all the possible species which could be involved in the pyrolysis of o-phenylene carbonate and sulfite. He disagrees with the explanation of DeJongh and Thomson[35] based on CNDO/2 energies of products only, and argues that the processes are controlled by the thermodynamics of breaking of the first bond [Eqs. (6.21) and (6.22)] with the subsequent cleavage of CO_2 or of SO showing a clear thermochemical preference.

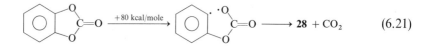

$$\text{(6.21)}$$

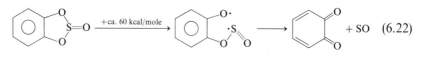

$$\text{(6.22)}$$

E. Lactones

The pyrolysis of β-lactones is considered in Chapter 8, Section II,D. Saturated γ-lactones are much more stable, and γ-butyrolactone undergoes only slight decomposition on pyrolysis at 590°.[38] ε-Caprolactones, however,

TABLE 6.3

Major Mass Spectral and Pyrolytic Processes in Three Tricyclic Esters

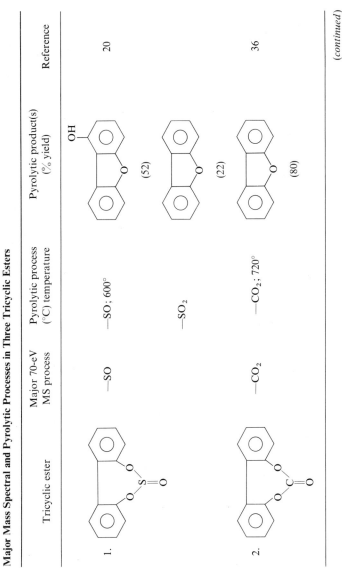

Tricyclic ester	Major 70-eV MS process	Pyrolytic process (°C) temperature	Pyrolytic product(s) (% yield)	Reference
1.	—SO	—SO; 600°	(52)	20
		—SO₂	(22)	
2.	—CO₂	—CO₂; 720°	(80)	36

(continued)

TABLE 6.3 (Continued)

Tricyclic ester	Major 70-eV MS process	Pyrolytic process (°C) temperature	Pyrolytic product(s) (% yield)	Reference
3. 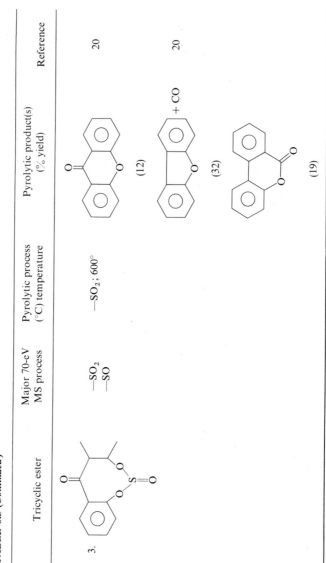	$-SO_2$ $-SO$	$-SO_2$; 600°	(see structures)	20 20

are conformationally sufficiently flexible to permit normal β-elimination, as shown by Bailey and Bird[38] for caprolactone and its ε-methyl derivative [Eqs. (6.23) and (6.24)].

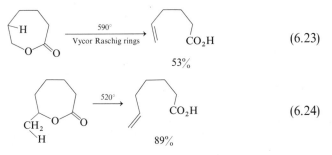

$$\begin{array}{cc} & (6.23) \\ & 53\% \end{array}$$

$$\begin{array}{cc} & (6.24) \\ & 89\% \end{array}$$

Wentrup and Müller[39] have compared the products of pyrolysis of phthalide (**30**) and of coumaranone (**31**). The pyrolysis of phthalide is directed by the ready breaking of the $ArCH_2$—OCO bond,[37] with subsequent loss of CO_2 and ring contraction to fulvenallene [Eq. (6.25)], whereas similar breaking of the $ArCH_2$—CO bond in **31** leads ultimately by bis-decarbonylation to fulvene, benzene, and other products [Eq. (6.26)]. The thermochemistry of these two reactions is discussed in detail,[37] and their activation energies are estimated to be 55–68 and 67 kcal/mole, respectively. Gilchrist and Pearson[40] have reported the smooth pyrolysis of the furanoid lactone **32** to give the well-stabilized diradical **33** which then undergoes ring cleavage to an acetylenic ketone [Eq. (6.27)].

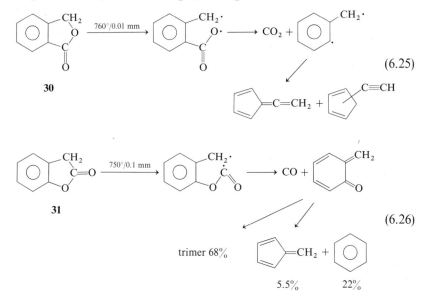

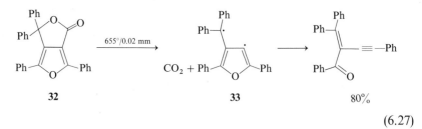

32   **33**   80%

$$(6.27)$$

Some β,γ-unsaturated lactones undergo decarbonylation with an ease which suggests that the reactions are synchronous chelotropic processes, as in the case of the synthetically interesting examples of Eqs. (6.28) and (6.29) reported by Skorianetz and Ohloff.[41]

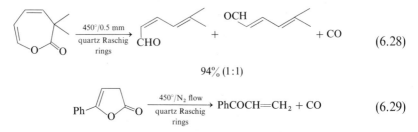

$$(6.28)$$

94% (1:1)

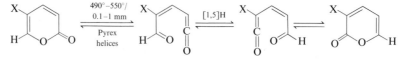

$$(6.29)$$

The pyrolysis of 2-pyrone at 900° and 2–5 mm N_2 gave furan (15%), and the further decarbonylation products propyne (41%) and allene (4%) which Brent et al.[42] showed were secondary products probably formed from furan. Coumarin on pyrolysis at 925° and 2–5 mm N_2 similarly gave benzofuran (63%) and some starting material. Pirkle and Turner[43] were able to throw some light on anomalies in the interpretation of the pyrolytic and mass spectral[44] fragmentations of substituted 2-pyrones by showing that many 3- and 5-substituted 2-pyrones are interconverted by pyrolysis at 490°–550° by a sequence of electrocyclic ring opening, 1,5 hydrogen migration, and electrocyclic ring closure [Eq. (6.30)]. Pyran-2-thiones and thiapyran-2-ones are similarly interconverted, and the nature of the rearrangement of 5-methyl-2-pyrone was confirmed with respect to oxygen by the use of ^{18}O-labeling. Pirkle et al.[45] have pointed out that the formation of furan from 2-pyrone at 900° may be due to decarbonylation of the intermediate ketene **34** to the carbene **35** which could cyclize to furan.

$X = CH_3$, Br, CH_3O, CH_3CO_2, or CF_3CO_2

$$(6.30)$$

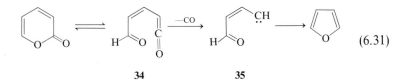

34   35

F. Lactams

2-Pyridone sublimes only with difficulty on a preparative scale because products of decomposition at 170°–210° tend to coat the sample and reduce its volatility, but Brent *et al.*[42] obtained pyrrole and a mixture of three butene nitriles [Eq. (6.32)] in a combined yield of 15% on pyrolysis of 2-pyridone at 960°–1020°. 2-Quinolone on flash vacuum pyrolysis at 850° and 0.3 mm gave only a trace of indole, and could be recovered in 86% yield.[46]

Brown and Butcher[47] found that pyrolysis of oxindole at 850° and 0.7 mm gave a tar and a complex mixture of liquid products including benzene, benzonitrile (26% yield), 2-vinylpyridine, aniline *o*-toluidine, and *o*-amino-styrene. This result was attributed to the formation of *o*-tolylnitrene on decarbonylation of oxindole, and the distribution of ^{13}C in benzonitrile formed from [3-^{13}C]oxindole [Eq. (6.33)] was consistent with the known rearrangements of arylnitrenes derived from aryl azides (see Chapter 5, Section III,E). This pyrolytic result suggested a reinterpretation of the 70 eV mass spectrum of oxindole, which shows prominent loss of CO and HCN. The carbon of the HCN might at first sight be expected to come from C-3 of oxindole, but the HCN lost from the molecular ion of [3-^{13}C]oxindole was found to be predominantly H[12]CN (ca. 70%), as expected by analogy with the pyrolytic process which forms [13]Ph[12]CN. The nitrene formed initially from 7-phenyloxindole was trapped by attack on the adjacent phenyl group to give 1-methylcarbazole [Eq. (6.34)] accompanied by some carbazole.

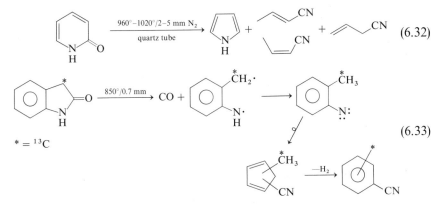

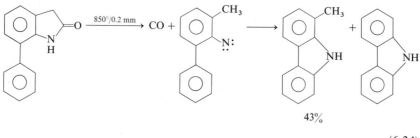

$$(6.34)$$

3,3-Dimethyloxindole on pyrolysis at 850° and 0.3 mm[46] gave 2-quinolone (52%) and indole (25%), but it is not clear whether 2-quinolone is formed by ring expansion of 3-methyleneoxindole (**38**), or from the diradical **36** by rearrangement and cyclization to **37** followed by elimination of methane [Eq. (6.35)]. Various mechanistic possibilities and related pyrolytic experiments are discussed by Brown and Butcher.[46]

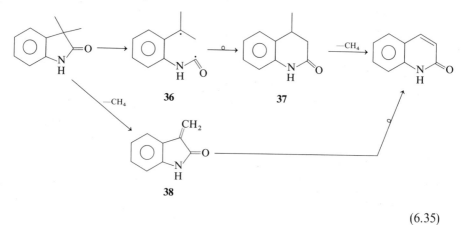

$$(6.35)$$

G.   Sultones and Sultams

DeJongh and Evenson[48,49] pyrolyzed the sultone naphth[1,8-*c,d*]-1,2-oxathiole 2,2-dioxide (**39**) at 650° in nitrogen and obtained 1-naphthol, traces of indene and naphthalene, and a *lactone*, naphtho[1,8-*b,c*]furan-2-one (**42**), in which SO_2 had been lost and CO added [Eq. (6.36)]. This remarkable result was explained[48] by the loss of SO_2 from **39** to give the diradical **40** or the cyclic species **41**. Abstration of hydrogen from another molecule by **40** gives 1-naphthol, and highly efficient trapping of CO by **40** or **41** gives the lactone **42**. In nitrogen the yields of these products were low, but addition of methanol vapor as hydrogen donor raised the yield of 1-naphthol to 22% and addition of carbon monoxide (200 mm) gave the lactone **42** in 79–95%

yield. The reality of the cyclic species **41** was shown by an experiment with the 6-methyl compound **43** which gave both possible methyl-1-naphthols in 24% combined yield with methanol as hydrogen donor [Eq. (6.37)].

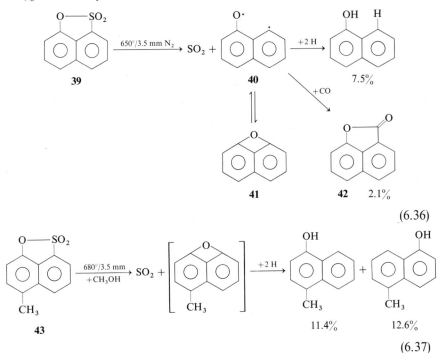

(6.36)

(6.37)

The sultam **44** similarly lost sulfur dioxide at 740° over a Chromel coil to give the diradical **45**, 1-naphthylnitrene, and ultimately the cyanoindenes [Eq. (6.38)] by rearrangement of this nitrene [compare Chapter 5,

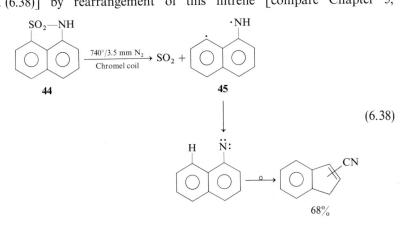

(6.38)

Section III,D; Eq. (5.61)]. The diradical from the corresponding N-phenyl derivative cyclized instead to 7H-benz[k,l]acridine.[49]

H.   Cyclic Anhydrides

The vigorous pyrolysis of five-membered cyclic anhydrides of aliphatic acids gives CO, CO_2, and hydrocarbons in good yield. Three examples studied by Rice and Murphy[50] are shown in Eqs. (6.39)–(6.41). The decomposition of succinic anhydride might be represented as the concerted process **46**, but it is quite probable that a diradical **47** or an equivalent dipolar intermediate is involved in its fragmentation leading to ethylene. The pyrolysis of several acyclic and cyclic anhydrides at 500° and atmospheric pressure with relatively long contact times has been studied by Brown and Ritchie,[51] who favor the intermediacy of the carboxyketene **48** in the pyrolysis of succinic anhydride because of the presence of both acrylic acid and acrolein in the liquid fraction of the pyrolyzate. It is possible that **47** fragments directly to ethylene at 900°, but undergoes some hydrogen transfer to give **48** at 500° under the conditions of Brown and Ritchie. Methylmaleic anhydride similarly gives propyne (58%) at 750° and 50 mm[50] and the pyrolysis of fluoromaleic anhydride through a packed Vycor tube at 650° and 5–7 mm provides a useful synthesis of fluoroacetylene.[52] The failure of Feist's anhydride to yield methylene-cyclopropene on pyrolysis is noted in Chapter 3, Section VI; and Eq. (3.58).

$$\xrightarrow[\substack{\text{quartz tube} \\ \text{CT 0.2 sec}}]{880°-900°/10 \text{ mm}} CH_2{=}CH_2 + CO + CO_2 \qquad (6.39)$$

$$\xrightarrow[\text{CT 0.08 sec}]{820°/10 \text{ mm}} HC{\equiv}CH + CO + CO_2 \qquad (6.40)$$

$$\xrightarrow[\text{CT 0.05 sec}]{900°/3-4 \text{ mm}} CH_2{=}C{=}CH_2 + CH_3C{\equiv}CH + CO + CO_2 \quad (6.41)$$

$$56\% \qquad\qquad 44\%$$

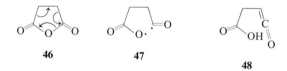

46   **47**

48

The anhydrides of both bicyclo[2.2.1]heptadiene- and bicyclo[2.2.1]hept-2-ene-2,3-dicarboxylic acids **49** and **50** yield rearranged products of pyrolysis. Morner et al.[53] obtained CO, CO_2, and fulvenallene on flash vacuum pyrolysis of **49** at 900°, but at 700° the intermediate benzocyclobutenone was formed; the probable pathway is shown in Eq. (6.42). Benzocyclobutenone and fulvenallene are also formed on pyrolysis of homophthalic anhydride at 530° and 2.3 mm N_2 over a Nichrome coil.[54] The anhydride **50** undergoes initial retro-Diels–Alder reaction at 550° to give ethylene and a cyclopentadienedicarboxylic anhydride **51**; Bloch[55] proposes that the final products, fulven-6-one and CO_2, are formed by fragmentation of the isomeric anhydride **52**.

4,4-Dimethylhomophthalic anhydride (**53**) at 560° over a Nichrome coil[54] gives *o*-isopropenylbenzaldehyde [Eq. (6.44)] via the same intermediate **54**, which is formed on pyrolysis of 3,3-dimethylindan-1,2-dione (see Section II,B; Table 6.1, entry 4).

Carbon suboxide is conveniently prepared by pyrolysis of diacetyltartaric anhydride (**55**)[56] and Crombie and co-workers[57] have summarized the most

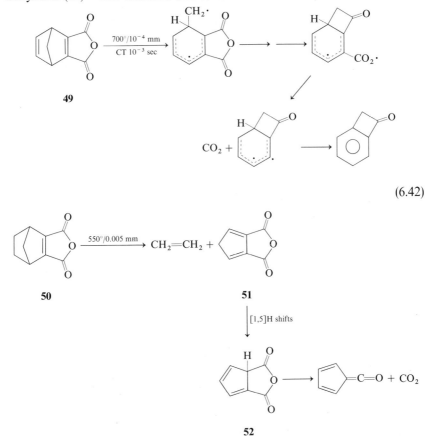

(6.42)

(6.43)

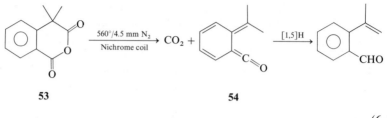

$$560°/4.5\ mm\ N_2\ (Nichrome\ coil) \rightarrow CO_2 + \quad \xrightarrow{[1,5]H}$$

53   **54**

(6.44)

satisfactory experimental procedures. They have carried out a careful [14]C- and ^{18}O-labeling study of the reaction, and they conclude that, after initial β-elimination of acetic acid leading to **56**, the reaction does not proceed through a symmetrical intermediate such as **57**, but must involve an un-symmetrical species such as **58**. The carbon atom lost on pyrolysis of acetoxymaleic anhydride (**56**) is C-1. Three possible pathways for the decomposition of **58** have been suggested; only one of these is shown in Eq. (6.45), and the nature of these final steps has not been clearly established. One alternative scheme leads from **58** to carbon dioxide, carbon suboxide, and acetaldehyde.

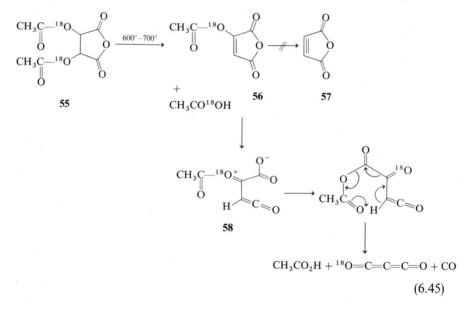

$$CH_3CO_2H + {}^{18}O{=}C{=}C{=}C{=}O + CO$$

(6.45)

I. Aromatic Cyclic Anhydrides

   The reactions of arynes formed by pyrolysis of aromatic anhydrides in the presence of high pressures of other aromatic compounds are discussed in Chapter 3, Section V. Cava and co-workers[58] and Brown, McOmie and co-workers[59] found that flash vacuum pyrolysis of phthalic anhydride was a simple method for the preparation of biphenylene by dimerization of benzyne. Both groups were guided to this method by consideration of the mass spectrum of phthalic anhydride, which shows stepwise fragmentation parallel to the pyrolytic processes of Eq. (6.46). No carbonyl compounds have been found in this high-temperature pyrolyzate (compare the pyrolysis of indanetrione in Section II,B) and attempts to detect benzocyclopropenone or the diradical **59** in the pyrolyzate stream by direct mass spectrometry have been unsuccessful,[53] although Fields and Meyerson[60] obtained some fluorenone by pyrolysis of a solution of phthalic anhydride in benzene. Small amounts of biphenyl, naphthalene, fluorene, and higher hydrocarbons are also formed in the reactions of Eq. (6.46). *o*-Sulfobenzoic anhydride (**60**) is an alternative pyrolytic precursor of benzyne at 690°–830°;[59,61] initial loss of SO_2 appears to produce the diradical **61**, and in the presence of benzene 3,4-benzocoumarin is formed.[61] The formation of biphenylene in 27% yield by gas-phase pyrolysis of phthaloyl peroxide at 600° was earlier reported by Wittig and Ebel.[62]

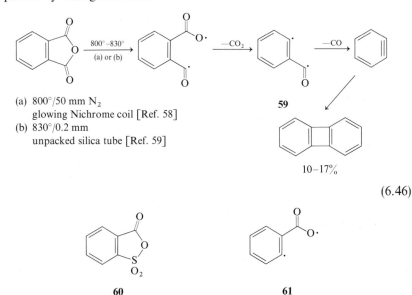

(a) 800°/50 mm N_2
   glowing Nichrome coil [Ref. 58]
(b) 830°/0.2 mm
   unpacked silica tube [Ref. 59]

59

10–17%

(6.46)

60   **61**

Several substituted biphenylenes have been prepared by the pyrolysis of substituted phthalic anhydrides, and examples are shown in Table 6.4. Yields are usually low except in the case of tetrachlorophthalic anhydride (entry 6) which, however, also gives much hexachlorobenzene. Tetrabromophthalic anhydride gives only hexabromobenzene (entry 7), reflecting the

TABLE 6.4

Preparation of Substituted Biphenylenes from Phthalic Anhydrides

Phthalic anhydride	Pyrolytic conditions	Substituted biphenylene (% yield)	Other products	Reference
1. 4-CH_3	850°/0.3 mm, N_2	2,6- and 2,7-$(CH_3)_2$	—	65
2. 4,5-$(CH_3)_2$	770°–790°/ 1.5 mm N_2	1,4,5,8-$(CH_3)_4$ (3)	—	59
3. 4-Cl	800°/0.1–1.5 mm	2,6- and/or 2,7-Cl_2 (15)	—	59
4. 4,5-Cl_2	800°/0.1 mm	2,3,6,7-Cl_4(6.5)	—	59
5. 3,6-Cl_2	800°/1 mm	1,4,5,8-Cl_4 (3.8)	—	59
6. 3,4,5,6-Cl_4	800° NiCr coil/ 15 mm N_2	1,2,3,4,5,6,7,8-Cl_8 (30)	C_6Cl_6 (34)	58
	800°/0.2 mm	(26)	(31)	59
7. 3,4,5,6-Br_4	NiCr coil; 15 mm N_2	—	C_6Br_6 (14)	58
8. 3,4,5,6-F_4	750°/0.6 mm	1,2,3,4,5,6,7,8-F_8 (23)	—	66
9. Unsubstituted +4-CH_3(9:1)	800°/0.3 mm N_2	Unsubstituted + dimethyl 2-CH_3(16)	—	65
10. (Ninhydrin) +3,4,5,6-Cl_4	800°/0.7 mm N_2	1,2,3,4-Cl_4 (31)	—	65

ease of breaking of C—Br bonds, whereas the tetrafluoro compound gives a reasonable yield of octafluorobiphenylene. Tetrafluorobenzyne has also been trapped with $CF_2{=}CF_2$ and CF_2.[63] Pyrolysis of mixtures of anhydrides or of ninhydrin and a substituted phthalic anhydride gives mixtures of biphenylenes containing the unsymmetrical compound formed by addition of the substituted benzyne to benzyne (entries 9 and 10). Tetraphenylphthalic anhydride on copyrolysis with aromatic compounds at 690° fails to give products formed by intermolecular trapping. Instead a complex mixture of not fully identified di- and triphenylbiphenylenes and other polycyclic hydrocarbons is formed, apparently by interaction of the phenyl-substituents with the reactive aryne positions.[64]

Although a few heterocyclic arynes have been trapped in the gas phase with other aromatic compounds (see Chapter 3, Section V,E), aza- and diazaarynes rearrange to acyclic nitriles under flash vacuum pyrolytic conditions, and thus azabiphenylenes cannot be prepared by this pyrolytic route. Examples of such ring opening reactions are shown in Eqs. (6.47)–(6.50). The primary products are probably isonitriles, but these and the isoelectronic methylenecarbenes [Eq. (6.50)] rearrange to nitriles and ethinyl groups.

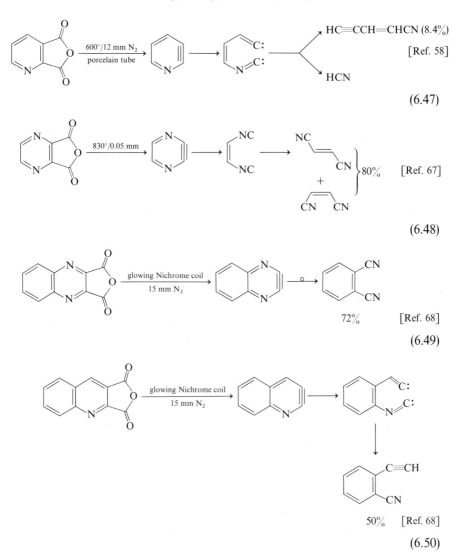

(6.47)

(6.48)

(6.49)

(6.50)

The pyrolysis of *N*-phenylpyrrole-3,4-dicarboxylic anhydride (**62**) at 650° and 12 mm N_2 over a Nichrome filament gives carbon dioxide and a highly rearranged product, furo[3,2-*c*]quinoline; this reaction is discussed in Chapter 9, Section II,C, and Eq. (9.25). The corresponding 2,3-dicarboxylic anhydride (**63**) also loses carbon dioxide under the same conditions and the intermediate pyrrolocyclopropenone or equivalent diradical resists decarbonylation.[69] Four colored dimers are obtained; two of these are the

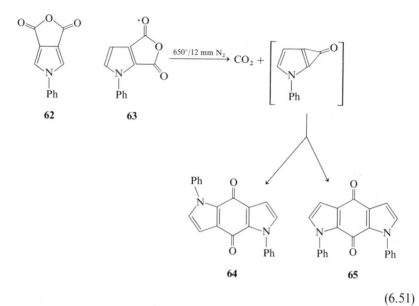

(6.51)

anthraquinone-like **64** and **65**, and two minor products are probably the corresponding phenanthrenequinone-like diones.

J.   Meldrum's Acid and Derivatives

The generation of carbenes from substituted ketenes formed by flash vacuum pyrolysis of derivatives of 2,2-dimethyl-1,3-dioxane-4,6-dione (Meldrum's acid) is discussed in Chapter 5, Sections II,B, and II,C, and the pyrolysis of o-alkylarylidene derivatives of Meldrum's acid to give 2-naphthols is described in Chapter 9, Section II,C, and Table 9.2. The chemistry of Meldrum's acid[70] and of its use for the pyrolytic generation of methylene-ketenes $R_2C=C=C=O$[71] have been reviewed.

Meldrum's acid (**66**) decomposes in a melt to give many products, but on flash vacuum pyrolysis at 430° it fragments cleanly to give acetone, carbon dioxide, and ketene [Eq. (6.52)].[72] It is not clear whether this reaction is concerted (**66**, arrows) or whether it proceeds through a diradical **67** or dipolar intermediate or through malonic anhydride (**68**). 5-Arylidene derivatives **70** (X = m- or p-H, CH_3, CH_3O, or Cl) can give the highly reactive arylideneketenes which can be collected at $-196°$ for infrared study, but which rapidly dimerize to symmetrical cyclobutane-1,3-diones above about $-100°$ [Eq. (6.53)].[72] Simple alkylidene derivatives such as **71** give alkyl-ideneketenes and the corresponding derivatives of cyclobutane-1,3-dione [Eq. (6.54)], but the behavior of other alkylidene derivatives is more complex, and cycloalkylidene derivatives tend to form cycloalkylidenecarbenes even

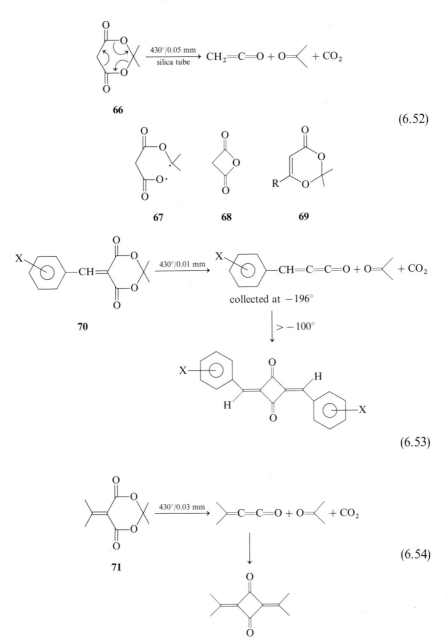

(6.52)

(6.53)

(6.54)

at the lower end of the temperature range 400°–600° [see Chapter 5, Section II,C, and Eqs. (5.17) and (5.18)]. The use of a retro-Diels–Alder reaction to produce the 5-methylene derivative of Meldrum's acid in the gas phase as a precursor of $CH_2{=}C{=}C{=}O$ itself is described in Chapter 8, Section III,B, Eq. (8.36).

The structurally somewhat related 1,3-dioxin-4-ones such as **69** undergo smooth thermal decomposition to acylketenes $RCOCH{=}C{=}O$ and acetone at temperatures as low as 120° in the condensed phase.[73]

III. PYROLYSIS OF CYCLIC SULFOXIDES AND SULFONES

A. Reactions Proceeding through Sulfenes $R_2C{=}SO_2$

J. F. King, P. de Mayo, and their co-workers at the University of Western Ontario have worked extensively on the thermal generation of sulfenes by a number of routes, and King[74] has provided a brief review. Sulfene $CH_2{=}SO_2$ can be generated pyrolytically from methanesulfonic anhydride at 650° and 0.004 mm[75] or from chlorosulfonylacetic acid[75] and under these mild conditions the sulfene survives. At higher temperatures desulfinylation occurs, as in the step following Cope rearrangement in Eq. (6.55), and following a 1,3 shift of chlorine in Eq. (6.56). At 930°–960° sulfene itself gives formaldehyde and sulfur monoxide [Eq. (6.57)]. Sulfene has been generated more efficiently from N-methylsulfonylphthalimide by pyrolysis through an unpacked quartz tube at 600° and 0.1 mm [Eq. (6.58)]. Phthalimide was obtained in quantitative yield, and the sulfene was readily trapped with ethylamine.[76]

In none of the reactions of Eqs. (6.55)–(6.58) was there any sign of decomposition of the sulfene to a carbene and sulfur dioxide, and King[74] estimates that this decomposition would be endothermic by 55–60 kcal/mole for $CH_2{=}SO_2$ itself. Such a decomposition has, however, been detected in the case of the diazosulfone **72**. Sarver et al.[79] decomposed this compound by

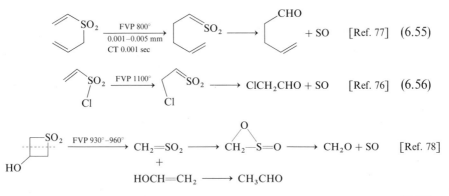

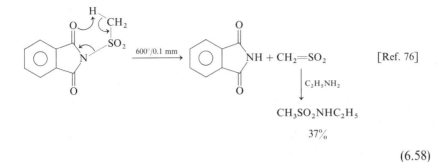

$$[\text{Ref. 76}]$$

$$CH_3SO_2NHC_2H_5$$

$$37\%$$

$$(6.58)$$

dropping it slowly onto a hot silica surface ($350°$–$600°$) at 10^{-4}–10^{-3} mm, and obtained 4-methylbenzophenone, formed by the usual loss of SO from **73**, and 2-methylfluorene, the characteristic product of rearrangement of *p*-tolylphenylcarbene **74**, formed by the unusual loss of SO_2 from **73**. Sulfur dioxide was identified as the major gaseous product [Eq. (6.59)].

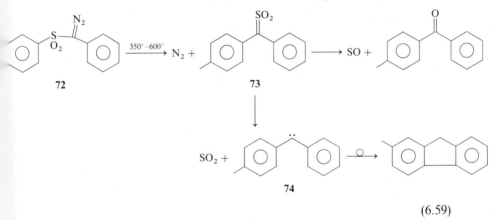

$$(6.59)$$

B. Cyclic Sulfoxides and Sulfones

A general review of the pyrolysis of cyclic sulfoxides and sulfones is to be found in the monograph by Block.[79a]

Ethylene episulfoxide decomposes slowly into ethylene and sulfur monoxide near $100°$, and the sulfur monoxide formed has been trapped in the presence of dienes.[80] Saito[81] has studied this decomposition by FVP using a 30-cm silica tube leading directly to a microwave spectrometer; he discusses whether the sulfur monoxide formed in this reaction [Eq. (6.60)] is initially in an excited singlet state or in the triplet ground state. The thermal decomposition of three-membered cyclic sulfoxides and sulfones is briefly reviewed by Braslavsky and Heicklen.[82]

$$\xrightarrow{580°/0.02 \text{ mm}} CH_2{=}CH_2 + SO \longrightarrow \longrightarrow S, S_2O \text{ and } SO_2 \quad (6.60)$$

Thietane 1,1-dioxide on flash vacuum pyrolysis at 950° loses sulfur dioxide and the 1,3-diradical then forms cyclopropane and propene [Eq. (6.61)].[83] Trost and co-workers[83a] have obtained mixtures of *cis*- and *trans*-dimethyl-cyclopropanes with some 2-pentenes from flow pyrolysis of the 2,4-dimethyl-thietane-1,1-dioxides at 350°. The very ready formation and dissociation of sulfolenes (2,5-dihydrothiophene 1,1-dioxides) into dienes and sulfur dioxide is a concerted chelotropic reaction which will not be considered here; it has been reviewed by Mock.[84] The bicyclic sulfolane **75** also decomposes smoothly at 120° to give 1,4-pentadiene and sulfur dioxide in a concerted $\sigma^2 s + \sigma^2 s + \sigma^2 s$ process.[85] The pyrolysis of simple sulfolanes (tetrahydro-thiophene 1,1-dioxides) to give alkenes and sulfur dioxide has been investigated by Bezmenova *et al.*[86] and by Mock *et al.*[87] It is clearly not concerted, as shown by the example of **76**, Eq. (6.62).[87] There is some low residual retention of stereochemistry, but the predominant loss of stereochemistry can be explained by internal rotation in a diradical such as **77**. Bridged cyclic sulfones generate 1,n-diradicals on flash pyrolysis, and these may fragment [Eq. (6.63)] or cyclize to bicyclic products [Eq. (6.64)].[88]

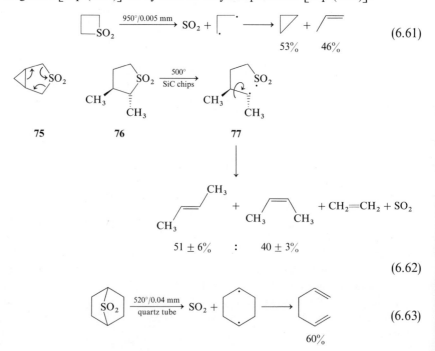

$$\xrightarrow{950°/0.005 \text{ mm}} SO_2 + \qquad \longrightarrow \qquad + \qquad\qquad (6.61)$$
$$53\% \qquad 46\%$$

75   **76**   **77**

$$51 \pm 6\% \quad : \quad 40 \pm 3\%$$

$$(6.62)$$

$$\xrightarrow[\text{quartz tube}]{520°/0.04 \text{ mm}} SO_2 + \qquad \longrightarrow \qquad\qquad (6.63)$$
$$60\%$$

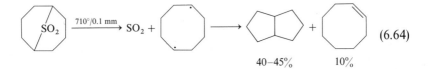

$$40\text{-}45\% \qquad 10\%$$

(6.64)

The bridged sulfone **78**, formally an adduct of sulfene and anthracene, does not undergo thermolysis to these components. On flash vacuum pyrolysis it loses sulfur dioxide and 5H-dibenzo[a,d]cycloheptene (**79**) is obtained in 89% yield, probably via an initial diradical; Hales et al.[89] also found that in an evacuated sealed tube at 300° 9-methylanthracene is formed instead. The course of these reactions is not well understood, but it appears possible that acid catalysis is involved at some stage of the sealed tube reaction.

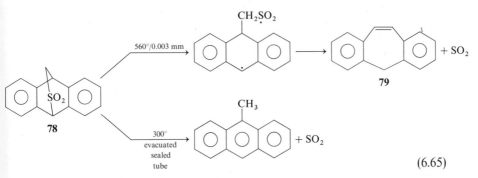

(6.65)

The pyrolysis of thiete 1,1-dioxide (**80**) has attracted much attention. When heated at 180°–190° with norbornene in a sealed degassed flask **80** slowly undergoes ring opening to vinylsulfene (**81**), which is trapped by Diels–Alder addition to norbornene.[90] On flash thermolysis of **80** at 390°–615° King and co-workers[83] obtained the isomeric sultine **82** in high yield [Eq. (6.66)]. At 950° acrolein was formed[78] by loss of sulfur monoxide from the vinylsulfene. The 2-phenyl derivative of **80** similarly isomerized at 455° and 0.01 mm to give the corresponding phenylsultine.

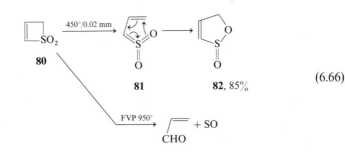

(6.66)

Hall and Smith[91] have obtained indene and 2H-1-benzopyran from flash thermolysis of 2H-1-benzothiapyran 1,1-dioxide (**83**). These products correspond to the loss of SO_2 and of SO, respectively, from **83**, but no details of yields or conditions appear to have been published.

83

C. Aromatic Cyclic Sulfones

In 1966 Fields and Meyerson[92] pyrolyzed dibenzothiophene 5,5-dioxide (**84**), hoping to obtain the 2,2'-biphenylylene diradical and biphenylene by loss of sulfur dioxide, but they found that the major products were dibenzofuran and dibenzothiophene in the ratio 6:1 [Eq. (6.67)]. The loss of sulfur monoxide in the formation of dibenzofuran was strikingly similar to some corresponding mass spectral processes. Losses of both SO and CO from the molecular ion are prominent in the mass spectrum of **84** and both processes require prior formation of a C—O bond. Similar consideration of the mass spectrum of the octafluoro derivative of **84** led Chambers and Cunningham[93] to the preparation of octafluorodibenzofuran in 72% yield by pyrolysis of the sulfone at 810° in a flow system. The intermediacy of sultines in such processes is suggested by the work of Hoffmann and Sieber[94] on pyrolysis of the highly strained sulfone naphtho[1,8-b,c]thietene 1,1-dioxide (**85**) which gave the sultine **86**.

van Tilborg and Plomp[95] have reported the products of pyrolysis of a range of 2,5-disubstituted and tetrasubstituted thiophene 1,1-dioxides at 800° and 0.01 mm. The dimethyl, di-t-butyl, and diphenyl compounds all

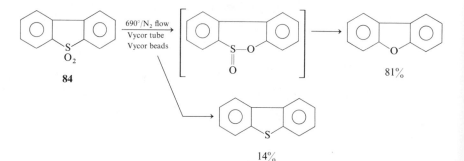

(6.67)

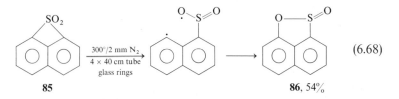

gave the corresponding furans in satisfactory yield [e.g., Eq. (6.69)], as did the dibenzo compound (89% yield of dibenzofuran). The tetraphenyl compound **87**, however, gave diphenylacetylene (75%) [Eq. (6.70)] with some tetraphenylfuran (2%), tetraphenylthiophene (0.5%), and 1,2,3-triphenylnaphthalene (2%). The last compound is a major product of pyrolysis of **87** in the condensed phase at 340°.[96] These authors[95] explain the formation of small amounts of thiophenes in such reactions by capture of elemental sulfur by butadiene-1,4-diyl radicals such as **88**, but it appears that there is no proof that the sulfur in the thiophene is not the sulfur of the original sulfone group. Flash pyrolysis of the iron tricarbonyl complex of 2,5-dimethylthiophene 1,1-dioxide at 640° and 0.01 mm gives the free 1,1-dioxide (70%), 2,5-dimethylthiophene (14%), and 2,5-dimethylfuran (8%), and the yield of the thiophene relative to that of the furan drops as the temperature is increased to 850°. This result appears to be taken as supporting the intermediacy of a butadiene-1,4-diyl or a coordinated cyclobutadiene in the formation of the thiophene.[95] It seems to the present author to indicate merely that at the lower temperature the iron tricarbonyl complex is capable of forming the thiophene and an iron oxide in competition with dissociation of 2,5-dimethylthiophene 1,1-dioxide from the complex; the composition of the

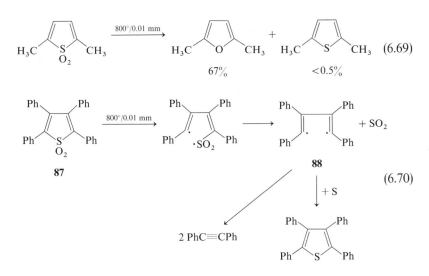

pyrolyzate formed from the complex at 850° approaches that from the free compound.

The long-sought benzothiete **91** was finally obtained by van Tilborg and Plomp[95,97] by flash vacuum pyrolysis of benzo[*b*]thiophene 1,1-dioxide at 1000° [Eq. (6.71)]. The yield of benzothiete was about 45%; benzene (8%) and benzothiophene (2%) were also formed but, surprisingly, only a trace of benzofuran was detected. The probable course of this remarkable reaction is summarized in Eq. (6.71). The authors suggest that the unusual radical rearrangement **89** → **90** probably involves transfer of the hydrogen atom via the sulfenyloxy radical. Benzothiete is a typical thermodynamically stable but strained molecule which is reactive in the condensed phase. It may be kept only for a few days at room temperature and it dimerizes above 100° to 6*H*,12*H*-dibenzo[*b,f*][1,5]dithiocin by initial rupture of the CH_2—S bond.

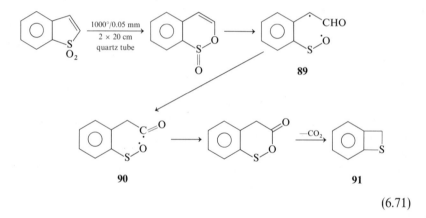

(6.71)

D. Preparation of Benzocyclobutenes from Sulfones

Benzocyclobutene (**92**) is a strained hydrocarbon which readily undergoes thermal ring opening to *o*-quinodimethane (**93**), a species which behaves as a highly reactive diene in cycloaddition reactions. There has recently been intense interest in the use of benzocyclobutenes in the synthesis of polycyclic compounds, and Oppolzer[98] has reviewed their use in intramolecular cyclo-additions of the type shown in Eq. (6.72). An example of the mutual trapping of *o*-quinodimethanes under flash pyrolytic conditions is the isomerization of the bis(benzocyclobutene) **94** to the [2,2,2](1,2,4)cyclophane (**95**) achieved by Aalbersberg and Vollhardt[99] [Eq. (6.73)].

The preparation of benzocyclobutene from 1,3-dihydroisothianaphthene 2,2-dioxide (**96**) was first reported by Cava and Deana,[100] who passed **96**

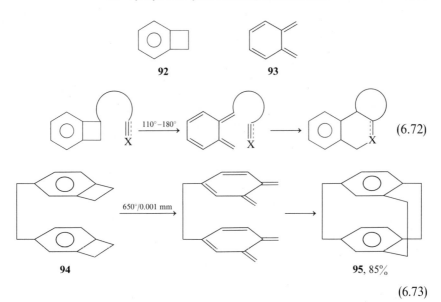

92   **93**

$$\xrightarrow{110°-180°}$$ (6.72)

X   X   X

$$\xrightarrow{650°/0.001 \text{ mm}}$$ $$\longrightarrow$$

94   **95, 85%**

(6.73)

over a glowing Nichrome coil at 670° and 2 mm N_2 and obtained **92** in 63% yield. It is uncertain whether this is a stepwise reaction, as shown in Eq. (6.74), or whether it should be regarded as a chelotropic elimination leading directly to *o*-quinodimethane and sulfur dioxide. The decomposition of 1,3-dihydronaphtho[2,3-*c*]thiophene 2,2-dioxide occurs readily even at 300° in diethyl phthalate[101] and forms naphtho[*b*]cyclobutene (**100**), whereas the decomposition of most such sulfones in solution leads to dimers and polymers. The sulfone route has been used to prepare systems with two fused cyclobutene rings such as **97**[102] and **98**,[103] and both naphthocyclobutenes **99**[104] and **100**.

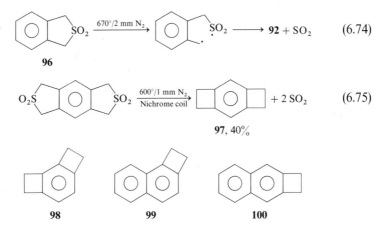

$$SO_2 \xrightarrow{670°/2 \text{ mm } N_2} SO_2 \longrightarrow 92 + SO_2 \qquad (6.74)$$

96

$$O_2S \qquad SO_2 \xrightarrow[\text{Nichrome coil}]{600°/1 \text{ mm } N_2} + 2 SO_2 \qquad (6.75)$$

97, 40%

98   **99**   **100**

Cava and Kuczkowski[105] have reported the preparation of 1,1′-spirobi-benzocyclobutene (**103**) in low yield by pyrolysis of the spirobisulfone **101** over a Nichrome coil. Some 2,2′-dimethylbenzophenone was also formed, apparently by addition of traces of water to the allenic intermediate **102**, but the major product (15% yield) was 5,11-di-*o*-tolyldibenzo[*a,e*]cyclooctate-traene (**104**) which collected on the walls of the apparatus above the coil. This must be formed by 4 + 4 dimerization of **102** followed by two 1,5 hydrogen shifts.

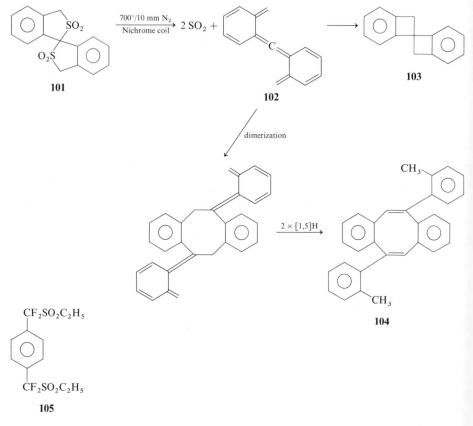

(6.76)

p-Xylylenes (*p*-quinodimethanes) have also been formed by pyrolysis of bisulfones. Thus *p*-CF$_2$=C$_6$H$_4$=CF$_2$ has been generated by pyrolysis of the bisulfone **105** at 750° with steam as diluent.[106] The dimer, 1,1,2,2,9,9,10,10-octafluoro[2,2]paracyclophane, was obtained in 9% yield.

E. Pyrolysis of Macrocyclic Sulfones and Bisulfones

The extension of the synthesis of cyclic hydrocarbons by pyrolysis of benzylic sulfones to the preparation of macrocyclic hydrocarbons was first made by Vögtle,[107] who obtained [2,2]metacyclophanes **106** in 15–20% yield as shown in Eq. (6.77). The reaction was found to be exclusively intra-molecular and presumably must occur stepwise, although no intermediate monosulfone could be isolated. Staab and Haenel[108] obtained [2,2](4,4′)bi-phenylophan (**108**) in 47% yield by pyrolysis of the bisulfone **107** at 500° and 0.1 mm. The sparingly volatile sample was sublimed from a zone heated at 300°–400° into a pyrolysis zone at 500°, and the products were collected on a cold finger immediately beyond the pyrolysis zone; a diagram of the simple apparatus is given.[108] Several other hydrocarbons closely related to **108** were prepared in the same way.

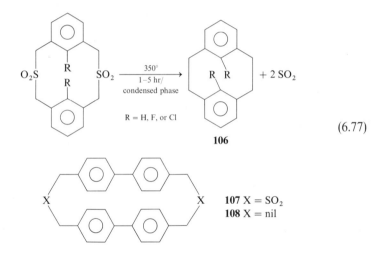

$$(6.77)$$

Vögtle's group has prepared many other macrocyclic hydrocarbons by this method, and three further examples are provided by the hydrocarbons **110, 112**, and **114**. The conditions noted below each structure are those required to effect pyrolysis of the sulfone to the hydrocarbon in the yield shown. The sulfones **109**[109] and **111**[110] are semi-benzylic and decomposition is probably initiated by homolysis of an $ArCH_2$—SO_2 bond, but some macrocyclic hydrocarbon **114** is obtained even from the non-benzylic sulfone **113**.[111] The sulfone **113** also gives the tetramethylene-bridged hydrocarbon **115** by loss of both ethylene and sulfur dioxide.

Vögtle and Neumann[111a] have reviewed the use of pyrolysis of macro-cyclic sulfones in cyclophane synthesis.

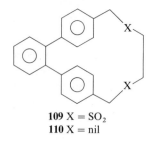

109 X = SO$_2$
110 X = nil

500°/0.1 mm (88%)

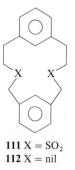

111 X = SO$_2$
112 X = nil

500°/10^{-3} mm (60%)

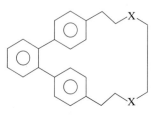

113 X = SO$_2$
114 X = nil

500°/10^{-6} mm (10%)

115

IV. FRAGMENTATION OF VARIOUS HETEROCYCLIC AND POLYHETEROCYCLIC COMPOUNDS

A. Introduction

This section deals with a quite heterogeneous collection of structures and processes, and any single system of organization by structure or by the nature of the stable fragment eliminated might obscure useful comparisons. I have chosen to group similar structures together where a common process is evident (e.g., loss of nitrogen from 1,2-diazaheterocycles with ring contraction) but in other areas pyrolytic reactions are grouped on the basis of an educated guess as to the nature of the first bond which would be broken in a stepwise process. Obviously such guesses are dangerous because little is known about the mechanism of many of these reactions but the resulting arrangement, though untidy, may present some points of interest.

B. Fragmentations with Breaking of C—O, C—S,
    and C—N Bonds

Flash vacuum pyrolysis of 1,3,4-oxadiazol-5-ones leads to loss of carbon dioxide, and formation of nitrilimines $R\bar{N}—\overset{+}{N}{\equiv}CR$. Reichen[112,113] has found that in the 4-phenyl series **116** the *N*-phenylnitrilimines formed at 450°–500° cyclize by intramolecular electrophilic substitution to give indazoles **117** in high yield, and that at higher temperatures hydrocarbons formed by further loss of nitrogen are obtained [Eq. (6.78)]. The related 2,5-diaryl-2*H*-tetrazoles behave similarly after loss of nitrogen.[113]

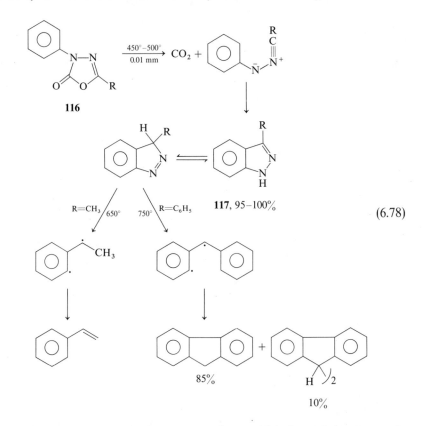

(6.78)

Two somewhat similar thermal reactions which have been run only in solution are the decomposition of isatoic anhydrides (**118**) and of 5-phenyl-1,3,4-oxathiazol-2-one (**120**). Isatoic anhydrides at 240°–290° form products derived from iminoketenes **119**[114] and the oxathiazolone **120**

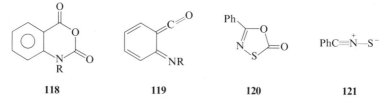

| 118 | 119 | 120 | 121 |

loses carbon dioxide at 130° to give benzonitrile sulfide (**121**), which can be trapped with acetylenedicarboxylic ester. At 165° **120** gives benzonitrile, sulfur, and carbon dioxide.[115]

There has been a limited amount of work on the pyrolysis of 1,3-dioxolanes in a static reactor[116] and in a Pyrex flow system[117] in the temperature range 455°–525°. One process leads to a carbonyl compound and products of re-arrangement of the C—C—O fragment [Eq. (6.79)], but alkenes and other products are also formed.

$$\text{(ring)} \longrightarrow R_2C{=}O + CH_3CH{=}O \qquad (6.79)$$

The highly reactive dienophile dicyanoacetylene is most satisfactorily pre-pared by flash vacuum pyrolysis of 4,5-dicyano-1,3-dithiol-2-one (**122**); Ciganek and Krespan[118] found this method to be much more satisfactory than the pyrolysis of 1,2-dichloro-1,2-dicyanoethylene or of tetracyanoeth-ylene. The corresponding 2-thione also gave dicyanoacetylene on pyrolysis, though in low yield (19%). The course of these reactions is unknown, but the authors point out that on flash pyrolysis of **122** at 500° the starting material is completely destroyed, but no dicyanoacetylene is obtained.

$$\text{(ring)} \xrightarrow{800°/1\ mm} NCC{\equiv}CCN + CS_2 + COS + S \qquad (6.80)$$

$$59–76\%$$

122

DeJongh and co-workers have made an extended study of the pyrolysis and the mass spectra of benzoheterazoline-2-thiones **123–125**[119] and of the corresponding 2-ones **126–128**.[120] In the case of the three thiones high-temperature pyrolysis led to primary loss of sulfur and formation of the parent benzoheterazoles; the other products were mainly formed by second-ary transformations of these parent compounds [Eqs. (6.81)–(6.83)] as shown by separate experiments. The formation of benzonitrile and derivatives is presumably due to the decomposition and rearrangement of *o*-substituted phenylisonitriles. The other products formed from the benzoxazoline-2-thione (**125**) but not shown in Eq. (6.83) were the 1- and 2-cyanonaphthalenes (12% and 15%, respectively) and naphthalene (7%); the origin of these is not

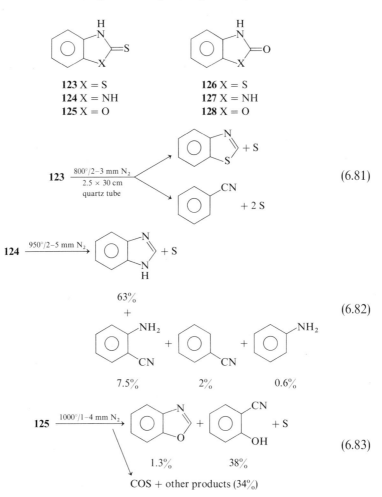

$$123 \xrightarrow[\substack{2.5 \times 30 \text{ cm} \\ \text{quartz tube}}]{800°/2-3 \text{ mm N}_2}$$

(6.81)

$$124 \xrightarrow{950°/2-5 \text{ mm N}_2}$$

(6.82)

$$125 \xrightarrow{1000°/1-4 \text{ mm N}_2}$$

(6.83)

yet completely clear, but initial loss of COS from **125** is required. This series of compounds again showed a close parallelism between the lowest energy pathways in their mass spectra and the major pyrolytic pathways of lowest energy.

The corresponding benzoheterazolin-2-ones[120] also required high temperatures for complete decomposition, and they gave more complex mixtures of products than the thiones **123** and **124**. The major products from benzothiazolin-2-one **(126)** were similar to the minor products from benzoxazoline-2-thione **(125)**, and loss of COS is required in each case [Eqs. (6.83) and (6.84)]. Benzimidazolin-2-one **(127)** fragmented mainly by loss of CO and the major products appear to be formed by dehydrogenation with ring opening of the intermediate *o*-quinonediimine [Eq. (6.85)].

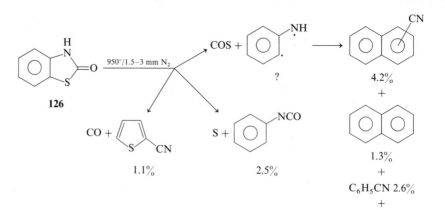

$$(6.84)$$

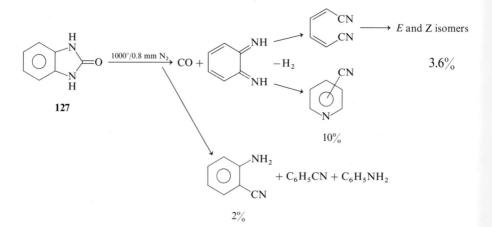

$$(6.85)$$

Benzoxazolin-2-one (**128**) underwent either two successive losses of CO, or loss of CO_2. It is clear that in this series there are many competing pathways, all with high initial energies of activation.

The introduction of *N*-phenyl substituents into some of these compounds permits the substantial diversion of pyrolytic intermediates towards products of internal cyclization [Eqs. (6.87)[121] and (6.88)[122]]. *N*-Phenyl-benzothiazoline-2-thione, however, mainly underwent phenyl migration [Eq. (6.89)].[122]

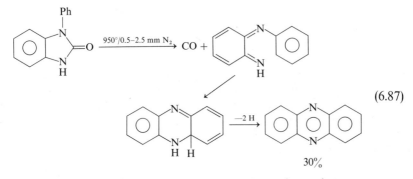

(6.87)

30%

+ other products

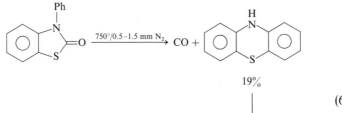

(6.88)

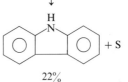

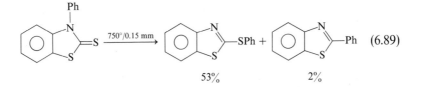

(6.89)

C. Ring Contraction of 1,2-Diaza Heterocycles with Loss of Nitrogen

Attempted pyrolytic extrusion of nitrogen from tetrafluoropyridazine at 815° led instead to isomerization to the corresponding pyrimidine and pyrazine and other products, as discussed in Chapter 9, Section VI. The azo group can, however, be extruded from tricyclic and higher fused systems such as benzo[c]cinnoline (129). MacBride[123] obtained biphenylene in 70% yield by pyrolysis of 129 at 870° [Eq. (6.90)] and octachlorobiphenylene was similarly formed in 80% yield at 700°. Barton and Walker[124] have obtained the hydrocarbon from double extrusion of nitrogen from the pentacyclic compound 130 in 9.5% yield by pyrolysis at 800° and 0.04 mm through a 1-m silica tube; products formed by single extrusion of nitrogen were also isolated. The reaction has been extended to the 1-aza and 3-aza derivatives of 129 to give the corresponding azabiphenylenes,[125] and to the preparation of the biquinoxalylene (131)[126] and of symmetrical diazabiphenylenes such as 132 [Eq. (6.91)]. The by-product 133 was shown to be a product of second-ary pyrolysis of 132.[127]

This approach thus provides a synthetic route to aza and polyaza bi-phenylenes which are not available via generation and dimerization of

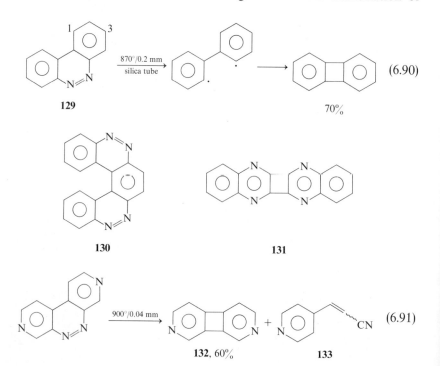

heteroarynes in the gas phase because of preferred ring fission [see Section II.I; Eqs. (6.47)–(6.50)].

Chambers and co-workers[128] have succeeded in obtaining products of extrusion of nitrogen and ring fission from a number of perhalogenated pyridazines and cinnolines, as shown in Eqs. (6.92)–(6.94).

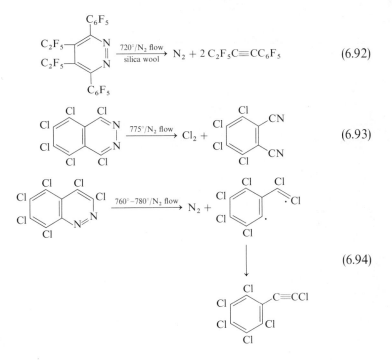

$$(6.92)$$

$$(6.93)$$

$$(6.94)$$

D. Fragmentation with Breaking of C—N and N—N Bonds in Triazoles, Triazines, Tetrazoles, and Some Benzo Derivatives

The formation of 1H-azirine intermediates on flash vacuum pyrolysis of substituted 1H-1,2,3-triazoles is discussed in Chapter 5, Section III,D. Gilchrist *et al.*[129] have examined the products of pyrolysis of all nine mono-, di-, and triphenyl-1,2,4-triazoles. These compounds were thermally rather stable compared to the 1,2,3-triazoles, but those phenylated on nitrogen underwent a 1,5 phenyl shift at high temperatures and the intermediates then lost nitrogen to give, ultimately, isoindoles. Even at 750°–800° conversions were low, but yields of diphenylisoindole corrected for recovery of starting material were high. Thus 1,3,5-triphenyl-1,2,4-triazole (**134**) gave 1,3-diphenylisoindole in 95% yield based on triazole consumed [Eq. (6.95)], the

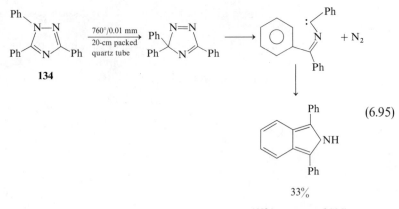

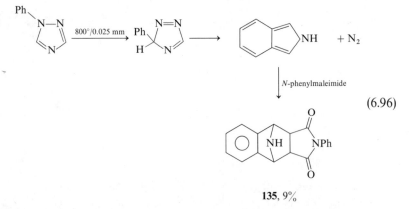

(6.95)

33%

(65% recovery of SM)

diphenyl-1,2,4-triazoles gave 1-phenylisoindole, and the 1-phenyl compound gave isoindole itself. The isoindole was collected on a surface coated with N-phenylmaleimide and isolated as the adduct **135** [Eq. (6.96)] in 9% yield.

(6.96)

135, 9%

3-Phenyl- (**136**) and 4-phenyl-1,2,4-triazole (**137**) behaved differently in that nitrogen was not lost as N_2, but products of ring opening and cleavage were formed instead [Eqs. (9.67) and (6.98)].[129]

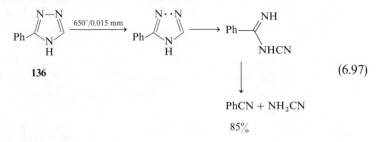

(6.97)

$PhCN + NH_2CN$

85%

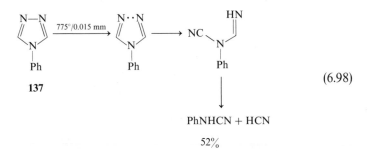

(6.98)

PhNHCN + HCN

52%

1-Substituted 1*H*-benzotriazoles pyrolyze to give intermediate 1,3-diradicals which can interact with aromatic and unsaturated substituents to give cyclic and rearranged products [Eqs. (6.99)–(6.101)]. The formation of 3-methylcarbazole (**138**), but no 2-methylcarbazole, in the reaction of Eq. (6.99) showed that under the conditions shown such 1,3-diradicals are not equilibrated via 1*H*-benzazirines.[130] This is an example of the well-known Graebe–Ullmann synthesis of carbazoles[131] in the gas phase rather than in solution.

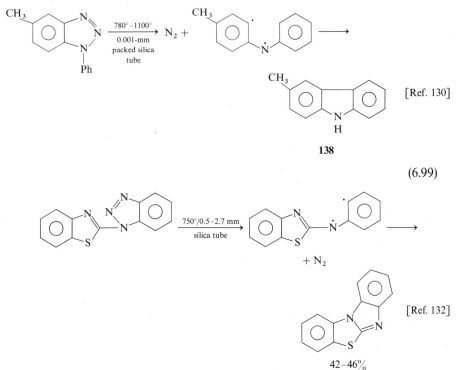

[Ref. 130]

138

(6.99)

[Ref. 132]

42–46%

(6.100)

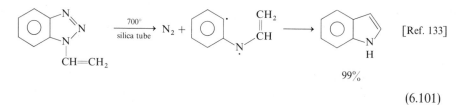

[Ref. 133]

99%

(6.101)

1,2,3-Triazines readily lose nitrogen on pyrolysis near 500°. The trimethyl compound fragmented to give 2-butyne, acetonitrile, and nitrogen [Eq. (6.102)], possibly via an azatetrahedrane or cyclopropylnitrene intermediate. Closs and Harrison[134] also found that the azide **139** gave the same products on flow pyrolysis at 300°. Seybold and co-workers[135] obtained tris(dimethyl-amino)azete (**141**), which is somewhat stabilized by push–pull interactions, by flash vacuum pyrolysis of the 1,2,3-triazine **140**. The azete **141** was too unstable to be purified, but it was fully characterized in solution. Some fragmentation leading to $(CH_3)_2NCN$ was also observed.

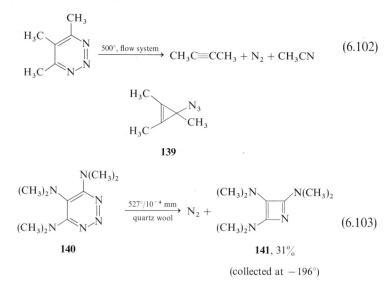

$$\xrightarrow{500°,\text{ flow system}} CH_3C{\equiv}CCH_3 + N_2 + CH_3CN \qquad (6.102)$$

139

$$\xrightarrow[\text{quartz wool}]{527°/10^{-4}\text{ mm}} N_2 +$$

140   **141**, 31%

(6.103)

(collected at −196°)

Rees, Storr, and co-workers[136,137] have made a thorough study of pyrolytic routes to benzazetes from 1,2,3- and 1,2,4-benzotriazines and related compounds. No general route was found because of the competing tendency of benzotriazines to fragment completely to nitrogen, a nitrile, and benzyne. This behavior is illustrated for 1,2,3-benzotriazine itself in Eq. (6.104); the 4-methyl, 4-phenyl, and 6-chloro-4-phenyl-derivatives behaved similarly and gave biphenylene (**142**) or derivatives on pyrolysis near 500°, and benzazetes could be obtained only under carefully controlled conditions in a few cases.

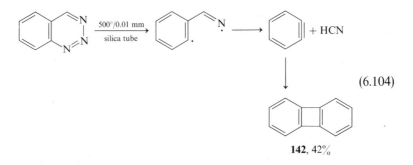

$$(6.104)$$

142, 42%

1,2,4-Benzotriazines required much higher pyrolytic temperatures [Eq. (6.105)]. Another approach by pyrolysis of the sulfoximide **143** led mainly to an acetylenic nitrile by ring fission, with minor formation of biphenylene [Eq. (6.106)]. The pyridotriazine **144** underwent hydrogen transfer in the initial diradical to give a nitrile [Eq. (6.107)], rather than elimination to give a 2,3-pyridyne.

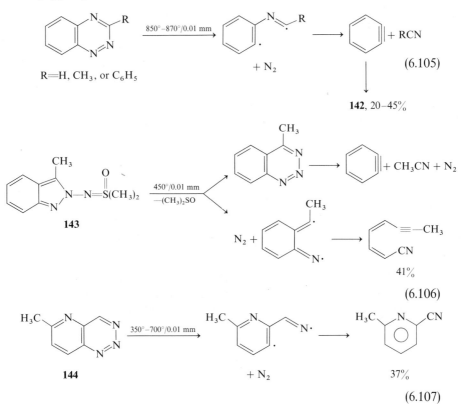

R=H, CH$_3$, or C$_6$H$_5$

$$(6.105)$$

142, 20–45%

41%

$$(6.106)$$

143

37%

144

$$(6.107)$$

2-Phenylbenzazete (**145**) was finally obtained[137] by careful pyrolysis of 4-phenyl-1,2,3-benzotriazine at 420°–450° and 0.001 mm with immediate collection of the products on a cold finger at −78° [Eq. (6.108)]. The red benzazete was accompanied by small amounts of biphenylene, 9-phenyl-acridine, benzonitrile, and starting material. It was characterized by the preparation of cycloadducts with highly reactive dienes at low temperatures; at room temperature a dimer (**146**)[138] is rapidly formed. The more stable 2-*p*-methoxyphenylbenzazete and 2-phenylnaphth[2,3-*b*]azete were obtained in the same way. Pyrolysis of 4-methyl-1,2,3-benzotriazine at 450° and 0.1 mm gave only biphenylene and 9-methylacridine, and only the 4-*t*-butyl compound of the alkyl compounds tried gave any indication of alkylbenzazete formation; on pyrolysis at 450° and 0.1 mm biphenylene (22%), benzonitrile (54%), and a dimer of 2-*t*-butylbenzazete (12%) of structure similar to **146** were isolated.[139] Some evidence for the formation of monomer was obtained by reactions with hydrazine or phenylhydrazine at a low temperature.

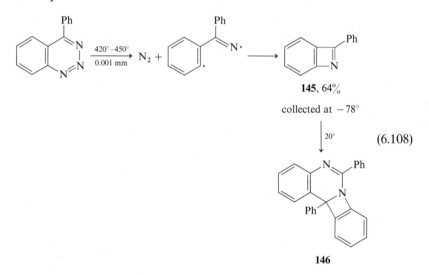

(6.108)

1,2,3-Benzotriazin-4-ones (**147**) decompose with loss of nitrogen on heating in inert solvents[114] to give the same intermediates **119** which can be generated from isatoic anhydrides **118** (see Section IV,B), but there seems

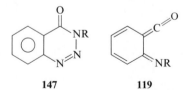

to be no report of any flow pyrolytic work. Gilchrist and co-workers[129] have pyrolyzed 1-phenyltetrazole and 3,5-diphenyl-1,3,4-oxadiazole and obtained the products of fragmentation shown in Eqs. (6.109) and (6.110).

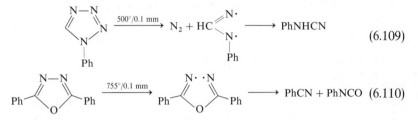

$$ (6.109) $$

$$ (6.110) $$

A spectacular series of rearrangements in the pyrolysis of 1-(2,6-dimethylphenyl)-5-phenyltetrazole (**148**) and several related compounds was discovered by Gilchrist *et al.*[140] Although a major product from **148** was the diarylcarbodiimide **149** formed by straightforward 1,2-phenyl migration,

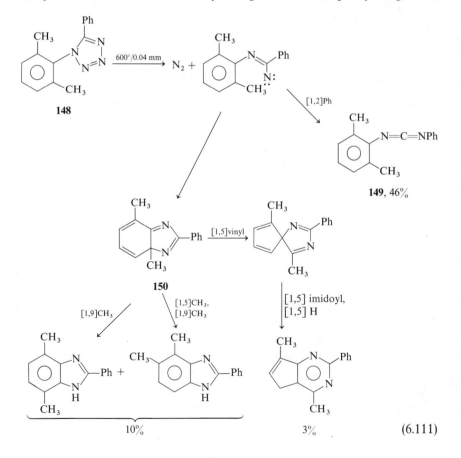

$$ (6.111) $$

other products were derived from 1,5 and 1,9 migrations of either methyl groups or ring bonds in a bicyclic intermediate **150** formed from an imidoyl-nitrene [Eq. (6.111)]. No 1,5 migration of a methyl group in **150** to nitrogen was detected, although this would have given an aromatic 1-methylbenzi-midazole directly.

E. Rearrangements and Fragmentations with Breaking of N—O and N—S Bonds

Many heterocycles containing N—O and N—S bonds rearrange or frag-ment thermally through the initial breaking of these bonds, which tend to be relatively weak compared with C—N, C—O, or C—S bonds. A typical example is the rearrangement of 3-hydroxybenzisoxazole (**151**) to 2-benzo-xazolinone (**152**) described by Kinstle and Darlarge.[141] The work of Ha-zeldine's group on adducts of nitrosotrifluoromethane also contains many examples of the fragmentation of heterocycles containing N—O bonds; examples are shown in Eqs. (6.113)[142] and (6.114).[143]

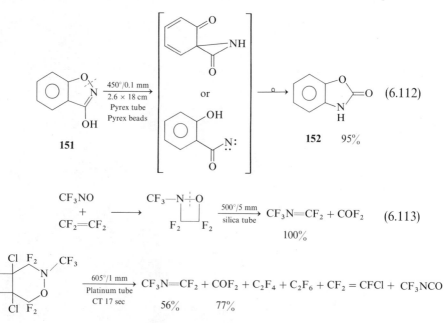

$$CF_3N=CF_2 + COF_2 \quad (6.113)$$

100%

$$CF_3N=CF_2 + COF_2 + C_2F_4 + C_2F_6 + CF_2=CFCl + CF_3NCO$$

56%   77%

$$(6.114)$$

Heterocyclic compounds containing the *O*-acyloxime structure $RCO_2N=C$ also fragment readily, and an example in the 1,2-oxazin-6-one series is noted in Chapter 3, Section III,E, and Eq. (3.24). David and

McOmie[144] showed that the 4-methyl and 4-phenyl derivatives of 1H-2,3-benzoxazin-1-one (**154**) are benzyne precursors which lead to biphenylene (**142**), whereas the isomeric 1H-2,4-benzoxazin-1-ones (**153**) withstand attempted pyrolysis at 800°. The fragmentation and subsequent rearrangements in solution of the 1,2,4-oxadiazolin-5-ones (**155**) and (**156**) and of the 1,2,4-oxadiazolin-3-one (**157**) have been reported by Boyer and co-workers.[145,146] Furazan N-oxides such as **158** dissociate initially to nitrile oxides $ArC\equiv N^+ — O^-$ in boiling diphenyl ether (257°),[147] and 3,5-diphenyl-1,2,4-oxadiazole gives benzonitrile and phenylisocyanate [Eq. (6.116)] when heated in an evacuated sealed tube.[148]

The fragmentation of 4-arylideneisoxazol-5-ones to carbon dioxide, a nitrile, and an arylidenecarbene is discussed in Chapter 5, Section II,C, and Eq. (5.19), and that of pyrido[2,3-*a*](1,2,4)oxadiazolone to 2-pyridylnitrene in Section III,F, and Eq. (5.88).

Flash vacuum pyrolysis of isothiazoles[149] and of 1,2,3-thiadiazoles gives thioketenes. Seybold and Heibl[150] have shown that the pyrolysis of thiadiazoles at 580° and 10^{-4} mm is the simplest and most convenient general

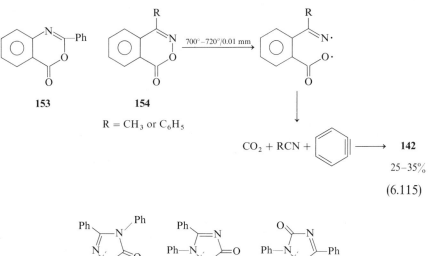

153 **154**

$R = CH_3$ or C_6H_5

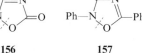

$CO_2 + RCN +$ ⟶ **142**

25–35%

(6.115)

155 **156** **157**

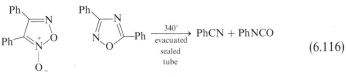

$\xrightarrow[\text{evacuated sealed tube}]{340°}$ PhCN + PhNCO

(6.116)

158

method for the preparation of thioketenes in high yield, and they have prepared a number of mono- and disubstituted thioketenes. Benzothiadiazole also gives a thioketene by ring contraction on flash vacuum pyrolysis [Eq. (6.119)],[150] whereas its decomposition in the condensed phase gives the dimeric compounds thianthrene (**159**), dibenzothiophene, dibenzo[*c,e*]-*o*-dithiin (**160**), and other products.[151]

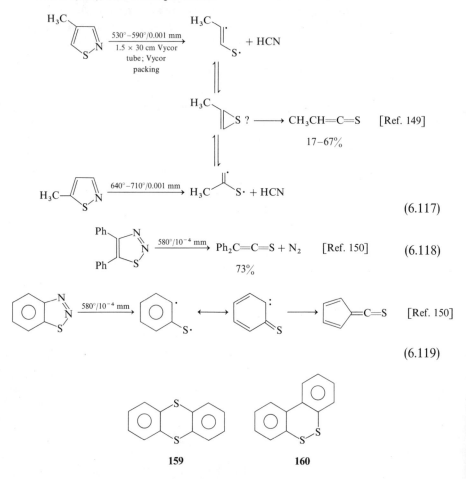

$$(6.117)$$

$$(6.118)$$

$$(6.119)$$

159   **160**

The mass spectrum of 3-phenyl-1,2-benzisothiazole-1,1-dioxide (**161**) shows a strong peak corresponding to the formation of the molecular ion of 2-phenylbenzazete by loss of sulfur dioxide, and Abramovitch and Wake[152] pyrolyzed **161** hoping that the thermal process would parallel the mass spectral process and give 2-phenylbenzazete, as did 4-phenyl-1,2,3-benzotriazine. This hope was not realized; **161** was stable to flow pyrolysis at 650° and pyrolysis at 850° gave instead benzonitrile and 2-phenylben-zoxazole (**162**), the latter requiring overall loss of sulfur monoxide. The course of these reactions is uncertain, and the intermediates shown in Eq. (6.120) are highly speculative. The 3-methyl derivative **164** unexpectedly gave 2-hydroxybenzonitrile [Eq. (6.121)] and a separate experiment showed that this could not be derived from the very stable 2-methylbenzoxazole (**165**). Davies *et al.*[153] have confirmed the behavior of the 3-phenyl derivative **161**, and have further found that an interchange of C and N is also shown by 3-phenylbenzisoxazole (**163**) and results in isomerization to the same product **162** at 800° and 0.1–1 mm. A reexamination[153] of the pyrolysis of 4-phenyl-1*H*-2,3-benzoxazin-1-one (**154**)[144] showed that at 625° and 0.1 mm formation of benzyne and biphenylene is accompanied by similar interchange of C and N giving 3-phenyl-1*H*-2,4-benzoxazin-1-one (**153**) in 11% yield.

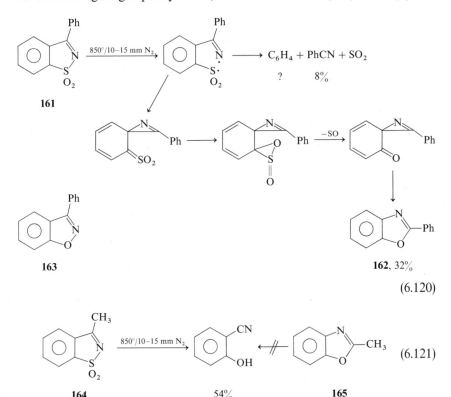

(6.120)

(6.121)

Ege and Beisiegel[154] have reported the decomposition of a number of ring-fused and substituted 1,2,5-thiadiazole 1,1-dioxides at 250°–330° in sealed tubes. The example of the diphenyl compound [Eq. (6.122)] suggests

$$\text{Ph}\underset{\text{Ph}}{\overset{}{\diagdown}}\!\!\!\!\!\!\!\!\!\!\!\!\!\! \overset{250°}{\underset{\text{sealed tube}}{\longrightarrow}} \ 2\ \text{PhCN} + \text{SO}_2 + \text{other products} \qquad (6.122)$$

59%

that a smoother fission could be achieved by flash vacuum pyrolysis; in a sealed tube the other products included 2,4,6-triphenyl-1,3,5-triazine and 2,4,5-triphenylimidazole.

F.   Extrusion of Sulfur, Tellurium, and Arsenic

The direct thermal extrusion of sulfur from cyclic thioethers is not usually a useful reaction in that conversions tend to be low, and the pyrolysis of sulfones is to be preferred. Telluroethers are more useful, however, and 1,3-dihydrobenzo[c]tellurophene gives benzocyclobutene in good yield on pyrolysis at 500° [Eq. (6.123)]. Cuthbertson and MacNicol[155] were also able to obtain naphtho[b]cyclobutene in good yield by a similar route.

$$\text{(structure)} \ \overset{500°/0.5\ mm\ He}{\underset{\text{quartz wool}}{\longrightarrow}} \ \text{(structure)} + \text{Te} \qquad (6.123)$$

Tou and co-workers[156] noted that the molecular ions of phenoxarsine derivatives tended to lose fragments AsX in the mass spectrometer, and they found that several such compounds yielded dibenzofuran when a methylene chloride solution was injected into a coupled pyrolysis–GLC apparatus [Eq. (6.124)]. This behavior is similar to that of phenothiazine, which shows some loss of sulfur at 750° [see Eq. (6.88)].

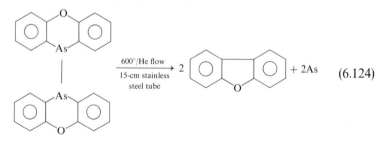

$$\overset{600°/\text{He flow}}{\underset{\substack{\text{15-cm stainless}\\\text{steel tube}}}{\longrightarrow}} \ 2 \ \text{(structure)} + 2\text{As} \qquad (6.124)$$

REFERENCES

1. Delles, F. M., Dodd, L. T., Lowden, L. F., Romano, F. J., and Daignault, L. G. (1969). *J. Am. Chem. Soc.* **91**, 7645.
2. de Mayo, P., and Verdun, D. L. (1970). *J. Am. Chem. Soc.* **92**, 6079.
3. Sato, T. (1965). *Tetrahedron* **21**, 2947.

4. Sato, T., Murata, K., Nishimura, A., Tsuchiya, T., and Wasada, N. (1967). *Tetrahedron* **23**, 1791.

5. Sato, T., and Obase, H. (1967). *Tetrahedron Lett.*, p. 1633.

6. Cookson, R. C., Hudec, J., and Williams, R. O. (1960). *Tetrahedron Lett.* (22) p. 29.

7. Tezuka, T., Yamashita, Y., and Mukai, T. (1976). *J. Am. Chem. Soc.* **98**, 6051.

8. Brown, R. F. C., Gream, G. E., Peters, D. E., and Solly, R. K. (1968). *Aust. J. Chem.* **21**, 2223.

9. Brown, R. F. C., and Solly, R. K. (1966). *Aust. J. Chem.* **19**, 1045.

10. Anderson, D. J., Horwell, D. C., Stanton, E., Gilchrist, T. L., and Rees, C. W. (1972). *J. Chem. Soc. Perkin Trans. 1*, p. 1317.

11. Cava, M. P., (1967). *In* "Aromaticity," Chem. Soc. Special Publ. No. 21, p. 168. The Chem. Soc., London.

12. Hedaya, E., and Kent, M. E. (1970). *J. Am. Chem. Soc.* **92**, 2149.

13. Brown, R. F. C., and Butcher, M. (1969). *Aust. J. Chem.* **22**, 1457.

14. Brown, R. F. C., and Butcher, M. (1970). *Aust. J. Chem.* **23**, 1907.

15. Ksander, G. M., and McMurry, J. E. (1976). *Tetrahedron Lett.*, p. 4691.

16. Brown, R. F. C., Gardner, D. V., McOmie, J. F. W., and Solly, R. K. (1967). *Aust. J. Chem.* **20**, 139.

17. Haddon, R. C. (1972). *Tetrahedron Lett.*, p. 3897.

18. Hageman, H. J., and Wiersum, U. E. (1973). *Chem. Br.* **9**, 206.

19. DeJongh, D. C., Van Fossen, R. Y., and Bourgeois, C. F. (1967). *Tetrahedron Lett.*, p. 271.

20. DeJongh, D. C., and Van Fossen, R. Y. (1972). *J. Org. Chem.* **37**, 1129.

21. DeJongh, D. C., and Brent, D. A. (1970). *J. Org. Chem.* **35**, 4204.

22. Chapman, O. L., and McIntosh, C. L. (1971). *J. Chem. Soc. Chem. Commun.*, p. 770.

23. DeJongh, D. C., Brent, D. A., and Van Fossen, R. Y. (1971). *J. Org. Chem.* **36**, 1469.

24. Hageman, H. J., and Wiersum, U. E. (1972). *Angew. Chem. Int. Ed. Engl.* **11**, 333.

25. Hageman, H. J., and Wiersum, U. E. (1971). *J. Chem. Soc. Chem. Commun.*, p. 497.

26. Mukai, T., Nakazawa, T., and Shishido, T. (1967). *Tetrahedron Lett.*, p. 2465.

27. Szabo, A., Suhr, H., and Venugopalan, M. (1977). *Justus Liebigs Ann. Chem.*, p. 747.

28. Schaden, G. (1977). *Angew. Chem. Int. Ed. Engl.* **16**, 50.

29. Price, C. C., and Berti, G. (1954). *J. Am. Chem. Soc.* **76**, 1211.

30. Coxon, J. M., Dansted, E., Hartshorn, M. P., and Richards, K. E. (1969). *Tetrahedron* **25**, 3307.

31. Coxon, J. M., Hartshorn, M. P., Little, G. R., and Maister, S. G. (1971). *J. Chem. Soc. Chem. Commun.*, p. 271.

32. Matlack, A. S., and Breslow, D. S. (1957). *J. Org. Chem.* **22**, 1723.

33. Dougherty, R. C. (1968). *J. Am. Chem. Soc.* **90**, 5780.

34. Dougherty, R. C. (1974). *Top. Curr. Chem.* **45**, 93.

35. DeJongh, D. C., and Thomson, M. L. (1972). *J. Org. Chem.* **37**, 1135.

36. DeJongh, D. C., Van Fossen, R. Y., and Dekovich, A. (1970). *Tetrahedron Lett.*, p. 5045.

37. Wentrup, C. (1973). *Tetrahedron Lett.*, p. 2919.

38. Bailey, W. J., and Bird, C. N. (1977). *J. Org. Chem.* **42**, 3895.

39. Wentrup, C., and Müller, P. (1973). *Tetrahedron Lett.*, p. 2915.

40. Gilchrist, T. L., and Pearson, D. P. J. (1976). *J. Chem. Soc. Perkin Trans. 1*, p. 1257.

41. Skorianetz, W., and Ohloff, G. (1975). *Helv. Chim. Acta* **58**, 1272.

42. Brent, D. A., Hribar, J. D., and DeJongh, D. C. (1970). *J. Org. Chem.* **35**, 135.

43. Pirkle, W. H., and Turner, W. V. (1975). *J. Org. Chem.* **40**, 1617.

44. Pirkle, W. H., and Turner, W. V. (1975). *J. Org. Chem.* **40**, 1644.

45. Pirkle, W. H., Seto, H., and Turner, W. V. (1970). *J. Am. Chem. Soc.* **92**, 6984.

46. Brown, R. F. C., and Butcher, M. (1973). *Aust. J. Chem.* **26**, 369.

47. Brown, R. F. C., and Butcher, M. (1972). *Aust. J. Chem.* **25**, 149.
48. DeJongh, D. C., and Evenson, G. N. (1971). *Tetrahedron Lett.*, p. 4093.
49. DeJongh, D. C., and Evenson, G. N. (1972). *J. Org. Chem.* **37**, 2152.
50. Rice, F. O., and Murphy, M. T. (1942). *J. Am. Chem. Soc.* **64**, 896.
51. Brown, A. L., and Ritchie, P. D. (1968). *J. Chem. Soc. (C)*, p. 2007.
52. Middleton, W. J., and Sharkey, W. H. (1959). *J. Am. Chem. Soc.* **81**, 803.
53. Morner, O. A., Lossing, F. P., Hedaya, E., and Kent, M. E. (1970). *Can. J. Chem.* **48**, 3606.
54. Spangler, R. J., Beckmann, B. G., and Kim, J. H. (1977). *J. Org. Chem.* **42**, 2989.
55. Bloch, R. (1978). *Tetrahedron Lett.*, p. 1071.
56. Ott, E. (1914). *Ber. Dtsch. Chem. Ges.* **47**, 2388.
57. Crombie, L., Gilbert, P. A., and Houghton, R. P. (1968). *J. Chem. Soc. (C)*, p. 130.
58. Cava, M. P., Mitchell, M. J., DeJongh, D. C., and Van Fossen, R. Y. (1966). *Tetrahedron Lett.*, p. 2947.
59. Brown, R. F. C., Gardner, D. V., McOmie, J. F. W., and Solly, R. K. (1967). *Aust. J. Chem.* **20**, 139.
60. Fields, E. K., and Meyerson, S. (1965). *Chem. Commun.*, p. 474.
61. Fields, E. K., and Meyerson, S. (1966). *Chem. Commun.*, p. 275.
62. Wittig, G., and Ebel, H. F. (1960). *Justus Liebigs Ann. Chem.* **650**, 20.
63. Plantonov, V. E., Senchenko, T. V., and Yakobsen, G. G. (1976). *Zh. Org. Khim.* **12**, 816.
64. Fields, E. K., and Meyerson, S. (1968). *In* "Advances in Physical Organic Chemistry" (V. Gold, ed.), Vol. 6, p. 1. Academic Press, New York.
65. Martineau, A., and DeJongh, D. C. (1977). *Can. J. Chem.* **55**, 34.
66. Gardner, D. V., McOmie, J. F. W., Albriktsen, P., and Harris, R. K., (1969). *J. Chem. Soc. (C)*, p. 1994.
67. Brown, R. F. C., Crow, W. D., and Solly, R. K. (1966). *Chem. Ind. (London)*, p. 343.
68. Cava, M. P., and Bravo, L. (1968). *Chem. Commun.*, p. 1538.
69. Cava, M. P., and Bravo, L. (1970). *Tetrahedron Lett.*, p. 4631.
70. McNab, H. (1978). *Chem. Soc. Rev.* **7**, 345.
71. Brown, R. F. C., and Eastwood, F. W. (1980). *In* "The Chemistry of Ketenes and Allenes" (S. Patai, ed.), p. 757. Wiley, New York.
72. Brown, R. F. C., Eastwood, F. W., and Harrington, K. J. (1974). *Aust. J. Chem.* **27**, 2373.
73. Jäger, G., and Wenzelburger, J. (1976). *Justus Liebigs Ann. Chem.*, p. 1689.
74. King, J. F. (1975). *Acc. Chem. Res.* **8**, 10.
75. King, J. F., Marty, R. A., de Mayo, P., and Verdun, D. L. (1971). *J. Am. Chem. Soc.* **93**, 6304.
76. Mijs, W. J., Reesink, J. B., and Wiersum, U. E. (1972). *J. Chem. Soc. Chem. Commun.*, p. 412.
77. King, J. F., and Harding, D. R. K. (1971). *J. Chem. Soc. Chem. Commun.*, p. 959.
78. McIntosh, C. L., and de Mayo, P. (1969). *J. Chem. Soc. Chem. Commun.*, p. 32.
79. Sarver, B. E., Jones, M., Jr., and Van Leusen, A. M. (1975). *J. Am. Chem. Soc.* **97**, 4771.
79a. Block, E. (1978). "Reactions of Organosulfur Compounds." Academic Press, New York.
80. Dodson, R. M., and Sauers, R. F. (1967). *Chem. Commun.*, p. 1189.
81. Saito, S. (1968). *Tetrahedron Lett.*, p. 4961.
82. Braslavsky, S., and Heicklen, J. (1977). *Chem. Rev.* **77**, 473.
83. King, J. F., de Mayo, P., McIntosh, C. L., Piers, K., and Smith, D. J. H. (1970). *Can. J. Chem.* **48**, 3704.
83a. Trost, B. M., Schinski, W. L., Chen F., and Mantz, I. B. (1971). *J. Am. Chem. Soc.* **93**, 676.
84. Mock, W. L. (1977). *In* "Pericyclic Reactions" (A. P. Marchand and R. E. Lehr, eds.), Vol. II, p. 141. Academic Press, New York.
85. Mock, W. L. (1970). *J. Am. Chem. Soc.* **92**, 6918.

86. Bezmenova, T. E., Gutyra, V. S., and Kamakin, N. M. (1965). *Ukr. Khim. Zh.* **30**, 948 (1964); *Chem. Abstr.* **62**, 2752.
87. Mock, W. L., Mehrota, I., and Anderko, J. A. (1975). *J. Org. Chem.* **40**, 1842.
88. Corey, E. J., and Block, E. (1969). *J. Org. Chem.* **34**, 1233.
89. Hales, N. J., Smith, D. J. H., and Swindles, M. E. (1976). *J. Chem. Soc. Chem. Commun.*, p. 981.
90. Dittmer, D. C., McCaskie, J. E., Babiarz, J. E., and Roggeri, M. V. (1977). *J. Org. Chem.* **42**, 1910.
91. Hall, C. R., and Smith, D. J. H. (1974). *Tetrahedron Lett.*, p. 3633.
92. Fields, E. K., and Meyerson, S. (1966). *J. Am. Chem. Soc.* **88**, 2836.
93. Chambers, R. D., and Cunningham, J. A. (1967). *Chem. Commun.*, p. 583.
94. Hoffmann, R. W., and Sieber, W. (1967). *Justus Liebigs Ann. Chem.* **703**, 96.
95. van Tilborg, W. J. M., and Plomp, R. (1977). *Rec. Trav. Chim. Pays-Bas* **96**, 282.
96. McOmie, J. F. W., and Bullimore, B. K. (1965). *Chem. Commun.*, p. 63.
97. van Tilborg, W. J. M., and Plomp, R. (1977). *J. Chem. Soc. Chem. Commun.*, p. 130.
98. Oppolzer, W. (1977). *Angew. Chem. Int. Ed. Engl.* **16**, 10.
99. Aalbersberg, W. G. L., and Vollhardt, K. P. C. (1979). *Tetrahedron Lett.* p. 1939.
100. Cava, M. P., and Deana, A. A. (1959). *J. Am. Chem. Soc.* **81**, 4266.
101. Cava, M. P., and Shirley, R. L. (1960). *J. Am. Chem. Soc.* **82**, 654.
102. Cava, M. P., Deana, A. A., and Muth, K. (1960). *J. Am. Chem. Soc.* **82**, 2524.
103. Giovannini, E., and Vuilleumier, H. (1977). *Helv. Chim. Acta* **60**, 1452.
104. Cava, M. P., Shirley, R. L., and Erickson, B. W. (1962). *J. Org. Chem.* **27**, 755.
105. Cava, M. P., and Kuczkowski, J. A. (1970). *J. Am. Chem. Soc.* **92**, 5800.
106. Chow, S. W., Pilato, L. A., and Wheelwright, W. L. (1970). *J. Org. Chem.* **35**, 20.
107. Vögtle, F. (1969). *Angew. Chem. Int. Ed. Engl.* **8**, 274.
108. Staab, H. A., and Haenel, M. (1973). *Chem. Ber.* **106**, 2190.
109. Grütze, J., and Vögtle, F. (1977). *Chem. Ber.* **110**, 1978.
110. Rossa, L., and Vögtle, F. (1977). *J. Chem. Res. (M)*, p. 3010.
111. Vögtle, F., and Rossa, L. (1977). *Tetrahedron Lett.*, p. 3577.
111a. Vögtle, F., and Neumann, P. (1973). *Synthesis*, p. 85.
112. Reichen, W. (1976). *Helv. Chim. Acta* **59**, 1636.
113. Wentrup, C., Damerius, A., and Reichen, W. (1978). *J. Org. Chem.* **43**, 2037.
114. Crabtree, H. E., Smalley, R. K., and Suschitzky, H. (1968). *J. Chem. Soc. (C)*, p. 2730.
115. Franz, J. E., and Black, L. L. (1970). *Tetrahedron Lett.*, p. 1381.
116. Guenther, W. B., and Walters, W. D. (1951). *J. Am. Chem. Soc.* **73**, 2127.
117. Neish, A. C., Haskell, V. C., and MacDonald, F. J. (1947). *Can. J. Res.* **25B**, 266.
118. Ciganek, E., and Krespan, C. G. (1968). *J. Org. Chem.* **33**, 541.
119. DeJongh, D. C., and Thomson, M. L. (1973). *J. Org. Chem.* **38**, 1356.
120. Thomson, M. L., and DeJongh, D. C. (1973). *Can. J. Chem.* **51**, 3313.
121. Lin, D. C. K., Thomson, M. L., and DeJongh, D. C. (1974). *Can. J. Chem.* **52**, 2359.
122. Lin, D. C. K., Thomson, M. L., and DeJongh, D. C. (1975). *Can. J. Chem.* **53**, 2293.
123. MacBride, J. A. H. (1972). *J. Chem. Soc. Chem. Commun.*, p. 1219.
124. Barton, J. W., and Walker, R. B. (1978). *Tetrahedron Lett.*, p. 1005.
125. Barton, J. W., and Walker, R. B. (1975). *Tetrahedron Lett.*, p. 569.
126. Kanoktanoporn, S., and MacBride, J. A. H. (1977). *Tetrahedron Lett.*, p. 1817.
127. MacBride, J. A. H. (1974). *J. Chem. Soc. Chem. Commun.*, p. 359.
128. Chambers, R. D., Clark, D. T., Holmes, T. F., Musgrave, W. R. K., and Ritchie, I. (1974). *J. Chem. Soc. Perkin Trans. 1*, p. 114.
129. Gilchrist, T. L., Rees, C. W., and Thomas, C. (1975). *J. Chem. Soc. Perkin Trans. 1*, p. 12.
130. Wentrup, C. (1972). *Helv. Chim. Acta* **55**, 1613.

131. Graebe, C., and Ullmann, F. (1896). *Justus Liebigs Ann. Chem.* **291**, 16.
132. Lin, D. C. K., and DeJongh, D. C. (1974). *J. Org. Chem.* **39**, 1780.
133. Lawrence, R., and Waight, E. S. (1970). *Org. Mass Spectrom.* **3**, 367.
134. Closs, G. L., and Harrison, A. M. (1972). *J. Org. Chem.* **37**, 1051.
135. Seybold, G., Jersak, U., and Gompper, R. (1973). *Angew. Chem. Int. Ed. Engl.* **12**, 847.
136. Adger, B. M., Keating, K., Rees, C. W., and Storr, R. C. (1975). *J. Chem. Soc. Perkin Trans. 1*, p. 41.
137. Adger, B. M., Rees, C. W., and Storr, R. C. (1975). *J. Chem. Soc. Perkin Trans. 1*, p. 45.
138. Rees, C. W., Storr, R. C., and Whittle, P. J. (1976). *J. Chem. Soc. Chem. Commun.*, p. 411.
139. Rees, C. W., Storr, R. C., and Whittle, P. J. (1976). *Tetrahedron Lett.*, p. 4647.
140. Gilchrist, T. L., Moody, C. J., and Rees, C. W. (1976). *J. Chem. Soc. Chem. Commun.*, p. 414.
141. Kinstle, T. H., and Darlarge, L. J. (1969). *J. Heterocycl. Chem.* **6**, 123.
142. Barr, D. A., and Hazeldine, R. N. (1955). *J. Chem. Soc.*, p. 1881.
143. Banks, R. E., Hazeldine, R. N., and Taylor D. R. (1965). *J. Chem. Soc.*, p. 978.
144. David, M. P., and McOmie, J. F. W. (1973). *Tetrahedron Lett.*, p. 1361.
145. Boyer, J. H., and Frints, P. J. A. (1970). *J. Heterocycl. Chem.* **7**, 59.
146. Boyer, J. H., and Ellis, P. S. (1977). *J. Chem. Soc. Chem. Commun.*, p. 489.
147. Chapman, J. A., Crosby, J., Cummings, C. A., Rennie, R. A. C., and Paton, R. M. (1976). *J. Chem. Soc. Chem. Commun.*, p. 240.
148. Cotter, J. L., and Knight, G. J. (1966). *Chem. Commun.*, p. 336.
149. Castillo, G. E., and Bertorello, H. E. (1978). *J. Chem. Soc. Perkin Trans. 1*, p. 325.
150. Seybold, G., and Heibl, C. (1975). *Angew. Chem. Int. Ed. Engl.* **14**, 248.
151. Benati, L., Montevecchi, P. C., and Zanardi, G. (1977). *J. Org. Chem.* **42**, 575.
152. Abramovitch, R. A., and Wake, S. (1977). *J. Chem. Soc. Chem. Commun.*, p. 673.
153. Davies, K. L., Storr, R. C., and Whittle, P. J. (1978). *J. Chem. Soc. Chem. Commun.*, p. 9.
154. Ege, G., and Beisiegel, E. (1974). *Synthesis*, p. 22.
155. Cuthbertson, E., and MacNicol, D. D. (1975). *Tetrahedron Lett.*, p. 1893.
156. Tou, J. C., Wang, C. S., and Alley, E. G. (1970). *Org. Mass Spectrom.* **3**, 747.

Chapter 7

Fragmentation of Acyclic Structures, and Related Cleavage of Cyclic Structures

I. DECARBOXYLATIONS AND DECARBONYLATIONS

A. Decarboxylation of Carboxylic Acids

Flow pyrolysis at high temperatures has been very little used for the decarboxylation of carboxylic acids. Most of the decarboxylations surveyed by Henecka[1] and by Clark[2] involve either uncatalyzed or catalyzed reactions in solution. However, the example of indole-4-carboxylic acid discussed below [Eq. (7.6)] suggests that flash pyrolytic methods may occasionally be useful for effecting smooth decarboxylation of otherwise stable acids. The following brief account gives some examples of decarboxylation in flow systems, and the major review by Smith and Kelly[3] includes systematic data on the gas-phase decarboxylation of unsaturated acids.

Bigley's group has made an extended study of the decarboxylation of unsaturated acids, starting in 1964 with kinetic and labeling studies of decarboxylations run in boiling phenanthrene.[4] The decarboxylation of β,γ-unsaturated acids is a concerted and largely nonpolar cyclic process[5] [Eq. (7.1)], as first suggested by Arnold and co-workers.[6] In liquid-phase or static gas-phase decarboxylations polymerization of the starting material or acid-catalyzed isomerization of the product may be a problem, but this

$$\text{(structure)} \longrightarrow \text{(structure)} + CO_2 \qquad (7.1)$$

can readily be avoided by flow pyrolysis or flash vacuum pyrolysis as in the examples of Eqs. (7.2)–(7.5). These examples also show that allenic and acetylenic acids decarboxylate readily by the cyclic process. Indeed, the rate of decarboxylation of the acetylenic acid $HC\equiv CCH_2CO_2H$ in the gas phase at 227° is 3.9 times faster than that of the olefinic acid $CH_2=CHCH_2CO_2H$, an acceleration attributed by Bigley and Weatherhead[7] to the lower energy of breaking of the first π bond in an alkyne (225 kJ/mole) compared to that in an alkene (264.4 kJ/mole). The acetylenic bond is probably accommodated in the cyclic transition state through the low frequency bending mode corresponding to the $C-C\equiv C$ bending mode of propyne at 336 cm^{-1}. The rate of decarboxylation of the allenic acid $CH_2=C=CHCO_2H$ could not be measured because it polymerized so readily, even in the flow apparatus used for kinetic measurements at 227° and 150 mm. When this allenic acid was heated in a sealed tube at 190° for 2 hr it gave acetone, formed by dimerization of the acid, 6-center rearrangement of the dimer to give acetoacetic acid and (by inference) butatrienone, and decarboxylation of the acetoacetic acid.[8]

As a single example of the decarboxylation of a relatively stable acid, Eq. (7.6) shows the decarboxylation of labeled indole-4-carboxylic acid, used

$$C_6H_5CH=CHC(CH_3)_2CO_2H \xrightarrow[\substack{\text{petroleum} \\ \text{diluent} \\ \text{glass helices}}]{420°/N_2 \text{ flow}} C_6H_5CH_2CH=C(CH_3)_2 + CO_2 \qquad [\text{Ref. 9}]$$

$$(7.2)$$

$$HC\equiv CCH_2CO_2H \xrightarrow[\text{CT 1.5 sec}]{427°/0.01 \text{ mm}} \underset{100\%}{CH_2=C=CH_2} + CO_2 \qquad [\text{Ref. 7}] \qquad (7.3)$$

$$CH_2=C=CHCO_2H \xrightarrow[\text{CT 1.5 sec}]{327°-387°/0.01 \text{ mm}} \underset{100\%}{CH_3C\equiv CH} + CO_2 \qquad [\text{Ref. 8}] \qquad (7.4)$$

$$CH_2=C=CHCH_2CO_2H \xrightarrow[\text{CT 1.5 sec}]{327°-387°/0.01 \text{ mm}} CH_2=CHCH=CH_2 + CO_2 \qquad [\text{Ref. 8}]$$

$$(7.5)$$

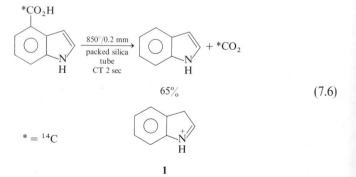

$$(7.6)$$

$* = {}^{14}C$

by Brown and Smith[10] to determine the fate of a ^{14}C label in the rearrangement of 2-quinolylnitrene [see Chapter 5; Eq. (5.89)]. It is uncertain whether the electron-rich heterocyclic ring participates in this decarboxylation; if it does this would imply the formation of a dipolar intermediate **1**.

B. Decarboxylation of Trimethylsilyl Esters with Silyl Migration

Trimethylsilyl esters of malonic acids, β-keto acids and some α-keto acids can undergo pyrolytic decarboxylation at high temperatures, with migration of a trimethylsilyl group to a new oxygen atom in a cyclic process. The pyrolysis of the malonic esters **2** and **3** was studied by Bloch and Denis,[11] who were able to detect the formation of the intermediate acetal **4**, $R=CH_3$, at 650°. At 700° clean pyrolysis to the ketene occurred by a further 1,3 migration of a silyl group.

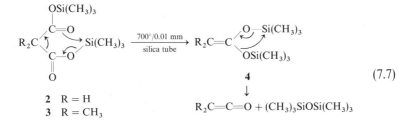

$$(7.7)$$

2 $R = H$
3 $R = CH_3$

$$R_2C=C=O + (CH_3)_3SiOSi(CH_3)_3$$

The trimethylsilyl ester of 1-acetylcyclopropane-1-carboxylic acid and of cyclopropane-1,1-dicarboxylic acid (**5**) showed similar initial 1,5 migration of one trimethylsilyl group to oxygen, but subsequent steps involved migration to carbon rather than to oxygen. Thus pyrolysis of the ester **5** failed to give the elusive ketene **6** [compare Chapter 5, Eq. (5.7)] but gave instead the C-silylated esters **7** and **8**.[11]

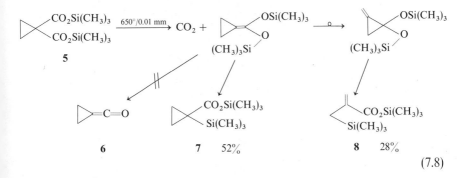

$$(7.8)$$

Bloch and co-workers[12] have found that even the ester **9** of a bridgehead β-keto acid will undergo this migration to give a bridgehead enol ether **10** [Eq. (7.9)]. This intermediate **10** is not isolable, but it undergoes a 1,3 hydrogen shift to give the isomeric enol ether **11** and the corresponding ketone formed by adventitious hydrolysis. Other minor products appear to be formed by a retro-Diels–Alder reaction of **10**. The occurrence of the 1,3 hydrogen shift has been proved by deuterium labeling as indicated in Eq. (7.9); the barrier to such a forbidden suprafacial shift must be reduced by relief of the strain energy (ca. 12 kcal/mol) associated with the bridgehead double bond in **10**.

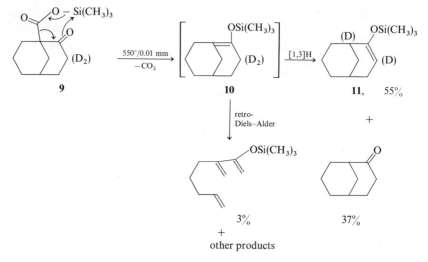

$$\tag{7.9}$$

The decarboxylation of the α-ketoester **12** by vigorous destructive distillation from a flask heated with a free flame [Eq. (7.10)] as reported by Brook

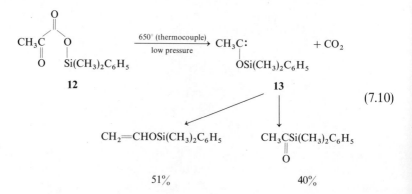

$$\tag{7.10}$$

and co-workers[13] is clearly closely related to the previous cases. The products are consistent with the formation of an intermediate silyloxycarbene **13**.

C. Decarboxylation of Allylic Esters through Rearrangement of the Enol

Allylic esters of acetoacetic acid and cyanoacetic acid can undergo a form of Claisen rearrangement through their enolic tautomers leading to a β-keto acid or a cyano acid which decarboxylates to an allyl substituted ketone or nitrile. The reaction occurs readily in the liquid phase at $150°–200°$ for acetoacetic esters (the Carroll reaction[14]) but the less readily enolized cyano-acetic ester requires more vigorous conditions and rearrangement is conveniently effected by flow pyrolysis. Kooyman and co-workers[15] found that the pyrolysis of allyl cyanoacetate to give carbon dioxide and allylacetonitrile [Eq. (7.11)] was catalyzed by addition of the vapor of acetic acid or of tertiary amines or, even more effectively, by a packing of glass wool; in all cases catalysis must promote the necessary enolization of the ester.

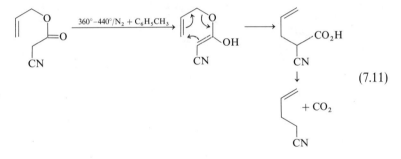

$$(7.11)$$

D. Decarbonylation of Carbonyl Compounds; Ketoesters and Vinyl Esters

The thermal decarbonylation of a range of simple aliphatic and aromatic aldehydes and ketones has been reviewed by Schubert and Kintner.[16] Aliphatic aldehydes RCHO decarbonylate to give hydrocarbons RH at about $500°$, but aromatic aldehydes require a higher temperature, $650°–700°$. Some of these reactions have been studied with flow reactors, but in most cases the partial pressure of reactant has been quite high, and radical chain reactions have predominated.

In this section we consider the decarbonylation of two somewhat more complex carbonyl systems. The first of these is the decarbonylation of glyoxylic esters derived from the condensation of oxalic ester with ketones. This decarbonylation is generally carried out in the liquid phase by heating the glyoxylic ester with a little powdered soft glass, as in the preparation of cyclohexanone-2-carboxylic ester.[17] The reaction succeeds under these conditions only for glyoxylic esters derived from acyclic or unstrained cyclic ketones; the failure of the cyclopentanone derivatives to decarbonylate has been attributed to their reluctance to form endocyclic enols.[18] Brown and Butcher[19] found, however, that the cyclopentanoneglyoxylic ester 14 afforded the corresponding 2-carboxylic ester in almost quantitative yield on flash vacuum pyrolysis at 500°–700° [Eq. (7.12)]. The carbon monoxide lost in such reactions is probably derived from the carbonyl group of the ester function, as suggested by the work of Banholzer and Schmid[20] on a related system, and a plausible intermediate 15 is shown in Eq. (7.12) for the present case. 1-Indanone-2-carboxylic ester could similarly be prepared without difficulty [Eq. (7.13)]. These pyrolyses followed a quite different course over a packing of Pyrex wool, which promoted an apparently internal alkylation of the ketone [e.g., Eq. (7.14)] with loss of both carbon dioxide and carbon monoxide.

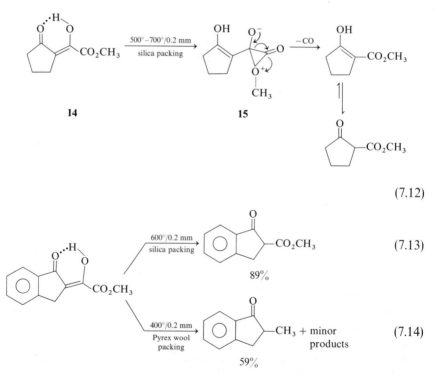

$$(7.12)$$

$$(7.13)$$

$$(7.14)$$

The second group of decarbonylations to be considered is that of the vinyl carboxylates. Young et al.[21] discovered a general thermal rearrangement of enol esters at 500° to give 1,3-dicarbonyl compounds, which can be obtained in 70–85% yield with the aid of recycling apparatus. The reaction is intramolecular, as shown by the negative results of crossover experiments. The reaction appears to be promoted by the surface effect of a packing, and a dipolar four-membered cyclic intermediate has been suggested.[21] Allan and co-workers[22] have found the 1,3-acyl shift to be reversible at 500°, with the 1,3-diketone strongly favored at equilibrium. The pyrolysis of isopropenyl benzoate[21] is shown in Eq. (7.15). In the case of unsubstituted vinyl esters the initial product is a formylketone, which readily decarbonylates to give a ketone. This form of the reaction has been extensively investigated by Ritchie and co-workers, who first observed it in studying the pyrolysis of ethylene dibenzoate.[23] The primary product from vinyl benzoate is formylacetophenone (16) which has been shown to decarbonylate readily at 550°, so that the pyrolysis of vinyl benzoate at 550° gives acetophenone as the major product [Eq. (7.16)],[23] accompanied by small amounts of benzoic acid and styrene. The decarbonylation is shown as concerted in Eq. (7.16), but there does not seem to be any evidence to rule out a radical chain mechanism. A closely related decarbonylation occurs in the pyrolysis of propargyl esters to give unsaturated ketones [see Chapter 9, Section V,D; Eq. (9.85)].

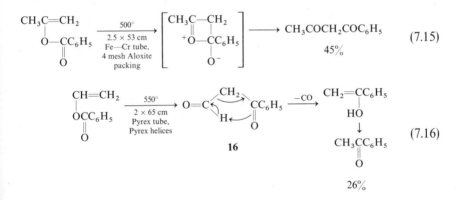

II. FISSION AND RING OPENING OF γ,δ-UNSATURATED ALCOHOLS

The thermal fission of γ,δ-unsaturated alcohols is a special case of the retro-ene reaction, and is included in Hoffmann's review[24] of ene reactions. The reaction has both historical and technical interest in the case of the destructive distillation of castor oil and ricinoleic acid, summarized by Hurd,[25] which provides a route to heptanal and undecylenic acid [Eq. (7.17)].

$$C_6H_{13}-\overset{\overset{\frown}{\underset{O-H}{}}}{}(CH_2)_7CO_2R \xrightarrow{\Delta} C_6H_{13}CHO + CH_2\!=\!CH(CH_2)_8CO_2R \tag{7.17}$$

R = H or glyceryl

Both fission of acyclic systems and ring opening of 2-(1'-alkenyl)cyclo-alkanols are mentioned in this section. The reaction is conveniently con-ducted in a flow system at atmospheric or reduced pressure and at about 500°. The operation of the cyclic mechanism first proposed by Arnold and Smolinsky[26] is clearly revealed by the pyrolysis of the deuterated alcohol **17** [Eq. (7.18)] which gives allylbenzene deuterated at C-1'.[27]

$$C_6H_5-\overset{}{\underset{D}{\|}}\overset{C_2H_5}{\underset{O}{\diagup}}C_2H_5 \xrightarrow[\substack{2 \times 30\ cm \\ glass\ tube; \\ glass\ helices}]{500°/N_2\ flow} C_6H_5CHDCH\!=\!CH_2 + (C_2H_5)_2CO \tag{7.18}$$

17

A selection of fission and ring opening reactions is shown in Table 7.1; the original papers contain many more examples. Both olefinic and acetylenic alcohols will undergo this fission, and the latter show relative accelerations $k_{\text{acet}}/k_{\text{olef}}$ ranging from 1.3 to 6.9 for different patterns of substitution.[28] Viola and co-workers[28] have discussed the effect of eclipsing of substituents in the planar transition state[29] for fission of alkynols of the type $R_2C(OH)\cdot CR_2C\!\equiv\!CR$ as compared with their conformational effects in the nonplanar transition state for the corresponding olefinic systems. Arnold's group[30,31] has used the pyrolysis of appropriately substituted cycloalkanols for the synthesis of long chain unsaturated aldehydes and ketones (see entries 3 and 4). Marvell and Rusay[32] have shown that the *trans*-alkenyl cycloalkanols yield predominantly or exclusively the *trans* alkenals or alkenones, whereas the *cis*-alkenyl cycloalkanols give mixtures of cis and trans isomers (compare entries 5 and 6). Spencer and Hill[33] found about 25% retention of optical activity in the fission of (R)-(−)-(E)-1,4-diphenyl-3-penten-1-ol (entry 7) and concluded that the transition state cannot be strictly chairlike, because this should have led to an excess of the (S) enantiomer of the alkene rather than the (R) enantiomer as observed.

Finally, a failure of the method is noted in entry 8. This acetylenic alcohol underwent acid–base catalyzed dehydration and electrocyclic ring closure when heated in an evacuated Pyrex tube;[34] it is amusing to speculate whether ring opening might have been achieved by flash pyrolysis with conditioned silica apparatus.

TABLE 7.1

Pyrolytic Fission of γ,δ-Unsaturated Alcohols

Unsaturated alcohol	Product(s)	Yield (%)	Conditions	Reference number
1. (cyclohexenyl–CH_2–CH(OH)–C_6H_{13})	(methylenecyclohexane) + $C_6H_{13}CHO$	78	500°; N_2 flow; glass helices	26
2. (C_4H_9–C≡C–CH(OH)–, HO)	$C_4H_9CH{=}C{=}CH_2$ + CH_2O	52	430°/10 mm; 2.2 × 45 cm Pyrex tube; Pyrex helices	28
3. (cyclopentyl with C_6H_{13}, OH)	(chain with C_7H_{15}, CHO)	55	500°; N_2 flow; Pyrex helices	30
4. (bicyclic, OH)	(CHO)	56	520°–530°; N_2 flow; Pyrex helices	31
5. (cyclohexane with vinyl, OH, CH₃)	(CH_3; $COCH_3$)	90		32
6. (cyclohexane with vinyl, OH, CH₃)	(CH_3; $COCH_3$) $E{:}Z = 54{:}46$	84	440°; N_2 flow; quartz packing; CT 1.5–2 min	32
7. (Ph–CH(OH)–CH₃ ··· C(Ph)=CH₂) (R)-(−) isomer	PhCHO (CH₃)(Ph)(H)C– (R)-(−) isomer in excess	62	465°; N_2 flow; 2 × 100 cm; Vycov tube; glass beads	33
8. (cyclohexyl–cyclohexene, HO, C≡CH)	(tricyclic with O)	100	350°; evacuated Pyrex tube; 2 hr	34

III. RETRO-ENE REACTIONS

A. Cleavage of Alkenes, Cycloalkenes, Cycloalka-1,2-Dienes, and Cycloalkynes

Hoffmann[24] noted in 1969 that there were no well-established examples of retro-ene reactions of simple acyclic alkenes [Eq. (7.19)], although it is hard to believe that such reactions do not occur in high-temperature industrial cracking of alkenes. Frech *et al.*[35] have analyzed the factors affecting pyrolysis of the methylpentenes, and they interpret the pyrolysis of 2-methyl-1-pentene [Eq. (7.20)] entirely in terms of radical and radical chain processes. However, it seems plausible that at least some of the isobutene formed in Eq. (7.20) should be the product of a retro-ene reaction. A secondary retro-ene reaction leading to loss of propene certainly follows the pyrolytic generation of 1,11-dodecadiene (see Table 7.2, entry 5).

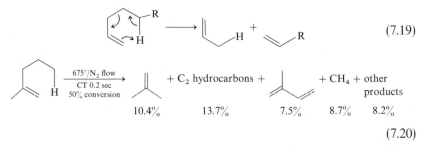

$$(7.19)$$

$$(7.20)$$

Transannular retro-ene reactions of medium ring cycloalkenes, cycloalka-1,2-dienes, and cycloalkynes are however of considerable synthetic interest for the preparation of medium chain hydrocarbons with terminal unsaturation at each end. Blomquist and Taussig[36] first discovered this ring cleavage in the attempted preparation of the cyclononenes by pyrolysis of cyclononyl acetate at 500°, which gave a pyrolyzate containing 70% of 1,8-nonadiene and only 30% of *cis*- and *trans*-cyclononene. *trans*-Cyclononene itself on pyrolysis at 500° gave 1,8-nonadiene in 66% yield, and the 12% of recovered cyclononene contained 93.4% of the cis isomer. It appears probable from the detailed results of Blomquist and Taussig that *cis*-cyclononene (**18**) does not readily undergo retro-ene cleavage except by initial isomerization to *trans*-cyclononene (**19**), which is conformationally ideal for cleavage to 1,8-nonadiene [Eq. (7.21)].

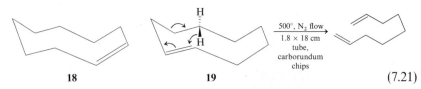

    18   **19** (7.21)

TABLE 7.2

Retro-ene Cleavage of Medium Ring Unsaturated Hydrocarbons

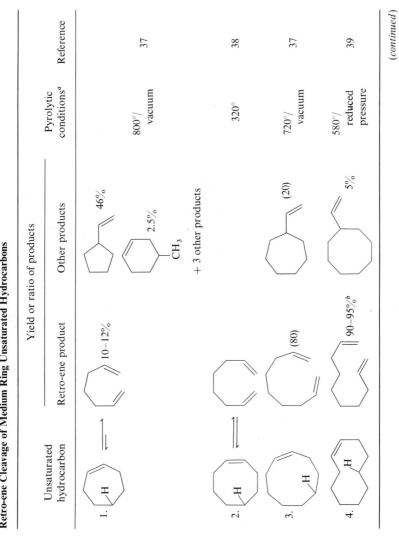

Unsaturated hydrocarbon	Yield or ratio of products		Pyrolytic conditions[a]	Reference
	Retro-ene product	Other products		
1.	10–12%	46%	800°/ vacuum	37
		2.5% + 3 other products		
2.			320°	38
3.	(80)	(20)	720°/ vacuum	37
4.	90–95%[b]	5%	580°/ reduced pressure	39

(*continued*)

TABLE 7.2 (Continued)

Unsaturated hydrocarbon	Yield or ratio of products		Pyrolytic conditions[a]	Reference
	Retro-ene product	Other products		
5.	$(CH_2)_8$ (34) + $(CH_2)_5$ (5)	Cyclododecenes (52) (3)	720°/ vacuum	37
6. $(CH_2)_8$	$(CH_2)_5$ (35) $^+CH_3CH=CH_2$	1,11-Dodecadiene (65)	720°/ vacuum	37
7.	(65)	(23) (12)	640°/ 0.3 mm	40
8.	(25)	Cyclononyne (54) (21)	600°/ 0.05 mm	41

[a] In most cases 17- to 30-cm packed tubes were used.
[b] The high yield was obtained by the use of a recycling apparatus, with removal of the lower-boiling diene.

Further examples of cleavage of medium ring hydrocarbons are shown in Table 7.2. *cis*-Cyclooctene (entry 2) undergoes reversible ring opening at temperatures as low as 320°. Cycloheptene (entry 1) is converted into 1,6-heptadiene at 800° and the reaction is partly reversible, but the major process (discussed below) is one of ring contraction to give vinylcyclopentane. At the same time 1,6-cycloheptadiene appears to undergo a unique ene reaction to give 4-methylcyclohexene [Eq. (7.22)]. At high temperatures the primary C_n dienes undergo secondary retro-ene reaction (entries 5 and 6) to give C_{n-3} dienes and propene. 1,2-Cyclononadiene and cyclononyne both undergo ene cleavage (entries 7 and 8), but cyclooctyne rearranges instead by ring contraction to cycloheptylidenecarbene (see Chapter 5, Section II,D).

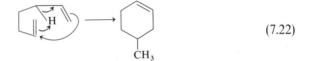

$$(7.22)$$

Ring contraction to give vinylcycloalkanes via intermediate allylic diradicals competes with ring cleavage of cycloalkenes, particularly at high temperatures (Table 7.2, entires 1, 3, 4, and 5), and is particularly prominent in cycloheptene [Eq. (7.23)] for which the transition state for retro-ene cleavage must be rather strained.

$$(7.23)$$

In the present discussion the cleavage reactions of cycloalkenes have been assumed to be concerted processes, but the results can also be rationalized in terms of diradical intermediates. Thus the cis–trans isomerization of cyclononene and its cleavage can be explained through generation and interconversion of 1,2 and 1,4 diradicals **20** and **21**,[36] and Crandall and Watkins[40]

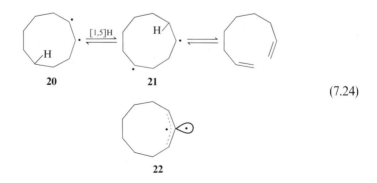

$$(7.24)$$

have similarly considered the diradical **22** as a possible intermediate in the cleavage of 1,2-cyclononadiene.

B.　Cleavage of Allyl and Vinyl Ethers, Acetals, and Thioethers

The range of heteroatom analogues of retro-ene reactions has been classified formally by Cookson and Wallis[42] and by Hoffmann;[24] it includes the fission of γ,δ-unsaturated alcohols described in Section II. In this section we consider the fragmentation of unsaturated ethers and related compounds to give alkenes and carbonyl or thiocarbonyl compounds.

Cookson and Wallis[42] investigated the products and rates of pyrolysis of some eighteen allylic ethers of the general type $R_2CHOCH(R^1)CR^2{=}CHR^3$ in a flow system at 450°–560°. The simplest compound, allyl ethyl ether, required a temperature of 560° for optimum formation of acetaldehyde [Eq. (7.25)] whereas the corresponding diphenylmethyl ether fragmented cleanly at 430°–540° to give benzophenone and propene [Eq. (7.26)]; the relative rates at 430°, corrected for the number of α-hydrogen atoms, were 0.09:1. Additional examples are shown in Eqs. (7.27) and (7.28). Some applications of this method include the preparation of terephthalaldehyde in 92% yield by pyrolysis of the bisether **23** at 540°,[43] the preparation of allyl esters from allyl acetals **24**,[44,45] and the preparation of many enol ethers from unsaturated acetals such as **25** at about 500°.[45]

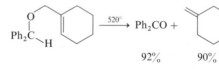

$$\xrightarrow[\text{silica granules}]{560°,\ \text{slow N}_2\ \text{flow}} \quad CH_3CHO + CH_2{=}CHCH_3 \qquad (7.25)$$

$$36\% \qquad\qquad 42\%$$

$$\text{Ph}_2\text{C}\diagdown_H \ (D) \quad \xrightarrow{540°} \quad Ph_2CO + CH_2{=}CHCH_3 \ (D) \qquad (7.26)$$

$$92\% \qquad\qquad 91\%$$

$$\text{Ph}_2\text{C}\diagdown_H \quad \xrightarrow{520°} \quad Ph_2CO + \qquad\qquad (7.27)$$

$$92\% \qquad\qquad 90\%$$

$$\text{Ph}-\text{CH}\diagdown_H \quad \xrightarrow{490°} \quad PhCHO + PhCH{=}CHCH_3 \qquad (7.28)$$

$$80\% \qquad\qquad 84\%$$

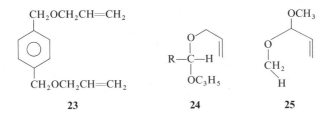

23   24   25

A spectacular example of stereochemical control through the application of ene and retro-ene reactions in synthesis [Eq. (7.29)] is due to Arigoni's group.[46] The chiral methyl compound **27** required for oxidative degradation to chiral acetic acid was prepared by consecutive intramolecular reactions of the chiral ether **26**.

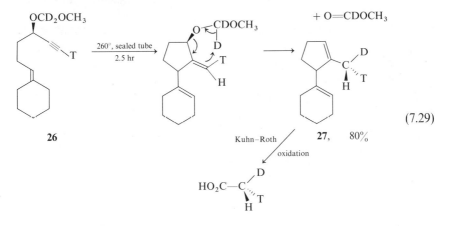

(7.29)

Appropriately substituted alkyl vinyl ethers also undergo fragmentation near 500°. Bailey and Di Pietro[47] found that methyl vinyl ether is stable to pyrolysis at 400° (99% recovery), whereas butyl vinyl ether is almost completely decomposed in one pass at 480° [Eq. (7.30)].

$$\underset{}{\overset{H}{\diagup}} \quad \xrightarrow[\substack{3\text{-mm Pyrex} \\ \text{helices}}]{480°} \quad \diagup\diagup \; + \; CH_3CHO$$

                                  71%   42%   (7.30)

All of the previous fragmentations are probably concerted 6-center reactions, but Viola and co-workers[48] have recently discovered two cases of 8-center decompositions which show activation parameters quite similar to those of the corresponding 6-center reactions; the activation energies are estimated to be only 2–3 kcal/mole higher. Both olefinic and acetylenic compounds show this fragmentation [Eqs. (7.31) and (7.32)], and the authors

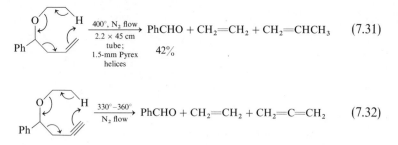

$$\text{(structure)} \xrightarrow[\substack{2.2 \times 45 \text{ cm} \\ \text{tube;} \\ 1.5\text{-mm Pyrex} \\ \text{helices}}]{400^\circ, \ N_2 \ \text{flow}} \text{PhCHO} + CH_2{=}CH_2 + CH_2{=}CHCH_3 \quad (7.31)$$

$$42\%$$

$$\text{(structure)} \xrightarrow[N_2 \ \text{flow}]{330^\circ - 360^\circ} \text{PhCHO} + CH_2{=}CH_2 + CH_2{=}C{=}CH_2 \quad (7.32)$$

consider that these are the first examples of a general class of fragmentations which may occur whenever alternative 6-center pathways are not available.

The retro-ene elimination of propene from allylic thioethers has been exploited by de Mayo's group[49] for the preparation of sensitive thiones and thioaldehydes. A simple example is the generation of cyclohexanethione by flash vacuum pyrolysis of allyl cyclohexyl sulphide [Eq. (7.33)]; other thiones such as thiocamphor, thiobenzaldehyde, thioacrolein, monothiobiacetyl, and dithiomalondialdehyde were prepared in a similar manner. The monomeric thiones were characterized by infrared spectroscopy at 77°K and were isolated as polymers or trimers.

$$\text{(structure)} \xrightarrow[\text{CT 1 msec}]{\text{FVP 650}^\circ} \text{(thione structure)} + CH_3CH{=}CH_2 \quad (7.33)$$

$$31\% \ \text{as trimer}$$

Finally, two 6-center reactions of unsaturated esters are formally closely related to the reactions of ethers considered above. Vernon and Waddington[50] have reported the decomposition of allylic formates in the gas phase to give high yields of alkenes and carbon dioxide [Eq. (7.34)], and Blackman *et al.*[51] have used microwave spectroscopy to detect the formation of methylene-ketene amongst the products of pyrolysis of diphenylmethyl propiolate [Eq. (7.35)].

$$\text{(structure)} \xrightarrow[\substack{\text{static} \\ \text{reactor}}]{332^\circ - 342^\circ} \text{(structure)} + CO_2 \quad (7.34)$$

$$\text{(structure)} \xrightarrow{560^\circ/0.05 \ \text{mm}} Ph_2CO + CH_2{=}C{=}C{=}O \quad (7.35)$$

C. Formation of Species with Si=C Bonds

The synthetic power of retro-ene reactions is well illustrated by the elegant synthesis of silacyclobutenes due to Block and Revelle[52] [Eq. (7.36)] and the adaptation of this approach to the generation and trapping with acetylene of 1-methylsilabenzene [Eq. (7.37)] reported by Barton and Burns.[53]

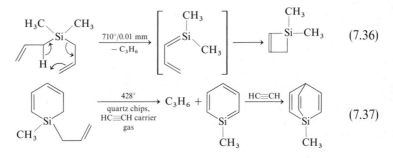

REFERENCES

1. Henecka, H. (1952). *In* "Methoden der Organischen Chemie (Houben-Weyl)" (E. Müller, ed.), Band VIII, Sauerstoffverbindungen III, p. 484. Thieme, Stuttgart.
2. Clark, L. W. (1969). *In* "The Chemistry of Carboxylic Acids and Esters" (S. Patai, ed.), p. 589. Wiley (Interscience), New York.
3. Smith, G. G., and Kelly, F. W. (1971). *In* "Progress in Physical Organic Chemistry" (A. Streitwieser and R. W. Taft, eds.) Vol. 8, p. 75. Wiley (Interscience), New York.
4. Bigley, D. B. (1964). *J. Chem. Soc.*, p. 3897.
5. Bigley, D. B., and Thurman, J. C. (1968). *J. Chem. Soc. B*, p. 436.
6. Arnold, R. T., Elmer, O. C., and Dodson, R. M. (1950). *J. Am. Chem. Soc.* **72**, 4359.
7. Bigley, D. B., and Weatherhead, R. H. (1976). *J. Chem. Soc. Perkin Trans. 2*, p. 592.
8. Bigley, D. B., and Weatherhead, R. H. (1976). *J. Chem. Soc. Perkin Trans. 2*, p. 704.
9. Bigley, D. B., and Thurman, J. C. (1965). *J. Chem. Soc.*, p. 6202.
10. Brown, R. F. C., and Smith, R. J. (1972). *Aust. J. Chem.* **25**, 607.
11. Bloch, R., and Denis, J. M. (1975). *J. Organomet. Chem.* **90**, C9.
12. Bloch, R., Boivin, F., and Bortolussi, M. (1976). *J. Chem. Soc. Chem. Commun.*, p. 371.
13. Brook, A. G., Harris, J. W., and Bassindale, A. R. (1975). *J. Organomet. Chem.* **99**, 379.
14. Carroll, M. F. (1941). *J. Chem. Soc.*, p. 507.
15. Kooyman, E. C., Louw, R., and De Tonkelaar, W. A. M. (1963). *Proc. Chem. Soc.*, p. 66.
16. Schubert, W. M., and Kintner, R. R. (1966). *In* "The Chemistry of the Carbonyl Group" (S. Patai, ed.), p. 695. Wiley (Interscience), New York.
17. Snyder, H. R., Brooks, L. A., and Shapiro, S. H. (1943). *In* "Organic Syntheses" (A. H. Blatt, ed.), Coll. Vol. II, p. 531. Wiley, New York.
18. Mayer, R. (1955). *Chem. Ber.* **88**, 1859.
19. Brown, R. F. C., and Butcher, M. (1971). *Aust. J. Chem.* **24**, 2421.
20. Banholzer, K., and Schmid, H. (1956). *Helv. Chim. Acta* **39**, 548.
21. Young, F. G., Frostick, F. C., Sanderson, J. J., and Hauser, C. R. (1950). *J. Am. Chem. Soc.* **72**, 3635.

22. Allan, R. J. P., McGee, J., and Ritchie, P. D. (1957). *J. Chem. Soc.*, p. 4700.
23. Allan, R. J. P., Forman, R. C., and Ritchie, P. D. (1955). *J. Chem. Soc.*, p. 2717.
24. Hoffmann, H. M. R. (1969). *Angew. Chem. Int. Ed. Engl.* **8**, 556.
25. Hurd, C. D. (1929). "The Pyrolysis of Carbon Compounds," pp. 164, 172. Chem. Catalog Co. (Tudor), New York.
26. Arnold, R. T., and Smolinsky, G. (1959). *J. Am. Chem. Soc.* **81**, 6443.
27. Arnold, R. T., and Smolinsky, G. (1960). *J. Org. Chem.* **25**, 129.
28. Viola, A., MacMillan, J. H., Proverb, R. J., and Yates, B. L. (1971). *J. Am. Chem. Soc.* **93**, 6967.
29. Viola, A., Proverb, R. J., Yates, B. L., and Larrahondo, J. (1973). *J. Am. Chem. Soc.* **95**, 3609.
30. Arnold, R. T., and Smolinsky, G. (1960). *J. Am. Chem. Soc.* **82**, 4918.
31. Arnold, R. T., and Metzger, G. (1961). *J. Org. Chem.* **26**, 5185.
32. Marvell, E. N., and Rusay, R. (1977). *J. Org. Chem.* **42**, 3336.
33. Spencer, H. K., and Hill, R. K. (1976). *J. Org. Chem.* **41**, 2485.
34. Voorhees, K. J., Smith, G. G., Arnold, R. T., Covington, R. R., and Mikolasek, D. G. (1969). *Tetrahedron Lett.*, p. 205.
35. Frech, K. J., Hoppstock, F. H., and Hutchings, D. A. (1976). *In* "Industrial and Laboratory Pyrolyses" (L. F. Allright and B. L. Crynes, eds.), ACS Symposium Series No. 32, p. 197. Amer. Chem. Soc., Washington, D.C.
36. Blomquist, A. T., and Taussig, P. R. (1957). *J. Am. Chem. Soc.* **79**, 3505.
37. Roth, W. R. (1966). Unpublished work quoted in *Chimia* **20**, 229.
38. Crandall, J. K., and Watkins, R. J. (1971). *J. Org. Chem.* **36**, 913.
39. Rienäcker, R. (1964). *Brennst.-Chem.* **45**, 206.
40. Crandall, J. K., and Watkins, R. J. (1970). *Tetrahedron Lett.*, p. 1251.
41. Baxter, G. J., and Brown, R. F. C. (1978). *Aust. J. Chem.* **31**, 327.
42. Cookson, R. C., and Wallis, S. R. (1966). *J. Chem. Soc. B*, p. 1245.
43. Beam, C. F., and Bailey, W. J. (1971). *J. Chem. Soc., C*, p. 2730.
44. Ho, T.-L., and Wong, C. M. (1975). *Synth. Commun.* **5**, 213.
45. Mutterer, F., Morgen, J. M., Biedermann, J. M., Fleury, F. P., and Weiss, F. (1970). *Tetrahedron* **26**, 477.
46. Townsend, C. A., Scholl, T., and Arigoni, D. (1975). *J. Chem. Soc. Chem. Commun.*, p.921.
47. Bailey, W. J., and J. Di Pietro, (1977). *J. Org. Chem.* **42**, 3899.
48. Viola, A., Madhavan, S., Proverb, R. J., Yates, B. L., and Larrahondo, J. (1974). *J. Chem. Soc. Chem. Commun.*, p. 842. ———
49. Giles, H. G., Marty, R. A., and de Mayo, P. (1976). *Can. J. Chem.* **54**, 537.
50. Vernon, J. M., and Waddington, D. J. (1969). *J. Chem. Soc. Chem. Commun.*, p. 623.
51. Blackman, G. L., Brown, R. D., Brown, R. F. C., Eastwood, F. W., McMullen, G. L., and Robertson, M. L. (1978). *Aust. J. Chem.* **31**, 209.
52. Block, E., and Revelle, L. K. (1978). *J. Am. Chem. Soc.* **100**, 1630.
53. Barton, T. J., and Burns, G. T. (1978). *J. Am. Chem. Soc.* **100**, 5246.

Chapter 8

Cleavage of
Carbocyclic Systems with
Related Heterocyclic Examples

I. INTRODUCTION

It is convenient to consider together the pyrolytic dissociation of cyclo-butanes to give two alkenes, and the retro-Diels–Alder reactions of various carbocyclic systems which give formal dienes and alkenes, because the synthetic use of both reactions may be rather similar. In both cases the carbocyclic skeleton can serve to protect a required alkene or diene system, and the removal of the protecting group can be accomplished by pyrolysis. Most of this chapter deals with carbocyclic systems, but it seems appropriate to mention some related $2 + 2$ retrogressions of β-lactones, β-lactams, oxetanes, thietanes, and dithietanes, and some retro-Diels–Alder reactions of heterocycles.

Interest in the detailed study of the pyrolysis of cyclobutanes has grown both from gas kinetics, where the simplicity of the substrates and the forma-tion of products in a clean unimolecular reaction is an obvious attraction, and from the theoretical considerations of Woodward and Hoffmann.[1] Thermal fission of a cyclobutane by a concerted and allowed $\sigma^2s + \sigma^2a$ process[1] would lead to retention of configuration in one alkene product, and inversion in the other [Eq. (8.1)].

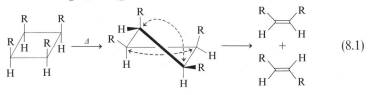

$$\qquad\qquad (8.1)$$

On the other hand, in a stepwise process proceeding by initial formation of a 1,4-diradical the stereochemical outcome will depend on the relative rates of radical fission of the first bond, rotations of the radical termini about single bonds in the 1,4-diradical, and fission of the second bond [Eq. (8.2)].

$$\text{(8.2)}$$

The behavior of most of a large number of cyclobutanes can best be explained in terms of stepwise diradical processes, and the variation in stereochemical outcome is usually ascribed to the effect of substituents on the lifetimes of individual 1,4-diradicals.

The pyrolysis of cyclobutanes has been covered in very detailed reviews of thermal reactions of alicyclic compounds, and much of the literature refers to kinetic experiments with static reactors. Consequently the pyrolysis of simple cyclobutanes is scarcely mentioned in this chapter. Unimolecular reactions of many hydrocarbons including cyclobutanes are reviewed by Frey[2] and by Frey and Walsh,[3] and an extensive collection of kinetic data is provided by Willcott *et al.*[4] More recent surveys covering the mechanisms of cyclobutane fission are those of Seebach,[5] Brown,[6] and of Becker and Brodsky.[7] All of these reviews give much kinetic and thermodynamic data, reflecting the intense interest in the detailed mechanisms of thermal reactions of cyclobutanes.

Most monographs and reviews on the Diels–Alder reaction naturally emphasize the synthetic value of the forward cycloaddition reaction. However Smith and Kelly[8] review retro-Diels–Alder reactions under homogeneous gas-phase conditions with emphasis on structure–reactivity relationships and temperatures required for appropriate reaction. A review by Kwart and King[9] is entirely devoted to the retro-Diels–Alder reaction, and a well-indexed monograph by Wollweber[10] contains a section covering the literature on retro-Diels–Alder reactions through 1971. Ripoll and the Rouessacs[10a] have extended review coverage of these reactions to the end of 1976.

At the end of this chapter some retrogressions of higher order are briefly mentioned.

II. CLEAVAGE OF CYCLOBUTANE DERIVATIVES

A. Cyclobutanes

Most work on the mechanism of pyrolysis of simple cyclobutanes[2-7] has involved the use of aged or conditioned static reactors in which the cyclo-

butane is decomposed under reduced pressure in the temperature range
$200°–450°$. The following few examples illustrate the conditions required for
cleavage in flow systems.

Baldwin and Ford[11] tested the stereochemical prediction of Woodward
and Hoffmann[1] by pyrolysis of 7,8-*cis*,exo-dideuteriobicyclo[4.2.0]octane in
both static and flow systems. With nitrogen as carrier in a flow system the
ethylene produced at $500°$ contained $62 \pm 3\%$ of the *trans*-dideuterioethylene
[Eq. (8.3)] and that produced at $450°$ and 1 mm in a sealed tube contained
$57 \pm 1\%$ of the trans isomer. If cleavage had been entirely concerted the
products of the $2 + 2$ process should have been cyclohexene (necessarily cis)
and 100% of *trans*-dideuterioethylene, whereas a stepwise diradical process
with rapid cleavage of the second bond should have given an excess of the
cis isomer. Baldwin and Ford comment[11] that "the reaction occurs in a
stereochemical sense incompatible with both the tetramethylene diradical
hypothesis and the completely stereoselective antarafacial elimination."

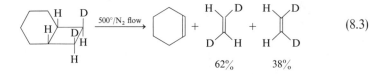

$$(8.3)$$

A quite different stereochemical result was found by Paquette and
Thompson[12] for the slow flow pyrolysis of a propelladiene [Eq. (8.4)]. The
deuterium-labeled methyl vinyl ether was formed with almost complete retention of stereochemistry and this is attributed to very fast cleavage of the
second bond in the intermediate diradical.

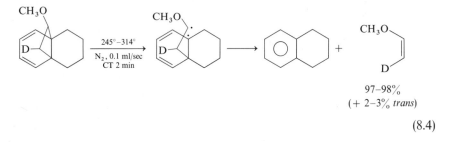

97–98%
(+ 2–3% *trans*)

$$(8.4)$$

In more strained systems cleavage may occur at lower temperatures, as
with the tricyclo[4.2.0.0^{2,5}]octanes studied by Martin and Eisenmann;[13]
1,4-diradicals are again invoked to explain the products [Eq. (8.5)].

The preparative value of the cleavage of cyclobutane rings is shown by the
work of Drysdale and co-workers[14] on the pyrolysis of various methylene-
cyclobutanes, which has provided a useful preparation[15] of an allenic ester,

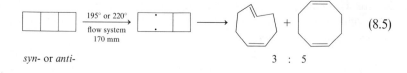

$$\text{(8.5)}$$

syn- or *anti*-   3 : 5

methyl 2,3-butadienoate [Eq. (8.6)], although some alternative cleavage and rearrangement also occurs. The starting material is derived from the cyclo-adduct of allene and maleic anhydride.

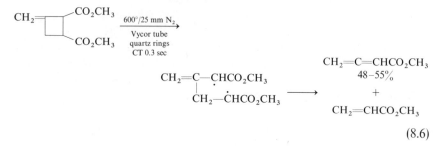

$$\text{(8.6)}$$

   The pinane group of terpenes provides a readily available class of cyclo-butanes which give pyrolytic reactions of commercial importance in the perfumery industry,[16] although the reactions are often much complicated by the formation of various secondary products. The pyrolysis of the terpene hydrocarbons was briefly reviewed by Banthorpe and Whittaker[17] in 1966, but there has been much recent work on functionalized pinanes.

   The simplest example of such pyrolyses is that of 6,6-dimethylnorpinane, studied by Pines and Hoffman.[18] The hydrocarbon was pyrolyzed at 500° by drip addition at 50–60 ml/hr to a vertical tube filled with copper punchings. The primary product, a 1,6-diene, was obtained in 50% yield, but was accompanied by an isopropenylmethylcyclopentane formed in a secondary ene reaction [Eq. (8.7)].

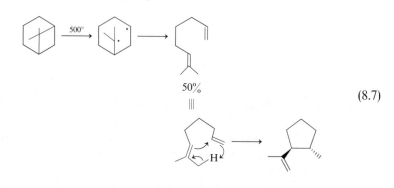

$$\text{(8.7)}$$

Similar reactions occur in the pyrolysis of pinane itself[19] and of many alcohols in the pinane series. For example the pyrolysis of 10α-pinan-2-ol at 580° studied by Coxon *et al.*[20] gives as the major product the 1,6-diene *R*-linalool (44%), together with a group of stereoisomeric 3-isopropenyl-cyclopentanols formed by ene reactions, and a methyl ketone formed by a complex sequence of ene reaction and β-hydroxyalkene fission from the alternative minor 1,6-diene [Eq. (8.8)]. Pinanols and pinenols with hydroxyl groups at C-2, C-3, and C-4 have been studied by this New Zealand group.[20,21,22]

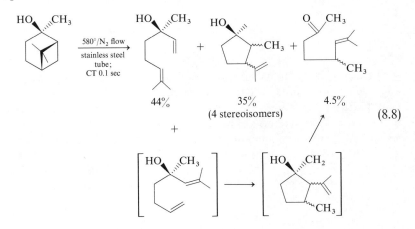

(8.8)

The commercially important pyrolyses of the α- and β-pinenes have been studied extensively,[16,17] particularly with the aim of improving the yields of the primary products ocimene and myrcene. Goldblatt and Palkin[23] showed that pyrolysis of α-pinene at 403° and 1 atm gives the conjugaged triene allo-ocimene (40%), (±)-dipentene [Eq. (8.9)], and other products, but the primary product, ocimene, can be obtained by pyrolysis of α-pinene at 5 mm

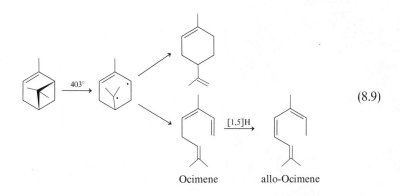

(8.9)

Ocimene   allo-Ocimene

over a Nichrome spiral under flow conditions which rapidly remove ocimene from the hot zone.[24]

β-Pinene on pyrolysis at 403° and 1 atm gives mainly myrcene (ca. 70%) and some (−)-limonene[23] [Eq. (8.10)]. Commercial processes for the production of myrcene have used higher temperatures (ca. 600°) and short contact times.[16]

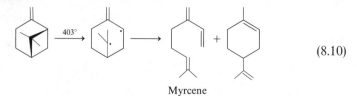

$$\qquad\qquad\qquad\qquad (8.10)$$

Myrcene

The effects of carbonyl groups at C-2, C-3, and C-4 on the pyrolysis of norpinane and pinane systems have been studied by Coxon et al.,[25] von Schulte-Elte et al.,[26] and by Mayer and Crandall.[27] Coxon et al. conclude[25] that the direction of fission of the cyclobutane ring is governed by the presence of adjacent substituents which can stabilize the 1,4-diradical, and that the effectiveness of these substituents is in the order exocyclic methylene > carbonyl > methyl > hydroxyl > hydrogen. The products of pyrolysis of nopinone[25,27] are shown in Eq. (8.11).

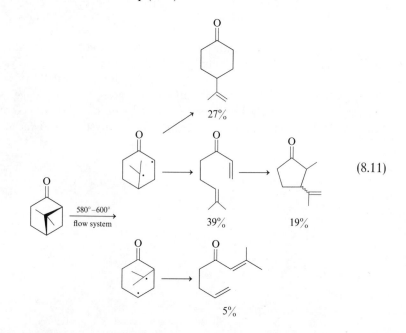

$$\qquad\qquad\qquad\qquad (8.11)$$

Pyrolytic cleavage of a cyclobutane ring was used as a structural tool by Büchi and Goldman[28] in the determination of the structure of carvonecamphor, the photocyclization product of carvone [Eq. (8.12)]. The direction of thermal cleavage is opposite to that of the photochemical formation of carvonecamphor, and must again be controlled by the interaction of the carbonyl group with the developing radical center during initial breaking of the bond *a—b.*

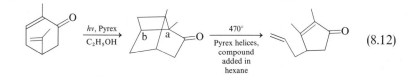

$$(8.12)$$

B. Silacyclobutanes

The decomposition of silacyclobutanes to give silaethenes and alkenes was first reported by Nametkin *et al.,*[29,29a] who showed that 1,1-dimethyl-1-silacyclobutane undergoes dissociation and recombination reactions at 600° [Eq. (8.13)].

$$(8.13)$$

The silaethene $(CH_3)_2Si{=}CH_2$ and ethylene generated by flash vacuum pyrolysis at 650° have been collected for infrared spectroscopy on a sodium chloride plate at $-196°$; Barton and McIntosh[30] assign strong absorption at 1407 cm^{-1} to the Si=C system. Recombination of the products to give the original silacyclobutane occurred above $-120°$.

The silaethenes are so reactive in dimerization and cycloaddition reactions that much of their chemistry has been studied by copyrolysis of silacyclobutanes with thermally stable trapping reagents such as benzophenone[31,32] or thiobenzophenone.[33] In most pyrolyses nitrogen at 1 atm is used as carrier gas and tube temperatures are 500°–600°. Golino and co-workers[31] generated silaethene, $H_2Si{=}CH_2$, by pyrolysis of silacyclobutane itself, and isolated products of its reactions with hexamethylcyclotrisiloxane and benzophenone [Eq. (8.14)] and other reagents.

In the pyrolysis of 2-substituted silacyclobutanes initial fission of the 2,3-bond is favored by the 2-substituent (CH_3 or C_6H_5) and the major product [Eq. (8.15)] is the more highly substituted silaethene.[32,34]

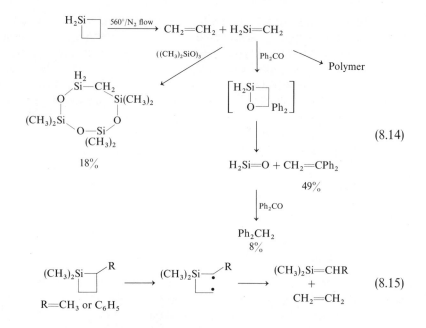

(8.14)

$$(CH_3)_2Si \underset{R=CH_3 \text{ or } C_6H_5}{\overset{R}{\longrightarrow}} \quad (CH_3)_2Si \overset{R}{\underset{\cdot}{\cdot}} \quad \longrightarrow \quad \begin{matrix} (CH_3)_2Si{=}CHR \\ + \\ CH_2{=}CH_2 \end{matrix}$$ (8.15)

C.   Cyclobutanones, Cyclobutane-1,3-Diones, and Cyclobutane-1,3-Dithiones

The thermal decomposition of simple cyclobutanones[7] has been studied mainly with static reactors over the temperature range 200°–450°. Cyclobutanone itself undergoes smooth cycloreversion to ethylene and ketene;[35,36,37] decarbonylation to cyclopropane via a diradical is a very minor process. The cycloreversion is probably a concerted $\sigma^2 a + \sigma^2 s$ process, the reverse of the $\pi^2 a + \pi^2 s$ cycloaddition of a ketene to an alkene.[1] In accord with this, Metcalfe et al.[37] have found 99% retention of stereochemistry in the formation of the 2-butenes from the 2,3-dimethylcyclobutanones.

Preferential fission of the 2–3 and 4–1 bonds of the cis compound [Eq. (8.16)] is considered[37] to be due to a conformational effect involving relief of methyl–methyl interactions in the passage of the distorted cyclobutane ring to the transition state. The corresponding ratio of products from *trans*-2,3-dimethylcyclobutanone was 52:48.

(8.16)

Homolytic α-cleavage of cyclobutanones is facilitated by α-alkylation, but the pyrolysis of 2,2,4,4-tetramethylcyclobutanone at 375° and 2–20 mm in a static system gave by this route only 6% of the decarbonylation product, 1,1,2,2-tetramethylcyclopropane; the major process was still cycloreversion to isobutene and dimethylketene.[38]

Rousseau and co-workers[39] attempted the pyrolytic synthesis of carbonyl-cyclopropane (dimethyleneketene, **2**) by flash vacuum pyrolysis of spiro [2,3] hexan-4-one (**1**), but a complex mixture of products including vinylketene, allene, and propyne was obtained [Eq. (8.17)]. Carbonylcyclopropane has been generated at 500° from a derivative of Meldrum's acid (see Chapter 5, Section II,B), but it decarbonylates to allene at higher temperatures. Thus it may be an intermediate in the pyrolysis of **1**, but the temperature is probably too high for its survival.

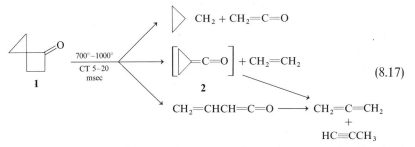

$$(8.17)$$

The cleavage of accessible cyclobutane-1,3-diones is a convenient method for the preparation of alkylketenes. Hanford and Sauer[40] describe a ketene lamp modified for the preparation of dimethylketene [Eq. (8.18)]. In the author's laboratory[41] conventional flash pyrolysis of tetramethylcyclo-butane-1,3-dione gave both dimethylketene and a product of radical decarbonylation, **3** [Eq. (8.19)].

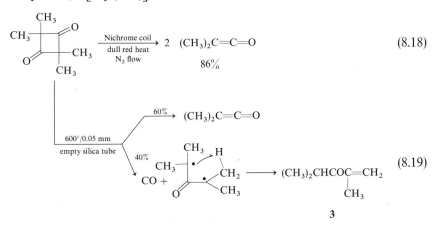

Dimethylthioketene has similarly been generated by flash vacuum pyrolysis of the corresponding 1,3-dithione, but Seybold[42] found that the reaction involves initial isomerization to a thietanethione. The thioketene was characterized by its infrared absorption measured at $-70°$; solutions could be kept for a time at $-80°$ but oligomerization occurred at room temperature.

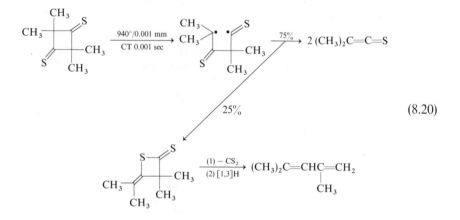

(8.20)

Thioketenes may also be prepared by pyrolysis of 1,2,3-thiadiazoles **4** and of 1,3-dithietane derivatives such as **5** and **6**.[43]

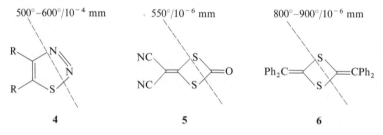

D. β-Lactones, β-Lactams, Oxetanes, Thietanes, and Dithietanes

Many simple β-lactones decompose readily to give carbon dioxide and an alkene on distillation or at quite low temperatures $(100°-250°)$; their decomposition has been reviewed by Etienne and Fischer.[44] 2-Oxetanone (β-propiolactone) itself decomposes smoothly at $200°-250°$ and 20 mm in a static gas phase system to give carbon dioxide and ethylene as the sole products of a first-order reaction.[45]

Hedaya and co-workers[46] used pyrolysis of a β-lactone ring in an approach to cyclobutadiene. Photo-α-pyrone (**7**) was subjected to flash vacuum pyrolysis with rapid quenching on a large Dewar. At 400° isomerization to α-pyrone was the major process, but at 835° some cyclobutadiene was generated [Eq. (8.21)] and was isolated as the dimer.

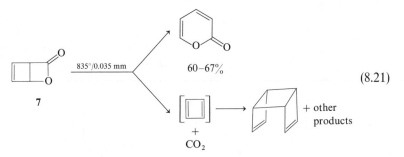

60–67%

7

+ other
products

+
CO_2

(8.21)

The pyrolysis of ketene dimers having the enol-β-lactone structure is of preparative importance. Diketene itself dissociates mainly to ketene [Eq. (8.22)] on conventional flow pyrolysis,[47] but Fitzpatrick[48] noted some fission to carbon dioxide and allene (8–18%) over a hot Nichrome filament.

$$CH_2 \quad \xrightarrow[\substack{\text{Pyrex packing} \\ \text{Vycor tube}}]{550°/N_2 \text{ flow}} \quad 2\,CH_2{=}C{=}O$$

46–55%

(8.22)

With more highly substituted dimers fission to carbon dioxide and an allene becomes the more important [Eq. (8.23)[49]] or the dominating process [Eq. (8.24)[50]].

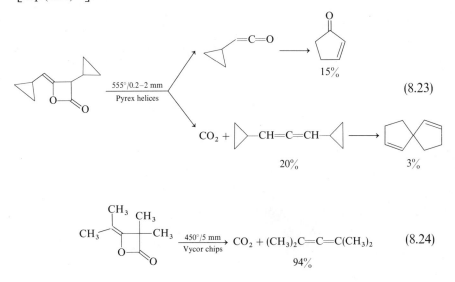

$$\xrightarrow[\text{Vycor chips}]{450°/5 \text{ mm}} CO_2 + (CH_3)_2C{=}C{=}C(CH_3)_2$$

94%

(8.24)

Volkova and co-workers[51a] have employed flash pyrolysis of azetidine at 400°–700° and 0.003 mm to generate ethylene and methanimine, CH_2=NH. The latter was characterized by mass spectrometry and low-temperature infrared spectrometry.

The pyrolysis of β-lactams has not been much used in synthesis, but Paquette *et al.*[51] have shown that simple β-lactams decompose smoothly on flow pyrolysis at 600° to give alkenes with almost complete retention of stereochemistry [Eq. (8.25)], consistent with a concerted $\sigma^2 s + \sigma^2 a$ process.

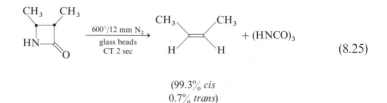

$$(8.25)$$

(99.3% *cis*
0.7% *trans*)

Brazlavsky and Heicklen[52] have recently reviewed the pyrolysis of oxetanes, which decompose to alkenes and carbonyl compounds at rates convenient for kinetic study between 400° and 500°. The balance of opinion favors a 1,4-diradical mechanism for this reaction, with only partial retention of stereochemistry in many cases [e.g. that of Eq. (8.26)[53]].

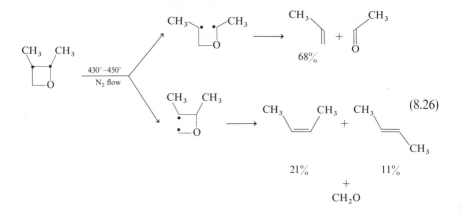

$$(8.26)$$

Jones and co-workers[54] have explored the synthetic value of the general photochemical/pyrolytic sequence of Eq. (8.27) for the production of long-chain enals related to insect pheromones. In the specific case of the bicyclic oxetane **8** pyrolysis gave *trans*-non-6-enal (**9**) in "fair yield" [Eq. (8.28)] and Rh(I)-catalyzed isomerization of **8** in solution was preferred.

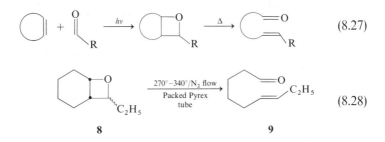

$$(8.27)$$

$$(8.28)$$

8   **9**

Although the photolysis of thietanes has been thoroughly explored[52] there appears to have been little work on their pyrolysis. Thietane 1,1-dioxides do not undergo thermal fission to sulfenes and alkenes (see Chapter 6, Section III,B). However the flash vacuum pyrolysis of thietane-1-oxide, 1,3-dithietane-1-oxide, and related compounds has been used for the generation of sulfines $R_2C{=}S{=}O$ (together with alkenes or thiocarbonyl compounds) for microwave[55] and photoelectron[56,56a] spectroscopic studies [Eqs. (8.29)–(8.31)].

$$\xrightarrow{\;>600°\;} CH_2{=}S{=}O + CH_2{=}CH_2 \qquad (8.29)$$

$$\xrightarrow{\;500°/0.025\ mm\;} CH_2{=}S{=}O + CH_2{=}S \qquad (8.30)$$

$$\xrightarrow{\;400°\;} (CH_3)_2C{=}S{=}O + (CH_3)_2C{=}S \qquad (8.31)$$

III. RETRO-DIELS–ALDER REACTIONS

A. Cyclohexene and Derivatives

Butadiene was a much less readily available laboratory starting material in the 1930s than is now the case, and much effort was devoted to refinement of its preparation by cracking of cyclohexene to butadiene and ethylene. Rice and co-workers[57] describe the preparation of butadiene with a simple flow apparatus which gave 40% conversion of cyclohexene in one pass and a 95%

yield of butadiene [Eq. (8.32)]. Rice and Murphy[58] extended the method to the preparation of butadiene, isoprene, and 2-phenylbutadiene from the appropriate 1- and 4-substituted cyclohexenes. Kistiakowsky and co-workers[59] decomposed cyclohexene vapor over a red-hot platinum wire and reported production of butadiene at 50 gm/hr.

$$\text{[structure]} \xrightarrow[\text{empty quartz tube}]{700°-800°/10-15\ mm} \text{||} + \text{[structure]} \qquad (8.32)$$

Retro-Diels–Alder reactions of cyclohexene derivatives have been used more recently for the preparation of unsaturated nitriles and esters. Ardis et al.[60] describe the preparation of vinylidenecyanide from 4,4-dicyanocyclo-hexene [Eq. (8.33)]; from the experimental directions there appears to be some loss in purification of this sensitive compound. Derivatives of butadiene-2,3-dicarboxylic acid are difficult to prepare by other than pyrol-ytic methods, and Cobb and Mahan[61] have modified methods reported in the patent literature to provide useful syntheses of the dimethyl ester and the 2,3-dinitrile [Eq. (8.34)]. In these experiments the starting materials were usually added as solutions in inert diluents of similar volatility (e.g., dimethyl phthalate, benzonitrile). Derivatives of butadiene-2,3-dicarboxylic acid may also be prepared by pyrolytic rearrangement of corresponding derivatives of cyclobutene-3,4-dicarboxylic acid (see Chapter 9, Section II,B).

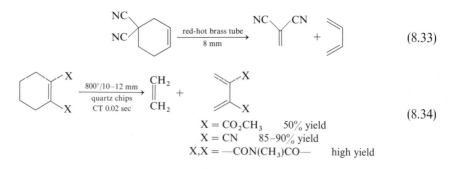

$$(8.33)$$

$$(8.34)$$

$$X = CO_2CH_3 \quad 50\%\ \text{yield}$$
$$X = CN \quad 85–90\%\ \text{yield}$$
$$X,X = -CON(CH_3)CO- \quad \text{high yield}$$

B.   Formal Adducts of Cyclopentadiene, Fulvene, and Furan

Diels–Alder retrogression of substituted norbornenes to cyclopentadiene and a substituted alkene often occurs smoothly in flow systems between 400° and 600°, and norbornenes and related systems can thus be used in synthesis as protected C=C bonds which can be unmasked on pyrolysis.

There are numerous examples in the literature,[8,9,10] and only a few typical experiments are mentioned below.

Haslouin and Rouessac[62] have prepared α-, β-, and γ-methylenebutyro-lactones in almost quantitative yield by pyrolysis of the appropriate nor-bornene spirolactones; the preparation of the γ-methylene compound is shown in Eq. (8.35). A similar approach was used by Brown and co-workers[63] for the generation of isopropylidene methylenemalonate in the gas phase as a precursor of methyleneketene [Eq. (8.36)]. Isopropylidene methylene-malonate (**11**) can be formed in solution and trapped as the adduct **10** with cyclopentadiene, but **11** cannot be isolated; methyleneketene has a half-life of about 16 sec at a pressure of 0.02 mm, and persists in the condensed phase only at −196° or below.

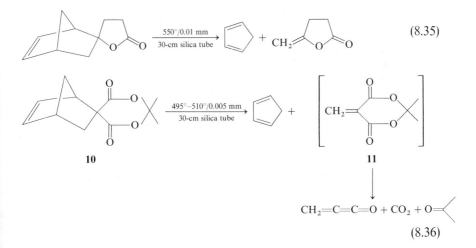

$$(8.35)$$

$$(8.36)$$

$$CH_2=C=C=O + CO_2 + O=\!\!<$$

The conditions used for the generation of some further terminal methylene compounds are shown in Table 8.1.

As an example of the method applied to the preparation of a thermo-dynamically less stable 1,2-disubstituted alkene we may note the preparation of *cis*-crotonaldehyde reported by Perrier and Rouessac[67] [Eq. (8.37)].

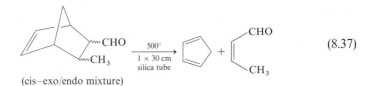

$$(8.37)$$

TABLE 8.1

Methylene Compounds from Pyrolysis of 2,2-Disubstituted Norborn-5-enes

Product, and position of attachment of former residue	Pyrolytic Conditions	Yield (%)	Reference
1. $CH_2{=}C\overset{R}{\underset{OSi(CH_3)_3}{}}$ R = allyl, n-butyl, vinyl or cyclopropyl	500°	100 for R = vinyl or cyclopropyl	64
2. $CH_2{=}\!\!\underset{O}{\bigtriangleup}$	600°/0.01 mm	70	65
3. $\underset{R'}{\overset{R}{\diagup}}\!\!\underset{O}{\overset{CH_2}{\diagup}}\!\!O$ R, R' = H, CH₃, C₆H₅, or n-C₄H₉	550°/0.01 mm	100	62
4. $\overset{CH_2}{\diagup}\underset{O}{}O$	550°/0.01 mm	100	62
5. $CH_2{=}\overset{R}{\underset{}{C}}COCH_2CO_2CH_3$ R = H or CH₃	600°/1–2 mm	80–90	66
6. $CH_2{=}\!\!\underset{S}{\diagup}$	FVP 600° microwave spectroscopy	High	66a

Highly reactive cyclic azocarbonyl compounds generated in solution by oxidation of the corresponding hydrazo compounds or hydrazides can be trapped with cyclopentadiene, and the azocarbonyl compounds can be re-generated as transient intermediates by flash vacuum pyrolysis of the adducts. In this way benzocyclobutenedione (**12**) can be obtained in high pyrolytic yield starting from phthalhydrazide[68] [Eq. (8.38)], and the indazolone adduct **13** ultimately gives benzyne and hence biphenylene.[68]

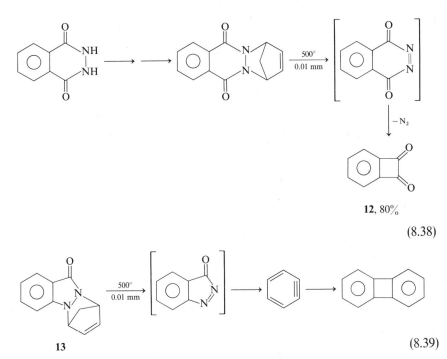

(8.38)

(8.39)

The conditions used for the pyrolytic preparation of a selection of cyclic and functionalized alkenes, some of which would be difficult to synthesize by conventional solution methods, are shown in Table 8.2.

Roth and his group[74] have provided an interesting comparison of the relative value of cyclopentadiene, 6,6-dimethylfulvene, furan, and benzene as stable formal diene fragments in the pyrolytic preparation of butatriene. 5,6-Dimethylenebicyclo[2.2.1]hept-2-ene (**14**) failed to undergo retro-Diels–Alder cleavage to cyclopentadiene and butatriene, and at 600° gave instead indan, formed by a complex sequence of rearrangements [Eq. (8.40)]. The

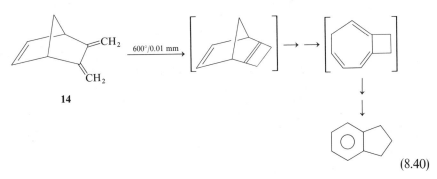

(8.40)

TABLE 8.2

Some Cyclic Alkenes Formed by Pyrolysis of Adducts with Cyclopentadiene

Product, and position of attachment of former residue	Pyrolytic conditions	Yield (%)	Reference
1.	550°/10 mm; quartz chips	28	69
2. CH$_2$ / CH$_2$	420°–460°/1 mm; CT 0.07 sec	100[a]	70
3. O / —CH$_3$ / CH$_3$	400°/N$_2$ flow; quartz chips	25	71
4. O O / O O	450°/0.1 mm; Pyrex chips	60	72
5. O / O / O	420°/10 mm	80–95	73

[a] No products other than this and cyclopentadiene were observed by nmr spectrometry.

dimethylfulvene adduct **15** decomposed in an irreproducible manner and was of no preparative interest, whereas the adducts **16** and **17** formed butatriene in good yield by cleavage to the aromatic systems furan and benzene (Table 8.3).

It is convenient at this point to mention the pyrolysis of homobasketene (**18**), which fragments quantitatively at 370° to give benzene and cyclo-

TABLE 8.3.

Pyrolytic Formation of Butatriene

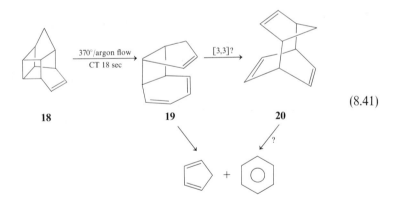

Starting material	Temperature (°C)	Yield of butatriene (%)
14 X = CH$_2$	600	0
15 X = C=C(CH$_3$)$_2$	500	1.2
16 X = O	580	ca. 80
		(+SM, 20)
17 X = —CH=CH—	620	ca. 100

pentadiene. Mauer and Grimme[75] have shown that the reaction involves initial retro-Diels–Alder opening to **19**, which can be trapped with maleic anhydride in static experiments. It is at present uncertain whether **19** then undergoes direct 2 + 2 cleavage to the products, or whether Cope rearrangement to **20** is followed by 4 + 4 cleavage.

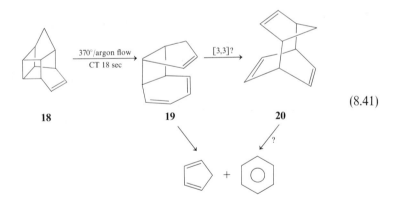

(8.41)

One of the most notable achievements of the modern flash vacuum pyrolytic method has been the generation and characterization of a simple pentalene, 1-methylpentalene, by de Mayo's group.[76] This success, which ended a long period of struggle to make pentalenes by conventional solution methods,[77] has been followed by the generation of pentalene itself in solution, its trapping with cyclopentadiene, and its regeneration by thermal dissociation of the cyclopentadiene adduct and of its dimers in the hands of Hafner's group.[78] The precursor **22** (exo/endo mixture) of 1-methyl-pentalene was prepared by base catalyzed cyclization of the ketone **21**, and the thermolysis was effected with a contact time of a few milliseconds with the flash vacuum thermolysis apparatus of King *et al.* (see Chapter 2, Fig. 2.12). 1-Methylpentalene dimerizes rapidly above $-150°$ to give **23** and minor isomers, and the monomer can be regenerated by pyrolysis at 700°. 1-Methylpentalene deposited on quartz or sodium chloride plates at $-196°$ was characterized by ultraviolet and infrared spectroscopy.

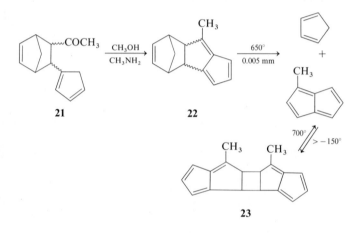

(8.42)

Attempted generation of benzocyclobutadiene by pyrolytic cleavage of cyclopentadiene from the exo or endo adducts **24** in the gas phase at 465° gave instead only the isomerization product **25**.[70] However, highly reactive species can be generated by retro-Diels–Alder reactions, as shown by the formation of diphenylbutadiyne in ca. 10% yield by pyrolysis of the phthal-azine derivative **26** at 550° and 0.05 mm. This reaction is considered by Gilchrist and co-workers[79] to involve intermediate formation of 3,6-di-phenyl-4-pyridazyne (**27**) with subsequent loss of nitrogen from **27** to form

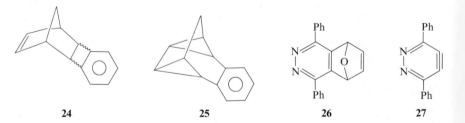

the diyne. Formation of the diyne was a minor process; the major process was isomerization of **26** to 1,4-diphenylphthalazin-5-ol (25%).

C. Adducts with 1,4-Bonded Aromatic Residues: Elimination of Benzene, Naphthalene, Anthracene, and Derivatives

Concerted 4 + 2 retrogressions usually proceed very smoothly when the formal 4π fragment which is eliminated forms part of a stable aromatic system. This principle has been employed very effectively by Vogel's group in syntheses of benzocyclopropene[80] [Eq. (8.43)] and naphthocyclopropene[81] [Eq. (8.44)] in which both products are aromatic.

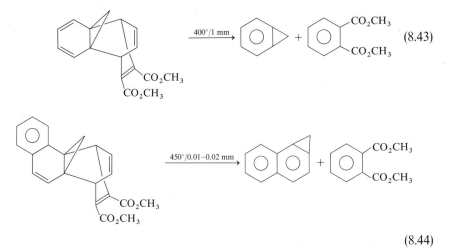

$$(8.43)$$

$$(8.44)$$

Oxirene (**25a**) is an unstable anti-aromatic molecule which at the time of writing has not been detected directly. Lewars and Morrison[82,83] attempted the generation of oxirene by retro-Diels–Alder fragmentation of its formal dimethyl phthalate and anthracene adducts **24a** and **26a** in the hope that the transition state for the formation of oxirene would be lowered by the thermodynamic stability of the incipient arene. This hope was not realized; in both cases pyrolysis through a quartz tube at 10^{-3}–10^{-4} mm gave mainly rearranged products [Eq. (8.45)] apparently formed by dipolar (or radical?) opening of the epoxide ring. Pyrolysis of **24a** in the temperature range $600°$–$800°$ also gave about 5% of ketene, but whether this was derived from rearrangement of oxirene or from isomerization to a bicyclo[2.2.2]octadienone followed by elimination of ketene is uncertain.

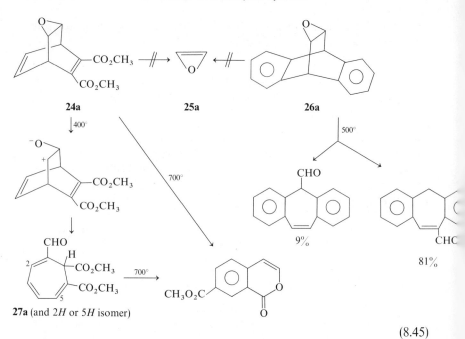

24a   **25a**   **26a**

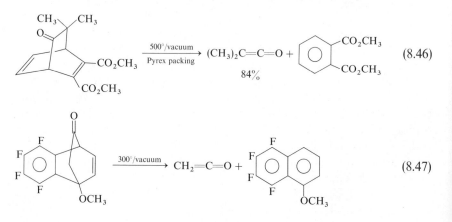

27a (and 2H or 5H isomer)

(8.45)

The formation of ketene from carbenaoxiran, generated by pyrolysis of a spiroepoxide related to **24a**, is discussed in Chapter 5, Section II,B.

Ketenes can be generated from bridged ketones of the benzene, naphthalene, and anthracene series [Eqs. (8.46),[84] (8.47),[85] and (8.48)[86]] although such reactions are more likely to be used for the preparation of the aromatic products than for the ketenes.

The pyrolysis of formal anthracene adducts is a valuable method for the preparation of very sensitive alkenes or of specifically labeled small mole-

(8.46)

(8.47)

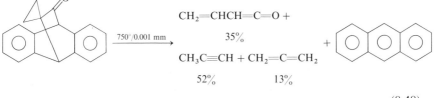

$$CH_2\text{==}CHCH\text{==}C\text{==}O +$$
$$35\%$$
$$CH_3C\text{≡}CH + CH_2\text{==}C\text{==}CH_2$$
$$52\% \qquad 13\%$$

(8.48)

cules. The first synthesis of pentatetraene, a cumulene with a half-life in 1% solution in $CDCl_3$ of 20 min at 40°, was achieved by Ripoll and Thuillin[87] by pyrolysis of an anthracene-bridged penta-1,2,4-triene [Eq. (8.49)]. Allenes and butatriene can be prepared by similar routes (see Table 8.4, entries 1, 2, and 3). Hart and Peiffer[88] found, however, that the diaza analogue $CH_3C_6H_4\text{—}N\text{==}C\text{==}C\text{==}N\text{—}C_6H_4CH_3$ could not be prepared from the corresponding anthracene-bridged bisketimine, which was stable to pyrolysis at 575° and 5–20 mm and was also resistant to photolysis.

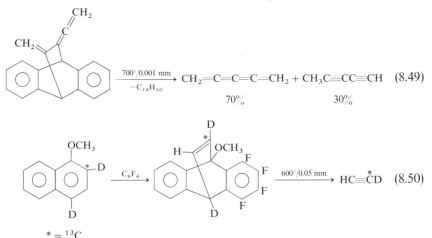

$$\xrightarrow[-C_{14}H_{10}]{700°/0.001\ mm} CH_2\text{==}C\text{==}C\text{==}C\text{==}CH_2 + CH_3C\text{≡}CC\text{≡}CH \quad (8.49)$$
$$70\% \qquad\qquad 30\%$$

$$\xrightarrow{C_6F_4} \qquad \xrightarrow{600°/0.05\ mm} HC\text{≡}\overset{*}{C}D \quad (8.50)$$

* = ^{13}C

For the preparation of doubly labeled acetylene [Eq. (8.50)] Brown and co-workers[89] exploited the work of Heaney's group[90] on the thermal

$$\xrightarrow[\substack{44\text{-cm silica tube;}\\ \text{silica wool}}]{550°\text{–}600°/0.001\ mm} C_2F_4 +$$

(8.51)

X = H, Y = CH_3, CF_3, CH_2Cl, or C_6H_5
X = Y = CH_3 or CF_3
X = CH_3 Y = CF_3

TABLE 8.4

Products of Pyrolysis of Formal Anthracene Adducts

Product or intermediate product and position of attachment of

former   residue

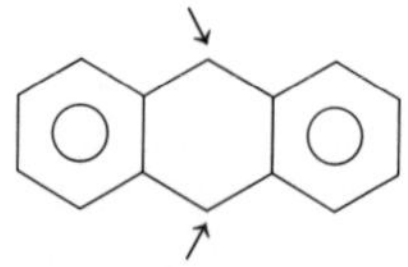

	Product or intermediate	Pyrolytic conditions	Yield (%)	Reference
1.	$R_2C{=}C{=}CHR$ $R = H$ or CH_3	$800°/10^{-3}$ mm	55–90	87
2.	$CH_2{=}C{=}C{=}CH_2$	$850°/10^{-3}$ mm	73	87
3.	$CH_2{=}C{=}C{=}CH_2$	$750°/10^{-3}$ mm	84	87
4.	$CH_2{=}C{=}C{=}C{=}CH_2$	$700°/10^{-3}$ mm	70	87
5. 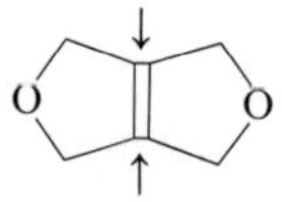		$600°/10^{-2}$ mm	100	
6. 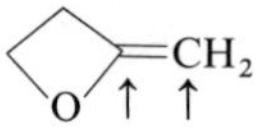	$=CH_2$	$340°–350°/N_2$ flow	10	92
7. 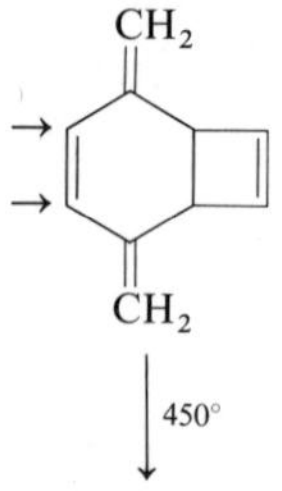	$CH_2 \dots CH_2$	$310°/0.2$ mm	19	93
	$\downarrow 450°$			
	$CH_2 \dots CH_2$	$450°/0.2$ mm	70–80	93

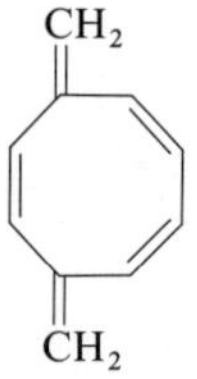

Table 8.4 (*Continued*)

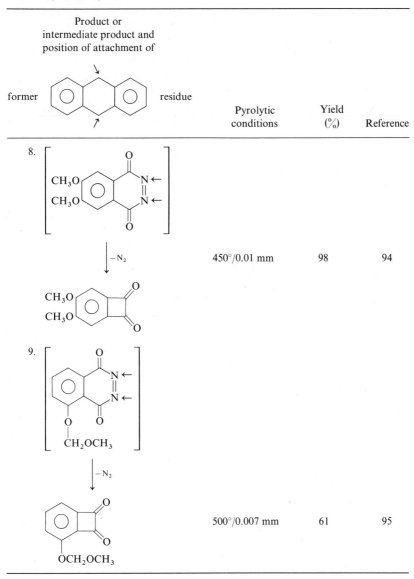

	Pyrolytic conditions	Yield (%)	Reference
8.			
	450°/0.01 mm	98	94
9.			
	500°/0.007 mm	61	95

elimination of acetylenes from benzo- and dibenzobarrelenes formed by addition of tetrafluorobenzyne to benzene and naphthalene derivatives.

A selection of compounds prepared mainly by flash vacuum pyrolysis of formal anthracene adducts is shown in Table 8.4. The generality of the cumulene synthesis is particularly noteworthy. The yield of 2-methylene-oxetane (entry 6) was low because extensive formation of rearranged adducts took place, probably in the condensed phase, during this pyrolysis at atmospheric pressure. The high yields of benzocyclobutenediones (e.g., entry 8) have encouraged the use of this method in natural product synthesis as in the case of entry 9, an intermediate required for the synthesis of naturally occurring anthraquinones and for an approach to the synthesis of the antibiotic adriamycin.

The extensive use of pyrolytic reactions in fluorocarbon chemistry has been reviewed by Plantonov and Yakobson.[96] Equation (8.51) shows the preparation of a range of tetrafluorobenzene derivatives by elimination of C_2F_4 from perfluorobicyclo[2.2.2]octadiene derivatives, reported by Anderson et al.[97]

There have been conflicting, if equally fascinating, reports of the products of pyrolysis of the heavily trifluoromethylated dihydrobenzvalene derivative **28**. Kobayashi et al.[98] reported cleavage to a cyclobutadiene and isolation of the syn dimer **29**, whereas Warrener's group[99] found in addition a less

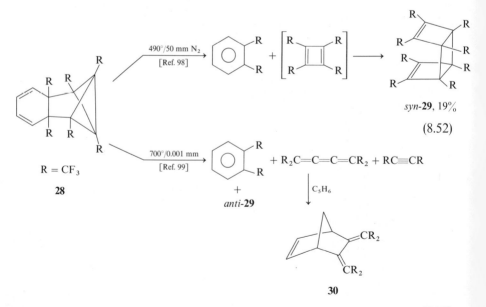

syn-**29**, 19%

(8.52)

anti-**29**

30

(8.53)

rational but well-established cleavage to tetrakis(trifluoromethyl)butatriene and some bis(trifluoromethyl)acetylene. The butatriene was characterized as its cyclopentadiene adduct **30**.

Retro-Diels–Alder reactions yielding aromatic products have also been used by Barton and Kilgour[100] [Eq. (8.54)] for the generation of tetramethyldisilene (detected by reaction with carbonyl compounds in the gas phase), and by Ashe and Friedman[101] for the introduction of substituents into arsabenzene [Eq. (8.55)].

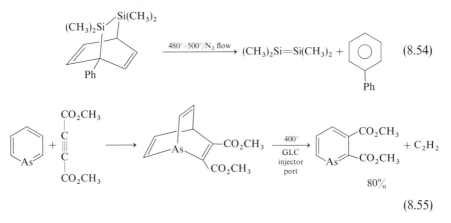

$$(8.54)$$

$$(8.55)$$

Finally, it would be a pity to omit the spectacular fragmentation of the ethylene acetal of quadricyclanone to benzene, carbon dioxide, and ethylene in almost quantitative yield, discovered by Lemal and co-workers.[102] This probably involves stepwise breaking of bonds and elimination of a cyclic dioxycarbene from an intermediate norbornadienone acetal.[103] Woodward and Hoffmann[1] have discussed orbital symmetry in such reactions considered as concerted chelotropic fragmentations.

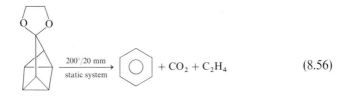

$$(8.56)$$

D.   Aromatic Compounds Which Lose 2π Components to Form *o*-Quinonoid Systems

We shall now examine a group of pyrolytic reactions of great synthetic value in which a stable benzene derivative is converted by loss of a 2π

component into a transient o-quinodimethane or an isobenzo heterocycle. The useful 2π fragments include ethylenes, formaldehyde, thioformaldehyde, formaldimine, and carbon dioxide. Much work in this area probably stems from the work of Maccoll's group on the correlation between formal retro-Diels–Alder reactions in the mass spectral fragmentation[104] and in the pyrolytic fragmentation[105] of tetralin and related heterocycles. Loudon and co-workers[105] showed that formation of o-quinodimethane and ultimately benzocyclobutene is a general major reaction of compounds 31 [Eq. (8.57)]. Isochroman (31, X = O) gives benzocyclobutene in 78% yield at 46% conversion in one pass, although many other minor products are formed in these pyrolyses at atmospheric pressure.

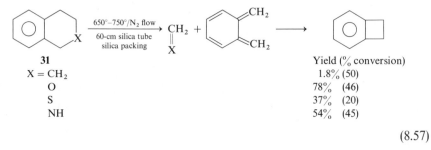

	31	Yield (% conversion)
X = CH$_2$		1.8% (50)
O		78% (46)
S		37% (20)
NH		54% (45)

$$(8.57)$$

Spangler and co-workers[106] have used 3-isochromanones and homophthalic anhydrides as starting materials for the preparation of benzocyclobutenes and benzocyclobutenones, by pyrolysis over a Nichrome filament in the apparatus shown in Chapter 2, Fig. 2.4. Benzocyclobutene itself can be prepared in high yield [Eq. (8.58)] and methoxy, o-dimethoxy, and methylenedioxy derivatives are obtained in 10–90% yield. Homophthalic anhydride gives both benzocyclobutenone and fulvenallene, a product of decarbonylation and ring contraction [Eq. (8.59)]. At higher temperatures homophthalic anhydride gives fulvenallene as the major product.[107] Dimethylhomophthalic anhydride (32) gives 2-isopropenylbenzaldehyde by a 1,5 hydrogen shift in the intermediate o-quinonoid ketene; the same intermediate and final product are obtained by decarbonylation of 3,3-dimethylindane-1,2-dione at 600° and 0.6 mm.[108]

Although many derivatives of isobenzofuran and isoindole are quite stable, the parent compounds 33 and 34 are sensitive to oxygen and require careful handling to avoid decomposition. Isobenzofuran was first prepared

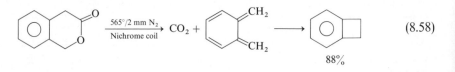

$$(8.58)$$

88%

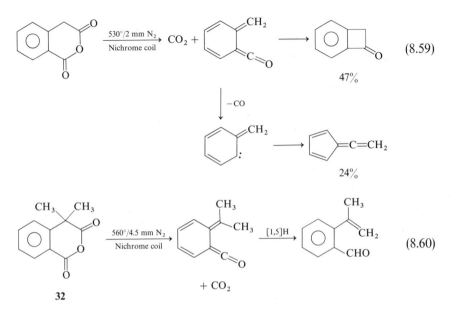

(8.59)

47%

24%

(8.60)

32

in a pure state by a mild thermal retrogression[109] and isoindole was first characterized spectroscopically at low temperatures in the pyrolyzate from a carbonate ester (see Chapter 3, Section IV,A). However both compounds and many derivatives are now freely available by the pyrolytic elimination of ethylene from bridged systems. Wiersum and Mijs[110] first prepared isobenzofuran in this way [Eq. (8.61)], and this was followed by the similar preparation of isoindole [Eq. (8.62)] due to Bornstein *et al.*[111]

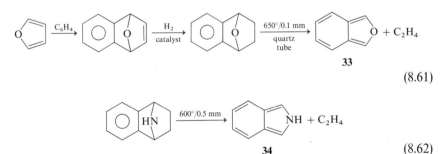

33

(8.61)

34

(8.62)

Substituted isobenzofurans,[112] furo[3,4-*c*]pyridine (**35**),[113] and poly-chlorinated and fluorinated isobenzofurans and isoindoles[114] have been prepared in the same way. A notable failure of the method has been found in the case of isobenzofulvene (**36**), which has been generated in solution[115] by elimination of hydrogen bromide from the mixture of bromides (**37**), but

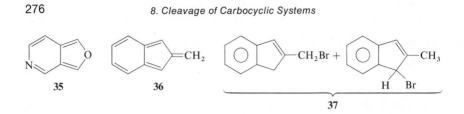

35   **36**   **37**

which cannot be prepared by pyrolytic elimination of ethylene from a
bridged system. Heaney and co-workers[114] found that in both the benzo
and the tetrafluorobenzo series [Eq. (8.63)] the major product **39** was produced by alternative recyclization of the diradical **38**; only a small proportion
lost ethylene to give ultimately the isopropenylindene **40** rather than an

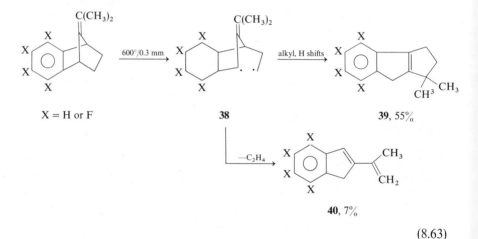

X = H or F   **38**   **39**, 55%

40, 7%

(8.63)

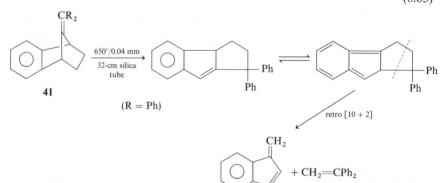

41   **42**

(R = Ph)

(8.64)

isobenzofulvene. Warrener's group[116] studied the pyrolysis of the bridged compounds **41** ($R = H$, CH_3, or C_6H_5) and found that the same 1,3 alkyl shift occurred initially, but that the diphenyl intermediate then underwent particularly smooth elimination of diphenylethylene to give benzofulvene (**42**), for which this reaction provides a useful preparative method.

IV. MISCELLANEOUS FRAGMENTATIONS

The highly reactive monomer *p*-xylylene (**43**) can be generated by pyrolytic dehydrogenation of *p*-xylene (see Chapter 4, Section II,A). *p*-Xylylene is of some industrial importance because the linear polymer poly-*p*-xylylene (Parylene) can be deposited directly from a pyrolyzate gas stream onto surfaces to form a tough protective coating.[117] The monomer is stable in dilute solution in hexane at $-190°$, and at $-78°$ has a half-life of 21 hr.[118] A very clean source of *p*-xylylene is the pyrolysis of [2,2]paracyclophane[117] [Eq. (8.65)], which has been used to generate **43** for static measurement of infrared,[119] ultraviolet,[119] and pmr spectra[120] at low temperatures, and for measurement of the He(II) photoelectron spectrum "on the fly."[121]

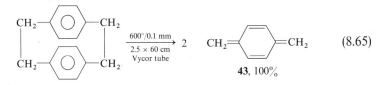

$$\text{(8.65)}$$

43, 100%

p-Xylylene is also formed in the remarkable $2 + 2 + 2$ fragmentation of a bridged Dewar benzene [Eq. (8.66)] investigated by Landheer and co-workers.[122]

$$\xrightarrow{\text{300°/0.5 mm N}_2} \quad CH_2=\!\!\!\bigcirc\!\!\!=CH_2 + C_2H_4 \qquad \text{(8.66)}$$

The modern literature of alicyclic chemistry abounds with examples of the fragmentation of unsaturated polycyclic systems, often proceeding by long sequences of rearrangements before fission. These examples often add little to our knowledge of primary pyrolytic processes, and only a single example, the tricyclotetradecatriene **44** studied by Gilbert and Walsh,[123] is shown in Eq. (8.67).

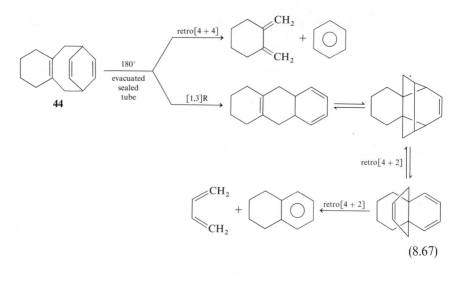

$$(8.67)$$

REFERENCES

1. Woodward, R. B., and Hoffmann, R. (1970). "The Conservation of Orbital Symmetry." Academic Press, New York.
2. Frey, H. M. (1966). *In* "Advances in Physical Organic Chemistry" (V. Gold, ed.), Vol. 4, p. 147. Academic Press, New York.
3. Frey, H. M., and Walsh, R. (1969). *Chem. Rev.* **69**, 103.
4. Willcott, M. R., Cargill, R. L., and Sears, A. B. (1972). *In* "Progress in Physical Organic Chemistry" (A. Streitwieser and R. W. Taft, eds.), Vol. 9, p. 25. Wiley (Interscience), New York.
5. Seebach, D. (1971). *In* "Methoden der Organischen Chemie (Houben-Weyl). Band IV/4. Isocyclische Vierring-Verbindungen" (E. Müller, ed.), p. 436. Thieme, Stuttgart.
6. Brown, J. M. (1973). *In* "International Review of Science." Organic Chemistry, Series One, Vol. 5, "Alicyclic Compounds" (W. Parker, ed.), p. 159. Butterworth, London.
7. Becker, D., and Brodsky, N. C. (1976). *In* "International Review of Science," Series Two, Vol. 5, "Alicyclic Compounds" (D. Ginsburg, ed.), p. 197. Butterworth, London.
8. Smith, G. G., and Kelly, F. W. (1971). *In* "Progress in Physical Organic Chemistry" (A. Streitwieser and R. W. Taft, eds.), Vol. 8, p. 75. Wiley (Interscience), New York.
9. Kwart, H., and King, K. (1968). *Chem. Rev.* **68**, 415.
10. Wollweber, H. (1972). "Diels-Alder Reaktion." Thieme, Stuttgart.
10a. Ripoll, J. L., Rouessac, A., and Rouessac, F. (1978). *Tetrahedron* **34**, 19.
11. Baldwin, J. E., and Ford, P. W. (1969). *J. Am. Chem. Soc.* **91**, 7192.
12. Paquette, L. A., and Thompson, G. L. (1972). *J. Am. Chem. Soc.* **94**, 7127.
13. Martin, H.-D., and Eisenmann, E. (1975). *Tetrahedron Lett.*, p. 661.
14. Drysdale, J. J., Stevenson, H. B., and Sharkey, W. H. (1959). *J. Am. Chem. Soc.* **81**, 4908.
15. Stevenson, H. B., and Sharkey, W. H. (1973). "Organic Syntheses" (H. E. Baumgarten, ed.), Coll. Vol. V, p. 734. Wiley, New York.
16. Sully, B. D. (1964). *Chem. and Ind. (London)*, p. 263.
17. Banthorpe, D. V., and Whittaker, D. (1966). *Q. Rev. Chem. Soc.* **20**, 373.

18. Pines, H., and Hoffman, N. E. (1954). *J. Am. Chem. Soc.* **76**, 4417.
19. Pines, H., Hoffman, N. E., and Ipatieff, V. N. (1954). *J. Am. Chem. Soc.* **76**, 4412.
20. Coxon, J. M., Garland, R. P., and Hartshorn, M. P. (1972). *Aust. J. Chem.* **25**, 353.
21. Coxon, J. M., Garland, R. P., and Hartshorn, M. P. (1972). *Aust. J. Chem.* **25**, 947.
22. Coxon, J. M., Garland, R. P., and Hartshorn, M. P. (1971). *Aust. J. Chem.* **24**, 1481.
23. Goldblatt, L. A., and Palkin, S. (1941). *J. Am. Chem. Soc.* **63**, 3517.
24. Hawkins, J. E., and Burris, W. A. (1959). *J. Org. Chem.* **24**, 1507.
25. Coxon, J. M., Garland, R. P., and Hartshorn, M. P. (1972). *Aust. J. Chem.* **25**, 2409.
26. von Schulte-Elte, K. H., Gadola, M., and Ohloff, G. (1971). *Helv. Chim. Acta* **54**, 1813.
27. Mayer, C. F., and Crandall, J. K. (1970). *J. Org. Chem.* **35**, 2688.
28. Büchi, G., and Goldman, I. M. (1957). *J. Am. Chem. Soc.* **79**, 4741.
29. Nametkin, N. S., Vdorin, V. M., Gusel'nikov, L. E., and Za'yalov, V. I. (1966). *Izvest. Akad. Nauk SSSR. Ser. khim.*, p. 584.
29a. Nametkin, N. S., Vdorin, V. M., Gusel'nikov, L. E., and Za'yalov, V. I. (1966). *Bull. Acad. Sci. USSR Divn. Chem. Sci.*, p. 563.
30. Barton, T. J.,and McIntosh, C. L. (1972). *J. Chem. Soc., Chem. Commun.*, p. 861
31. Golino, C. M., Bush, R. D., and Sommer, L. H. (1975). *J. Am. Chem. Soc.* **97**, 7371.
32. Golino, C. M., Bush, R. D., On, P., and Sommer, L. H. (1975). *J. Am. Chem. Soc.* **97**, 1957.
33. Sommer, L. H., and McLick, J. (1975). *J. Organomet. Chem.* **101**. 171.
34. Valkovich, P. B., Ito, T. I., and Weber, W. P. (1974). *J. Org. Chem.* **39**, 3543.
35. Das, M. N., Kern, F., Coyle, T. D., and Walters, W. D. (1954). *J. Am. Chem. Soc.* **76**, 6271.
36. Blades, A. T. (1969). *Can. J. Chem.* **47**, 615.
37. Metcalfe, J., Carless, H. A. J., and Lee, E. K. C. (1972). *J. Am. Chem. Soc.* **94**, 7235.
38. Frey, H. M., and Hopf, H. (1973). *J. Chem. Soc. Perkin Trans. 2*, p. 2016.
39. Rousseau, G., Bloch, R., Le Perchec, P., and Conia, J. M. (1973). *J. Chem. Soc. Chem. Commun.*, p. 795.
40. Hanford, W. E., and Sauer, J. C. (1946). *In* "Organic Reactions" (R. Adams, ed.), Vol. 3, p. 108. Wiley, New York.
41. Baxter, G. J. (1977). Ph.D. Thesis, Monash University.
42. Seybold, G. (1974). *Tetrahedron Lett.*, p. 555.
43. Seybold, G., and Heibl, C. (1977). *Chem. Ber.* **110**, 1225.
44. Etienne, Y., and Fischer, N. (1964). *In* "Heterocyclic Compounds with Three- and Four-Membered Rings" (A. Weissberger, ed.), Part 2, p. 729. Wiley (Interscience), New York.
45. Jones, T. L., and Wellington, C. A. (1969). *J. Am. Chem. Soc.* **91**, 7743.
46. Hedaya, E., Miller, R. D., McNeil, M. D., D'Angelo, P. F., and Schissel, P. (1969). *J. Am. Chem. Soc.* **91**, 1875.
47. Andreades, S., and Carlson, H. D. (1973). *In* "Organic Syntheses" (H. E. Baumgarten, ed.), Coll. Vol. V, p. 679. Wiley, New York.
48. Fitzpatrick, J. T. (1947). *J. Am. Chem. Soc.* **69**, 2236.
49. Berkowitz, W. F., and Ozorio, A. A. (1975). *J. Org. Chem.* **40**, 527.
50. Martin, J. C. (1964). U.S. Patent No. 3,131,234; *Chem. Abstr.* **61**, 2969.
51. Paquette, L. A., Wyvratt, M. J., and Allen, G. R. (1970). *J. Am. Chem. Soc.* **92**, 1763.
51a. Volkova, V. V., Gusel'nikov, L. E., Perchenko, V. N., Zaikin, V. G., Eremina, E. I., and Nametkin, N. S. (1978). *Tetrahedron Lett.*, p. 577.
52. Braslavsky, S., and Heicklen, J. (1977). *Chem. Rev.* **77**, 473.
53. Carless, H. A. J. (1974). *Tetrahedron Lett.*, p. 3425.
54. Jones, G., Acquadro, M. A., and Carmody, M. A. (1975). *J. Chem. Soc. Chem. Commun.*, p. 206.
55. Block, E., Penn, R. E., Olsen, R. J., and Sherwin, P. F. (1976). *J. Am. Chem. Soc.* **98**, 1264.
56. Bock, H., Solouki, B., Mohmand, S., Block, E., and Revelle, L. K. (1977). *J. Chem. Soc. Chem. Commun.*, p. 287.

56a. Block, E., Bock, H., Mohmand, S., Rosmus, P., and Solouki, B. (1976). *Angew. Chem. Int. Ed. Engl.* **15**, 383.

57. Rice, F. O., Ruoff, P. M., and Rodowskas, E. L. (1938). *J. Am. Chem. Soc.* **60**, 955.

58. Rice, F. O., and Murphy, M. T. (1944). *J. Am. Chem. Soc.* **66**, 765.

59. Kistiakowsky, G. B., Ruhoff, J. R., Smith, H. A., and Vaughan, W. E. (1936). *J. Am. Chem. Soc.* **58**, 146.

60. Ardis, A. E., Averill, S. J., Gilbert, H., Muller, F. F., Schmidt, R. F., Stewart, F. D., Trumbull, H. L. (1950). *J. Am. Chem. Soc.* **72**, 3127.

61. Cobb, R. L., and Mahan, J. E. (1977). *J. Org. Chem.* **42**, 2829.

62. Haslouin, J., and Rouessac, F. (1976). *Tetrahedron Lett.*, p. 4651.

63. Brown, R. F. C., Eastwood, F. W., and McMullen, G. L. (1977). *Aust. J. Chem.* **30**, 179.

64. Haslouin, J., and Rouessac, F. (1976). *Bull. Soc. Chim. Fr.*, p. 1122.

65. Haslouin, J., and Rouessac, F. (1973). *C. R. Acad. Sci., Ser. C.* **276**, 1691.

66. Stork, G., and Guthikonda, R. N. (1972). *Tetrahedron Lett.*, p. 2755.

66a. Block, E., Penn, R. E., Ennis, M. D., Owens, T. A., and Yu, S-L. (1978). *J. Am. Chem. Soc.* **100**, 7436.

67. Perrier, M., and Rouessac, F. (1977). *Nouv. J. Chim.* **1**, 367.

68. Forster, D. L., Gilchrist, T. L., Rees, C. W., and Stanton, E. (1971). *J. Chem. Soc. Chem. Commun.*, p. 695.

69. Katz, T. J., Carnahan, J. C., and Boecke, R. (1967). *J. Org. Chem.* **32**, 1301.

70. Martin, H.-D., Kagabu, S., and Schiwek, H. J. (1975). *Tetrahedron Lett.*, p. 3311.

71. De Selms, R. C., and Delay, F. (1973). *J. Am. Chem. Soc.* **95**, 274.

72. Kasai, M., Funamizu, M., Oda, M., and Kitahara, Y. (1977). *J. Chem. Soc. Perkin Trans. 1*, p. 1660.

73. Alder, K., Flock, F. H., and Beumling, H. (1960). *Chem. Ber* **93**, 1896

74. Roth, W. R., Humbert, H., Wegener, G., Erker, G., and Exner, H.-D. (1975). *Chem. Ber.* **108**, 1655.

75. Mauer, W., and Grimme, W. (1976). *Tetrahedron Lett.*, p. 1835.

76. Bloch, R., Marty, R. A., and de Mayo, P. (1972). *Bull. Soc. Chim. Fr.*, p. 2031.

77. Bergmann, E. D. (1959). *In* "Non-Benzenoid Aromatic Compounds" (D. Ginsburg, ed.), p. 141. Wiley (Interscience), New York.

78. Donges, R., Hafner, K., and Lindner, H. J. (1976). *Tetrahedron Lett.*, p. 1345.

79. Gilchrist, T. L., Gymer, G. E., and Rees, C. W. (1975). *J. Chem. Soc. Perkin Trans. 1*, p. 1747.

80. Vogel, E., Grimme, W., and Korte, S. (1965). *Tetrahedron Lett.*, p. 3625.

81. Tanimoto, S., Schafer, R., Ippen, J., and Vogel, E. (1976). *Angew. Chem. Int. Ed. Engl.* **15**, 613.

82. Lewars, E. G., and Morrison, G. (1977). *Can. J. Chem.* **55**, 966.

83. Lewars, E. G., and Morrison, G. (1977). *Can. J. Chem.* **55**, 975.

84. Alder, K., Flock, F. H., and Lessenich, H. (1957). *Chem. Ber.* **90**, 1709.

85. Buxton, P. C., Hales, N. J., Hankinson, B., Heaney, H., Ley, S. V., and Sharma, R. P. (1974). *J. Chem. Soc. Perkin Trans. 1*, p. 2681.

86. Ripoll, J. L. (1977). *Tetrahedron* **33**, 389.

87. Ripoll, J. L., and Thuillier, (1977). *Tetrahedron* **33**, 1333.

88. Hart, H., and Peiffer, R. W. (1974). *J. Chem. Soc. Chem. Commun.*, p. 126.

89. Brown, R. F. C., Eastwood, F. W., and Jackman, G. P. (1978). *Aust. J. Chem.* **31**, 579.

90. Brewer, J. P. N., Eckhard, I. F., Heaney, H., and Marples, B. A. (1968). *J. Chem. Soc. C*, p. 664.

91. Ripoll, J. L. (1974). *Tetrahedron Lett.*, p. 1665.

92. Hudrlik, P. F., Hudrlik, A. M., and Wan, C.-N. (1975). *J. Org. Chem.* **40**, 1116.

93. Oda, M., Fukaza, N., and Kitahara, Y. (1977). *Tetrahedron Lett.*, p. 3277.
94. McOmie, J. F. W., and Perry, D. H. (1973). *J. Chem. Soc. Chem. Commun.*, p. 248.
95. Jung, M. E., and Lowe, J. A. (1977). *J. Org. Chem.* **42**, 2371.
96. Plantonov, V. B., and Yakobson, G. G. (1976). *Synthesis*, p. 374.
97. Anderson, L. P., Feast, W. J., and Musgrave, W. K. R. (1969). *J. Chem. Soc. C*, p. 211.
98. Kobayashi, Y., Kumadaki, I., Ohsawa, A., Hanzawa, Y., and Honda, M. (1975). *Tetrahedron Lett.*, p. 3819.
99. Warrener, R. N., Nunn, E. E., and Paddon-Row, M. N. (1976). *Tetrahedron Lett.*, p. 2639.
100. Barton, T. J., and Kilgour, J. A. (1974). *J. Am. Chem. Soc.* **96**, 2278.
101. Ashe, A. J., and Friedman, H. S. (1977). *Tetrahedron Lett.*, p. 1283.
102. Lemal, D. M., Lovald, R. A., and Harrington, R. W. (1965). *Tetrahedron Lett.*, p. 2779.
103. Lemal, D. M., Gosselink, E. P., and McGregor, S. D. (1966). *J. Am. Chem. Soc.* **88**, 582.
104. Loudon, A. G., Maccoll, A., and Wong, S. K. (1970). *J. Chem. Soc. B*, p. 1727.
105. Loudon, A. G., Maccoll, A., and Wong, S. K. (1970). *J. Chem. Soc. B*, p. 1733.
106. Spangler, R. J., Beckmann, B. G., and Kim, J. H. (1977). *J. Org. Chem.* **42**, 2989.
107. Wiersum, U. E., and Nieuwenhuis, T. (1973). *Tetrahedron Lett.*, p. 2581.
108. Brown, R. F. C., and Butcher, M. (1969). *Aust. J. Chem.* **22**, 1457.
109. Warrener, R. N. (1971). *J. Am. Chem. Soc.* **93**, 2346.
110. Wiersum, U. E., and Mijs, W. J. (1972). *J. Chem. Soc. Chem. Commun.*, p. 347.
111. Bornstein, J., Remy, D. E., and Shields, J. E. (1972). *J. Chem. Soc. Chem. Commun.*, p. 1149.
112. Chacko, E., Sardella, D. J., and Bornstein, J. (1976). *Tetrahedron Lett.*, p. 2507.
113. Wiersum, U. E., Eldred, C. D., Vrijhof, P., and van der Plas, H. C. (1977). *Tetrahedron Lett.*, p. 1741.
114. Heaney, H., Ley, S. V., Price, A. P., and Sharma, R. P. (1972). *Tetrahedron Lett.*, p. 3067.
115. Warrener, R. N., Collin, G. J., Hutchison, G. I., and Paddow-Row, M. N. (1976). *J. Chem. Soc. Chem. Commun.*, p. 373.
116. Warrener, R. N., Gell, K. I., and Paddon-Row, M. N. (1977). *Tetrahedron Lett.*, p. 53.
117. Gorham, W. F. (1966). *J. Polym. Sci. Part A-1* **4**, 3027.
118. Errede, L. A., and Landrum, B. F. (1957). *J. Am. Chem. Soc.* **79**, 4952.
119. Pearson, J. M., Six, H. A., Williams, D. J., and Levy, M. (1971). *J. Am. Chem. Soc.* **93**, 5034.
120. Williams, D. J., Pearson, J. M., and Levy, M. (1970). *J. Am. Chem. Soc.* **92**, 1436.
121. Koenig, T., Wielesek, R., Snell, W., and Balle, T. (1975). *J. Am. Chem. Soc.* **97**, 3225.
122. Landheer, I. J., de Wolf, W. H., and Bickelhaupt, F. (1974). *Tetrahedron Lett.*, p. 2813.
123. Gilbert, A., and Walsh, R. (1976). *J. Am. Chem. Soc.* **98**, 1606.

Chapter 9

Rearrangements without Fragmentation

I. INTRODUCTION

The wide scope of this chapter leads to some difficulty in the selection and classification of examples of high-temperature rearrangements; the pyrolysis of quite simple molecules may involve several steps, each of which would place the example in a different category. Broadly, we begin with electrocyclic reactions and cycloadditions involving mainly four or six electrons, then move to reactions which are considered to proceed through diradical intermediates, and to sigmatropic rearrangements, and finally we consider some isomerizations of heterocyclic rings. In this general account there is little space for critical mechanistic discussion of so much material, but in most sections references are given to specialized reviews. There is a bias toward the selection of simple molecules as examples, and for this reason some most interesting work, such as that of Paquette's group on thermal rearrangements and degenerate rearrangements of complex alicyclic hydrocarbons, is seriously underrepresented.

II. REARRANGEMENTS WITH CONCERTED FOUR- OR SIX-ELECTRON STEPS

A. Azirines and Cyclopropenes

Pyrolysis of 3-phenyl-1-azirines commonly leads to the formation of substituted vinylnitrene intermediates by carbon-nitrogen bond fission. β-Styrylnitrene itself rearranges to a 1:1 mixture of phenylacetonitrile and indole, a reaction first observed by Isomura and co-workers[1] who studied the decomposition of 3-phenyl-1-azirine in boiling hexadecane [Eq. (9.1)]. In the absence of a hydrogen atom adjacent to the nitrene function indoles may be formed in high yield[2] [Eq. (9.2)].

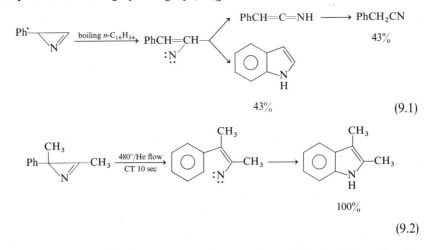

$$(9.1)$$

$$(9.2)$$

The mode of fission of the azirine ring may be completely changed by the presence of 2-dimethylamino or 2-phenyl substituents so that carbon-carbon bond fission now occurs. Demoulin and co-workers[3] found that 2-dimethyl-amino-3,3-dialkylazirines undergo smooth thermal isomerization to syn-thetically valable azadienes on flow pyrolysis at 340°–400° [Eq. (9.3)].

$$(9.3)$$

These reactions probably form dimethylaminocarbene intermediates such as **1** with subsequent 1,4 migration of hydrogen to form the azadiene. Similar processes were first proposed by Wendling and Bergman[4] to explain the formation of styrene, via an azadiene and an azetine, on pyrolysis of 3-methyl-2-phenyl-1-azirine. Wendling and Bergman[4] have further investi-gated the pyrolysis of a number of 2- and 3-phenyl-1-azirines. 3,3-Dimethyl-2-phenyl-1-azirine (**2**) at 472° forms a mixture of products [Eq. (9.4)] through electrocyclic ring closure of the intermediate azadiene **3** to an azetine **4** which can undergo fragmentation to styrene and acetonitrile; **2** also frag-ments directly to benzonitrile. At a higher temperature an alternative minor pathway is six-electron cyclization of the azadiene **3** to a dihydroisoquinoline **5** [Eq. (9.5)].

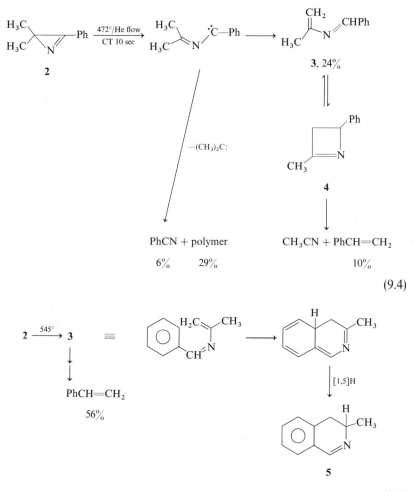

(9.4)

(9.5)

The kinetics of the pyrolysis of simple cyclopropenes have been studied mainly in static reactors, and mechanistic discussion[5,6] has centered on the intermediacy of diradicals or vinylcarbenes, or the possibility of concerted mechanisms, in their ring opening. 3-Methylcyclopropene (6) undergoes ring opening at 470°–500° to give 1-butyne as the major and butadiene as the minor product [Eq. (9.6)]. In this series Srinavasan[5] has noted a change in activation energy in passing from two methyl substituents to three or four, with a corresponding suppression of alkyne formation in favor of diene formation, and has suggested that in these cases formation of the major diene may be concerted as in 7. York et al.[6] have satisfactorily accounted for the racemization and the ring opening of optically active 1,3-diethyl-cyclopropene (**8**) in terms of the vinylcarbene intermediates **9** and **10**. Clearly much of the mechanistic discussion in these papers is relevant to the ring opening of 1-azirines as well.

Thermal reactions of small-ring hydrocarbons including cyclopropenes have been reviewed by Frey,[7] Brown,[8] Becker and Brodsky,[9] and by Closs.[10]

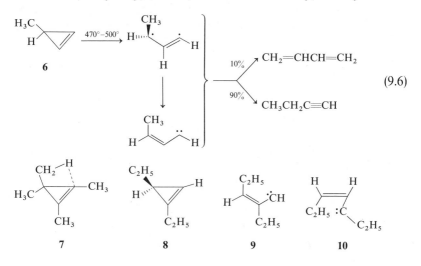

$$CH_2=CHCH=CH_2 \quad 10\%$$
$$CH_3CH_2C{\equiv}CH \quad 90\%$$

(9.6)

B. Ring Opening of Cyclobutenes and Four-Electron Ring Closures

The preparative value of flash vacuum pyrolysis in the ring opening of cyclobutenes is illustrated by the work of Belluš and Weis[11] on derivatives **11** and **12** of cyclobutene-1,2-dicarboxylic acid. Heating of such compounds in solution either does not effect ring opening or leads to dimeric compounds, but the corresponding butadiene derivatives can be obtained in high yield by gas-phase pyrolysis of the esters or nitriles [Eq. (9.7)]. The method has been extended by Dowd and Kang[12] to the preparation of butadiene-2,3-dicarboxylic acid itself by pyrolysis of the crude bistrimethylsilyl ester **13** at 420° and 10^{-5} mm and hydrolysis of the product ester with wet chloroform.

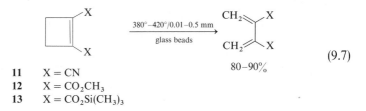

(9.7)

Many examples of the opening of cyclobutene rings in gas-phase pyrolysis involve cyclobutenes fused to further rings; in these systems the otherwise preferred conrotatory mode of opening may be disfavored by the ring-size of the product, or the initial stereochemical outcome may be changed through secondary isomerization. Willner and Rabinovitz[13] found that pyrolysis of 2,3:5,6-dibenzobicyclo[5.2.0]non-8-ene (**14**) at 290° did give the primary product of conrotation, the *cis,trans*-nonatetraene **15**, but this was accompained by the cis,cis compound **16** formed by secondary isomerization. At 400° the sole product was the trans-fused bicyclononene **17**, the product to be expected of conrotatory cyclization of **16** [Eq. (9.8)].

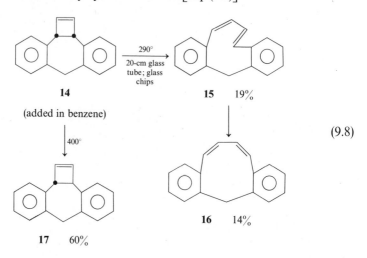

$$(9.8)$$

Paquette and Stowell[14] have achieved selective ring opening of one cyclo-butene ring in both the exo,exo and the exo,endo isomers of a tetracyclo-dodecatriene [Eq. (9.9)]; they suggest that disrotatory opening of the endo ring of **18** may be assisted by interaction of the developing sp^2 orbitals with the π lobes of the ethylene bridge. A further possibility is that both cyclo-butene rings open, but that the thermodynamically stable product **19** is reformed.[14]

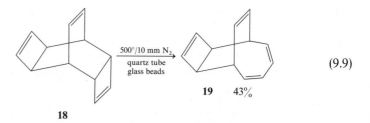

$$(9.9)$$

In Eq. (9.10)–(9.13) the initial and final products of ring opening of a selection of fused cyclobutenes are shown without further comment.

To close this section we consider two unusual electrocyclic ring closures. The first of these is the isomerization of hexakis(pentafluoroethyl)benzene

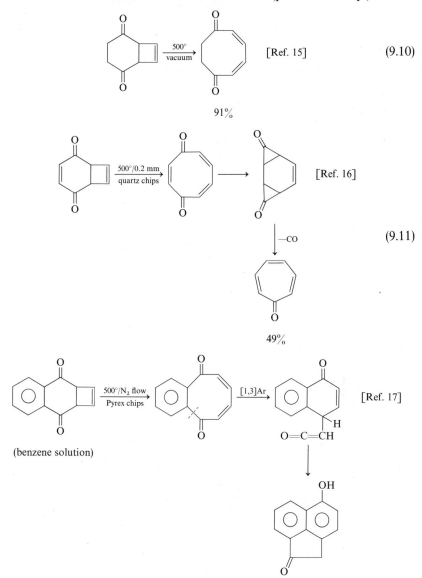

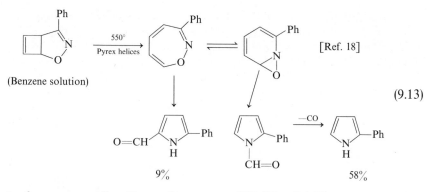

(Benzene solution)

[Ref. 18]

(9.13)

9%   58%

to the corresponding Dewar benzene at 400° [Eq. (9.14)], a process which Clifton *et al.*[19] suggest is favored by relief of the restriction of rotation of the —CF_2CF_3 groups in passing from the severely crowded benzene derivative to the Dewar benzene at 400°. At 140° the balance between this steric effect and the aromatic stabilization of the benzene favors isomerization of the Dewar benzene back to the benzene.

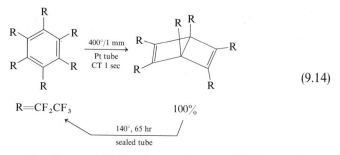

(9.14)

The second example, discovered by Wieser and Berndt,[20] is the cyclization of the nitrosoalkene **20** to a 4*H*-1,2-oxazete **21** at 220°, a temperature only slightly below that at which the oxazete fragments to di-*t*-butylketone and hydrogen cyanide.

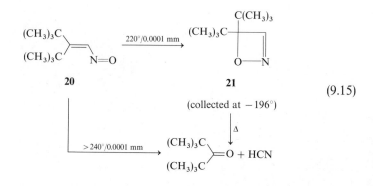

(9.15)

C. Electrocyclic Reactions of Six-Electron Systems

The formation of dihydroaromatic or aromatic systems by concerted cyclization of formal 6π intermediates is quite common at high temperatures, and we have already seen one example in the cyclization of a phenylazadiene to a dihydroisoquinoline [Section II,A, Eq. (9.5)]. Another common process, which has been encountered in other chapters as well, is electrocyclic ring-contraction of a larger ring to a smaller, often six-membered, ring with subsequent rearrangement or fragmentation. An example of the last process is the pyrolytic conversion of 2-hydroxycycloocta-2,3,5-trienone (**22**) to phenol [Eq. (9.16)] reported by Kitahara and co-workers.[21]

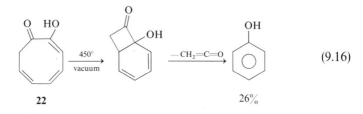

$$(9.16)$$

22   26%

The synthesis of dihydrobenzene and benzene derivatives from acyclic intermediates is illustrated by the extensive work of Schiess and co-workers[22,23,24] on the cyclization of heptatrienals **25** generated by pyrolysis of either bicyclo[3.2.0]hept-2-en-7-ones (**23**) or 8-oxabicyclo[5.1.0]octa-2,4-dienes (**24**) [Eq. (9.17)].

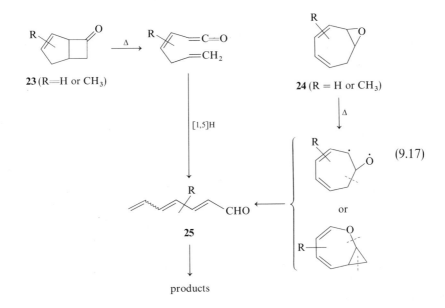

$$(9.17)$$

The single example of the formation of dihydrobenzaldehydes and benzal-dehyde from the unsubstituted bicycloheptenone **23** is shown in Eq. (9.18). The primary product of cyclization, **26**, was not detected in this experiment, although it was formed in 13% yield on pyrolysis of the epoxide **24** at 315°.

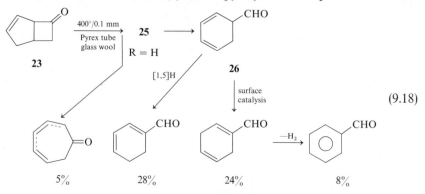

$$ (9.18) $$

Similar electrocyclic reactions can be used to convert 1-arylbutadienes into dihydronaphthalenes and naphthalenes, or related heterocyclic systems. Yoshida et al.[25] generated the arylbutadienes in the gas phase from the butenyl acetates **27** and obtained dihydrobicyclic compounds **28** and fully aromatized compounds **29** in 20–56 and 4–80% yields, respectively [Eq. (9.19)]. 1,2-Dihydronaphthalene (30%) and naphthalene (60%) were obtained from 1-phenylbutadiene at 620°.

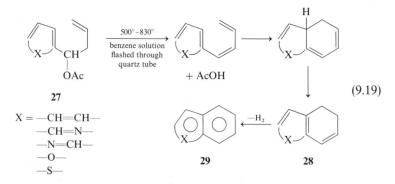

$$ (9.19) $$

This approach has been refined and extended by Weber's group[26-29] which has used preformed arylbutadienes and has found the best conditions for preparation of the dihydro products with little or no secondary aromatization. Thus pyrolysis of *trans*-1-phenylbutadiene in a stream of nitrogen at 450° gave 30% of starting material, 8% of its cis isomer, 57% of 1,2-dihydronaphthalene, and 4% of naphthalene. The course of this cyclization has been confirmed[26] by an experiment with the 2′,4′,6′-trideutero derivative **30** which gave the 2,5,7-trideutero compound **31** by a 1,5 shift of deuterium [Eq. (9.20)].

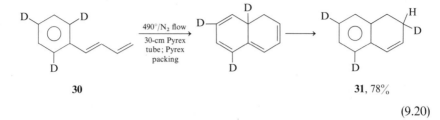

30   **31**, 78%

(9.20)

These reactions have been applied to a number of substituted and hetero-cyclic cases, and the preparative value of the method is shown by the entries of Table 9.1. Many of these dihydro compounds are not readily accessible by other routes. The preparation of 7,8-dihydroisoquinoline from the 1-(4-pyridyl)butadiene (entry 8) is particularly noteworthy because under the conditions used by the Japanese workers[25] this dihydro compound (55% yield) was accompanied by 21% of isoquinoline, and at 830° *quinoline* was obtained as well. This skeletal rearrangement to quinoline is believed to involve the dihydro compounds [Eq. (9.21)].

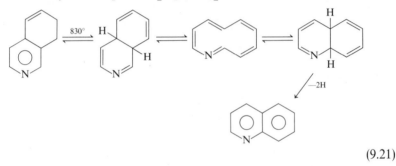

(9.21)

Related cyclizations in which the dihydropyridine ring of a dihydro-isoquinoline is formed have been observed by Gilchrist and co-workers[30] as a minor process in the pyrolysis of 1-methyl-4,5-diaryl-1,2,3-triazoles at 600° and 0.01 mm.

An alternative approach to 1,2-dihydronaphthalenes is by cyclization of the 6π (or 10π) *o*-xylylene intermediate generated from a 1-vinylbenzocyclo-butene **32** [Eq. (9.22)], although DeCamp *et al.*[31] found that a *cis*-methyl group in **32** led to the formation of 1-(*o*-tolyl)butadiene instead, by an antarafacial 1,7-hydrogen shift in the intermediate *o*-xylylene.

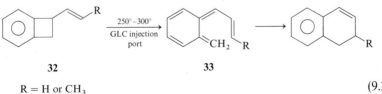

32   **33**

R = H or CH₃   (9.22)

TABLE 9.1

Preparation of Some Dihydroaromatic Compounds by Ring Closure of 1-Arylbutadienes and Derivatives

Starting material	Major product	Yield (%)	Conditions	Reference
1.		57	450°; N$_2$ flow	26
2. 4-CH$_3$	1-CH$_3$	60	450°; N$_2$ flow	26
3. 1-CH$_3$ ($E \rightarrow Z$ mixture)	4-CH$_3$	86	475°; N$_2$ flow	26
4. 2′,4′,6′-(CH$_3$)$_3$	2,5,7-(CH$_3$)$_3$[a]	25	470°; N$_2$ flow	26
5. 4′-OCH$_3$-1-CH$_3$	7-OCH$_3$-4-CH$_3$	96	439°; N$_2$ flow; 250-cm quartz spiral	27

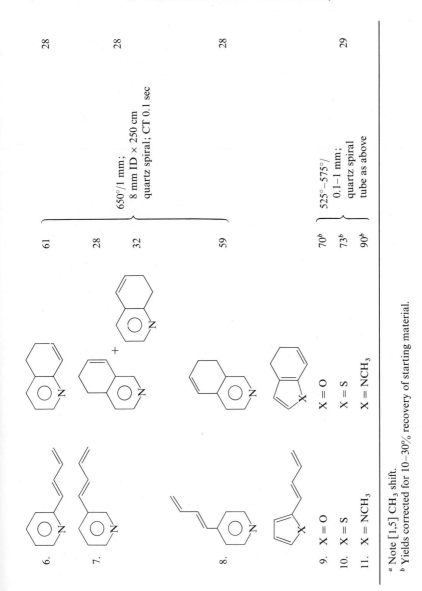

6.   61

7.   28

      28

      32

      650°/1 mm;
      8 mm ID × 250 cm
      quartz spiral; CT 0.1 sec

8.   59   28

9. X = O   70[b]
10. X = S   73[b]   525°–575°/
11. X = NCH₃ 90[b]   0.1–1 mm;
                       quartz spiral
                       tube as above   29

[a] Note [1,5] CH₃ shift.
[b] Yields corrected for 10–30% recovery of starting material.

The synthetic value of such reactions is somewhat limited by the poor yields of the starting materials **32** obtained by 1,2-addition of benzyne to dienes. However, intermediates similar to **33** can be approached from another direction, as shown by the work of Schmid's group[32] on the thermal rearrangement of *o*-methylarylallenes in solution [Eq. (9.23)].

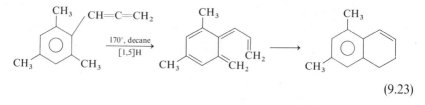

$$(9.23)$$

A pathway of the same formal type is also the basis of the general synthesis of 2-naphthols discovered by Brown and McMullen[33] and exploited in later papers (see Table 9.2). Condensation of *o*-alkylbenzaldehydes with Meldrum's acid (**34**) and pyrolysis of the arylmethylene derivatives gives 2-naphthols in high yield, and *m*-substituted phenols have been prepared in a similar way (Table 9.2, entries 3 and 7). The 1,5 hydrogen shift in the type reaction [Eq. (9.24)] occurs in a transient substituted methyleneketene **35**.[34]

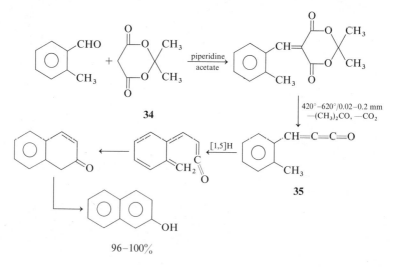

$$(9.24)$$

Cyclization of a phenyliminoketene is involved in an unexpected mode of pyrolytic decomposition of *N*-phenylpyrrole-3,4-dicarboxylic anhydride (**36**). Most aromatic cyclic anhydrides undergo loss of both carbon dioxide and carbon monoxide to form arynes (see Chapter 6, Section II,I), but Cava and

TABLE 9.2

Preparation of 2-Naphthols, Phenols, and Related Compounds by Application of Sequence Similar to that of Equation (9.24)

Carbonyl compound which was condensed with Meldrum's acid and the derivative pyrolyzed	Phenolic or other product	Yield (%)	Conditions of pyrolysis[a]	Reference
1.		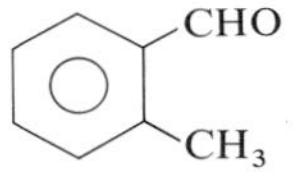 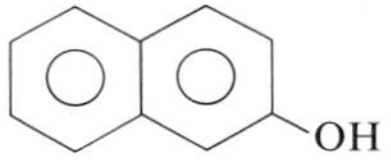 96–100	420°–620°/ 0.02–0.2 mm	33
2.		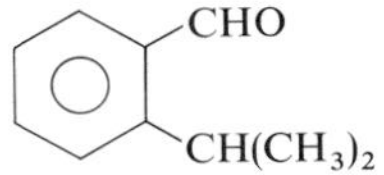 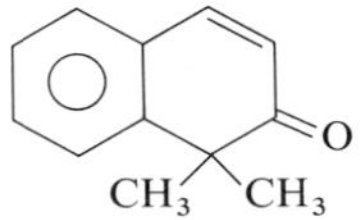 high	420°/ 0.02 mm	33
3.		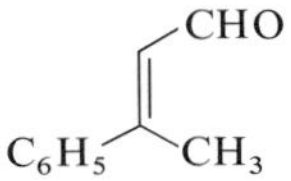 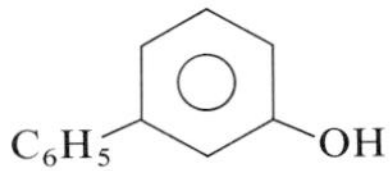 57	430°/ 0.01 mm	33
4.		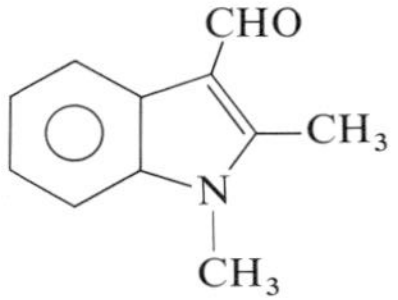 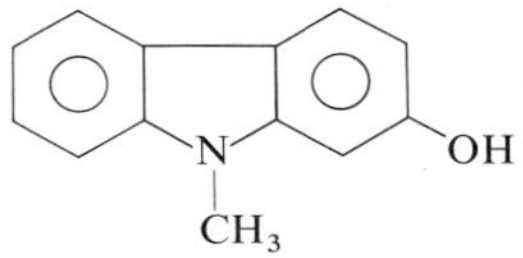 88	500°/ 0.05 mm	35
5.		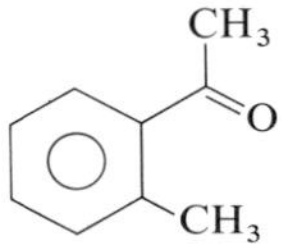 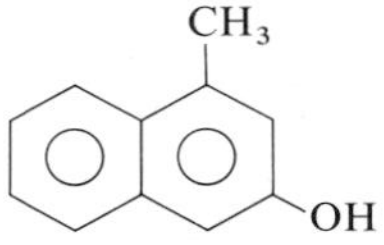 96	500°/ 0.05 mm	36

(continued)

TABLE 9.2 (*continued*)

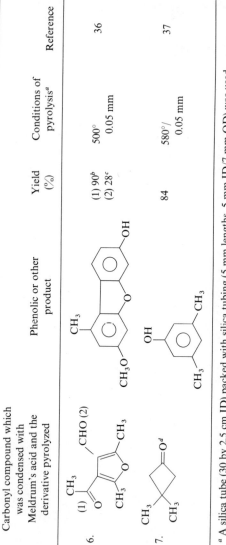

Carbonyl compound which was condensed with Meldrum's acid and the derivative pyrolyzed	Phenolic or other product	Yield (%)	Conditions of pyrolysis[a]	Reference
6.		(1) 90[b] (2) 28[c]	500° 0.05 mm	36
7.		84	580°/ 0.05 mm	37

[a] A silica tube (30 by 2.5 cm ID) packed with silica tubing (5 mm lengths, 5 mm ID/7 mm OD) was used.
[b] Double application of the sequence. The first benzofuranol was methylated, formylated at C-3, and condensed again with Meldrum's acid.
[c] The low yield was due to difficulty in sublimation of the intermediate.
[d] Complex pathway probably involving generation of transient $(CH_3)_2C=CHC(CH_3)=C=C=O$.

Bravo[38] found that **36** lost carbon dioxide at 650° to give an intermediate **37** or **38** which underwent ring fission to an ethinylphenyliminoketene **39**. This ketene then cyclized to the 4-quinolone **40**; a second cyclization led to the ultimate product, furo[3,2-*c*]quinoline (**41**).

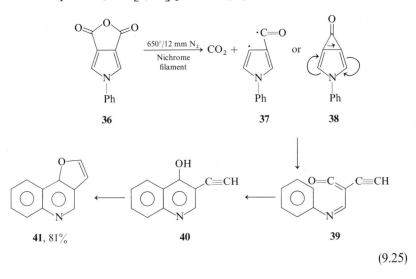

$$(9.25)$$

D. Rearrangements Initiated by Internal Diels–Alder Additions

This class of rearrangements includes many reactions in which intramolecular $[4 + 2]$ cycloaddition at a high temperature leads to a transient intermediate (usually highly strained) which then decomposes by a different pathway.

Paquette's group has encountered many such reactions in the chemistry of propellatrienes,[39] and the example[40] shown in Eq. (9.26) is typical. Similar reactions are involved in the conversion of bullvalene (**42**) to *cis*-9,10-dihydronaphthalene at 350°,[41] and in the conversion of 8-methoxy-7-azabicyclo[4.2.2]deca-2,4,7,9-tetraene (**43**) at 600° and 12 mm over glass beads to a mixture of quinoline and methoxyquinolines.[42]

Nomura *et al.*[43] have prepared the unstable endo-9-chloro-*cis*-bicyclo-[4.3.0]nona-2,4,7-triene (**45**) by flash vacuum pyrolysis of the chlorosulfite **44**. The endo-chloro compound **45** is highly unstable with respect to the exo compound and isomerizes rapidly in the condensed phase at room temperature, but a pyrolyzate containing 65% of the endo compound could be collected at −196°. The course of the rearrangement is complex, and it is

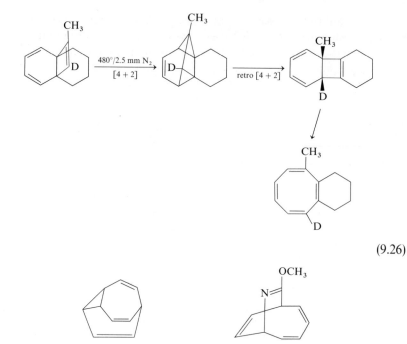

$$(9.26)$$

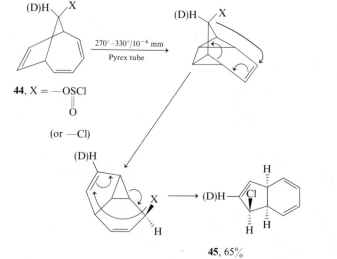

not clear whether the migrating group is —Cl or —OSOCl, but the first key step is a $[4 + 2]$ cycloaddition [Eq. (9.27)].

$$(9.27)$$

Another important group of rearrangements occurs through intramolecular Diels–Alder reactions of dienone intermediates produced in the Claisen rearrangement of allyl or propargyl phenyl ethers. This group includes the rearrangements of allyl pentafluorophenyl ether reported by Brooke[44] [Eq. (9.28)] and by Brooke and Hall,[45] and the rearrangements of aryl propargyl ethers studied by Trahanovsky and co-workers.[46,47,48] Phenyl propargyl ether on pyrolysis at 460°[46] gives a little starting material (14%) and the major products indan-2-one and benzocyclobutene [Eq. (9.29)]. The *o*-, *m*-, and *p*-tolyl ethers behave similarly at 480°.[47]

The 4-pyridyl ether **47** gives both of the cyclobutapyridines **48** and **49** [Eq. (9.30) rather than **49** alone as expected by strict analogy with Eq. (9.29). Riemann and Trahanovsky[48] have rationalized this result in terms of decarbonylation of the tricyclic ketone corresponding to **46** in the carbocyclic series, with rebonding of the resulting diradical followed by several possible rearrangements which might lead to carbenes capable of forming **48** and **49**.

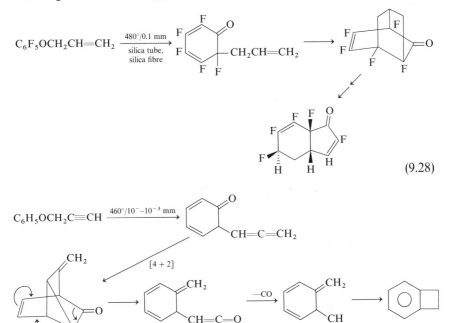

(9.28)

(9.29)

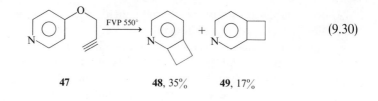

$$\textbf{47} \qquad\qquad\qquad \textbf{48}, 35\% \qquad \textbf{49}, 17\%$$

E. Intramolecular Ene Reactions and the Cyclization of ω-Unsaturated Ketones

The ene reaction, a 2π, $2\sigma + 2\pi$ counterpart of the Diels–Alder reaction [see Eq. (9.31)], has been reviewed comprehensively by Hoffmann.[49] We are concerned only with the intramolecular form of the reaction, and we have already seen some examples of such ene reactions in the further high-temperature reactions of 1,6-dienes produced by pyrolysis of pinanes and related compounds (see Chapter 8, Section II,A). Some further simple examples are shown in Eqs. (9.32)–(9.34).

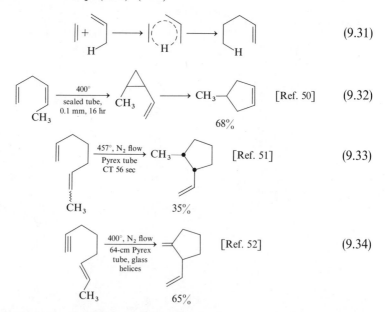

A particularly ready cyclization in a more rigid system was encountered by Bloch and Bortolussi[53] in the flash vacuum pyrolysis of the bis(methylene-cyclopropane) system **50**. The exo,exo, exo,endo, and endo,endo isomers were interconverted through diradicals of trimethylenemethane type, but the endo,endo isomer (**51**) finally underwent an irreversible ene reaction to give the cage compound **52** in almost quantitative yield.

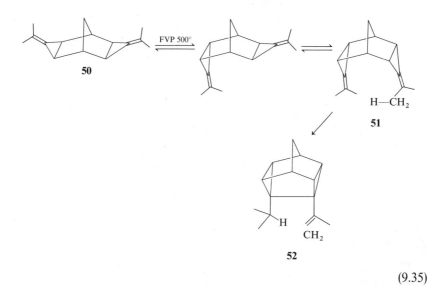

50

FVP 500°

51

H—CH₂

52

(9.35)

The cyclization of ω-unsaturated ketones studied extensively by Conia's group, and reviewed by Conia and Le Perchec,[54] constitutes a special case of the intramolecular ene reaction in which the 2π, 2σ component is an enol formed reversibly from the ketone. The reaction is an extremely versatile synthetic tool for the construction of cyclic and polycyclic systems which is certain to be used frequently with growing awareness of its power.

Because the reaction has been reviewed so comprehensively[54] only two examples are given below. Reaction is usually effected by heating the ketone in a sealed glass tube or in an evacuated glass bulb; enolization is probably promoted by the glass surface, so that in most cases a static reactor with long time of contact with glass is desirable. However, flow methods have been used, particularly on a substantial preparative scale,[54] with a result indistinguishable from that obtained with the static system[55] in the case of oct-7-en-2-one [Eq. (9.36)].

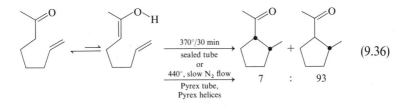

370°/30 min
sealed tube
or
440°, slow N₂ flow
Pyrex tube,
Pyrex helices

7 : 93

(9.36)

There are some cases, however, for which vapor-phase conditions are essential. Thus 4-allylcyclohexanone underwent double-bond migration and resinification when heated in a sealed tube, whereas cyclization to endo-7-methylbicyclo[3.2.1]octan-2-one (53) occurred smoothly in the vapor phase.[56]

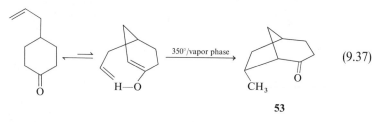

(9.37)

53

III.   REARRANGEMENTS WITH DIRADICAL OR DIPOLAR INTERMEDIATES

A.   Rearrangements of Epoxides

Thermal and acid-catalyzed rearrangements of epoxides have been reviewed by Rosowsky,[57] but the emphasis of the review is on catalyzed reactions. Hudrlik et al.[58] give a useful short list of references to kinetic and preparative studies of the thermal decomposition of simple epoxides, in which the major reaction is often rearrangement to a carbonyl compound.

Although epoxides are very susceptible to rearrangement under the influence of mild electrophiles, the results of most gas-phase studies suggest that diradical intermediates are involved. A typical example of such a rearrangement under flow conditions is that of norbornene oxide reported by Garin.[59] Many products were formed, but the major suggested pathways are shown in Eq. (9.38). The origin of cyclopentenone is uncertain; it may be formed as shown, or by a retro-Diels–Alder cleavage of the enol of norcamphor formed by hydrogen migration in the diradical **54**.

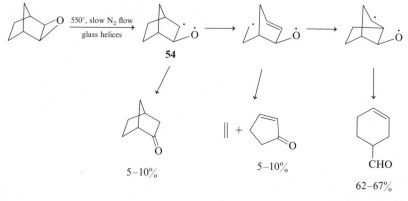

54

5–10%

5–10%

CHO

62–67%

(9.38)

A preparatively useful rearrangement is that of α,β-epoxysilanes to silyl enol ethers[58] which can be effected by flash vacuum pyrolysis in 66–77% yield for a range of such compounds. The simplest case is shown in Eq. (9.39); the authors propose initial heterolytic fission of the bond α to silicon, with migration of β-hydrogen in forming the final product.

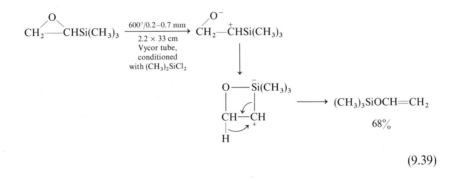

$$(9.39)$$

We have already seen an example of the fission of a cyclic epoxide with allylic unsaturation in the case of 8-oxabicyclo[5.1.0]octa-2,4-diene (see Section II,C); the pathways which may be involved in the formation of heptatrienal have been discussed in detail.[24] Schiess and Radimerski[60] have made an intensive study of the pyrolysis of 3,4-epoxycyclopentene which first gives *cis*-penta-2,4-dienal (**56**), probably via the diradical **55**. The components of the complex mixture of products were shown to vary with temperature over the range 290°–600° because of cis–trans isomerizations and a dominating isomerization of the cis aldehyde **56** to the ketene **57** by a 1,5 shift of the aldehydic hydrogen. The two ketenes **57** and **58** were identified by trapping with methanol at $-60°$.

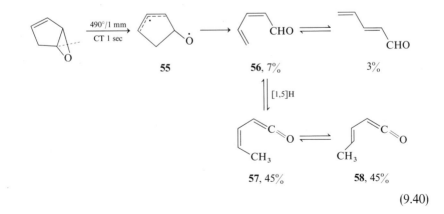

$$(9.40)$$

Pyrolysis of spiroepoxides may lead to ring expansion by alkyl migration, as shown in Eqs. (9.41)–(9.43).

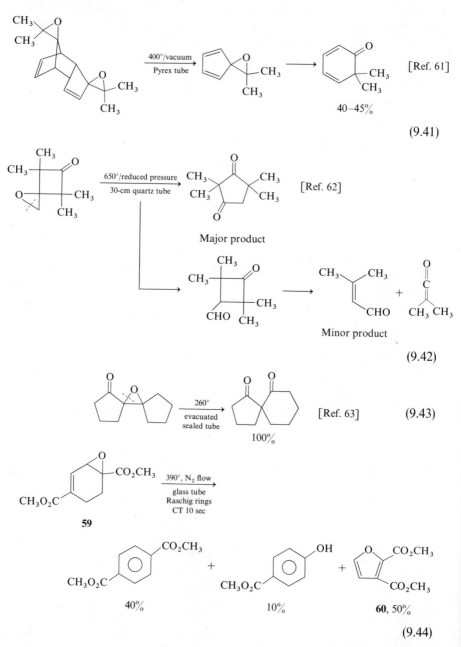

(9.41)

(9.42)

(9.43)

(9.44)

A dipolar intermediate **61** was suggested by Eberbach and Carre[64] to explain the formation of the furan 2,3-diester **60** as a product of pyrolysis of the epoxide **59**. Collapse of the intermediate **61** would form the bicyclic compound **62**, which could then lose ethylene by fission of the cyclobutane ring [Eq. (9.45)].

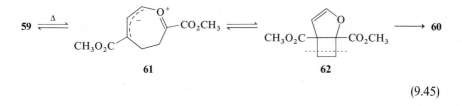

$$(9.45)$$

B. Diradical Rearrangements of Hydrocarbons and Carbonyl Compounds

Rearrangements occurring through thermal fission of methylenecyclopropanes to trimethylenemethane diradicals are quite common in the chemistry of small-ring hydrocarbons. A typical example is the pyrolysis of the bis(cyclopropylidene) **63**, reported by Crandall and co-workers,[65] which led to the isomeric methylenespiropentanes **65** and **67** [Eq. (9.46)] via the two possible trimethylenemethanes **64** and **66**. At a higher temperature (510°) **67**

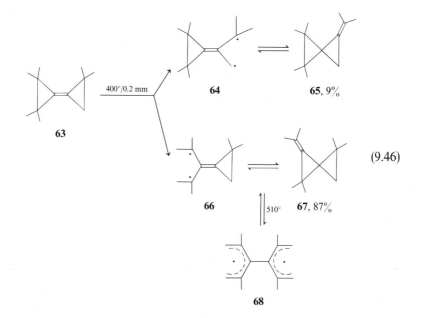

$$(9.46)$$

underwent further reactions through the bisallyl diradical **68**. Dolbier *et al.*[66] have made a quantitative study of the thermal reorganizations of such compounds in the vapor phase; they propose that the intermediate trimethylenemethane diradicals are nonplanar.

The isomerization of the tricyclooctene **69** into the tetracyclooctane **70** observed by Bloch and co-workers[67] similarly appears to involve interception of a trimethylenemethane diradical by the neighboring double bond [Eq. (9.47)]. At 600° a further 1,5 hydrogen shift and a thermal reorganization of the intermediate semibullvalene **71** leads to isopropylcyclooctatetraene [Eq. (9.48)].

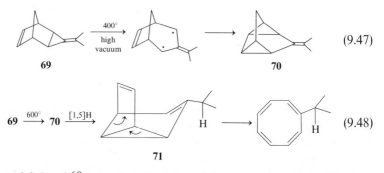

$$ (9.47) $$

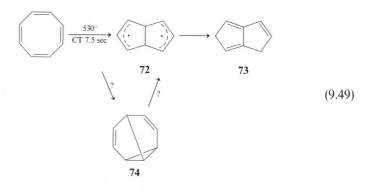

$$ (9.48) $$

Jones and Schwab[68] have described a useful preparation of the dihydropentalene **73** by flow pyrolysis of cyclooctatetraene. The product **73** is probably formed by a series of 1,5 hydrogen shifts in the diradical **72**, which may be generated directly from cyclooctatetraene or through semibullvalene (**74**). The optimum temperature range is fairly narrow, and at high temperatures much benzene and styrene are formed.

$$ (9.49) $$

Another group of hydrocarbon rearrangements proceeds through 1,3 alkyl shifts with formal intermediacy of allylic diradicals; a spectrum of mechanisms, ranging from continuous diradical transition states[69] to discrete diradicals with significant lifetimes, is possible for such reactions. Examples[70,71] are shown in Eqs. (9.50)–(9.52), with the position of bond-breaking marked [for one of two equivalent forward reactions in the case of Eq. (9.52)]. In the last case Bishop and co-workers[71] estimate an activation energy of 60–65 kcal/mole for the degenerate rearrangement of bicyclo [3.3.1]nona-2,6-diene (**75**), consistent with a transition state **76** in which a perpendicular allylic radical migrates across a planar allylic radical.

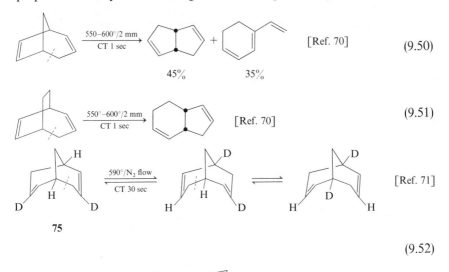

45%   35%   [Ref. 70]   (9.50)

[Ref. 70]   (9.51)

75

[Ref. 71]

(9.52)

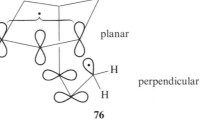

planar

perpendicular

76

A similar migration of a tertiary radical center must occur in the pyrolysis of β-himachalene (**77**) which gives a secondary product of dehydrogenation, cuparene (**78**), with 20% retention of optical activity.[72]

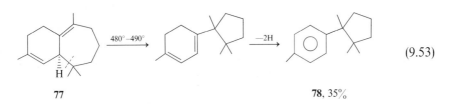

480°–490°   —2H   (9.53)

77   **78**, 35%

1,3 Shifts of both ketoalkyl radicals $RCOCH_2\cdot$ and of acyl radicals may occur in the pyrolysis of unsaturated ketones. Crandall and co-workers[73] found that cyclooct-4-enone underwent ring contraction to 3-vinylcyclohexanone by either photochemical or pyrolytic β-bond cleavage [Eq. (9.54)]. A group of synthetically valuable rearrangements of 2-alkenylcyclobutanones occuring under static vapor phase conditions at $280°-360°$ has been reported by Bertrand and co-workers.[74] In these conditions a 2-alkylidenecyclo-butanone such as **79** is equilibrated with the required 2-alkenylcyclobutanone **80** so that either an α,β- or a β,δ-unsaturated cyclobutanone may be used as starting material [Eq. (9.55)].

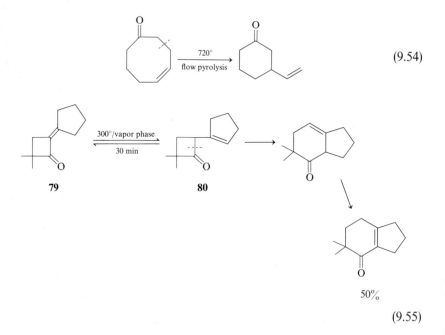

$$(9.54)$$

79   **80**

50%

$$(9.55)$$

The degenerate rearrangement of deuterated bicyclo[3.2.1]oct-2-en-7-one (**81**) has been investigated quantitatively by Berson's group;[75] the activation parameters (ΔH 51 ± 2 kcal/mole; ΔS 1.8 ± 3.4 eu) are consistent with either concerted rearrangement or the intermediacy of an allyl–acyl radical pair. A related reaction is the remarkable rearrangement of 7-isopropylidene-bicyclo[2.2.1]heptane-2,3-dione (**82**) to the dione **83**, in which migration of the ketoacyl radical occurs on flash vacuum pyrolysis without significant decarbonylation. Indeed Berson and Potter[76] obtained an almost quantitative yield of **83**, which was isolated as its enol acetate in 71% yield after purifi-

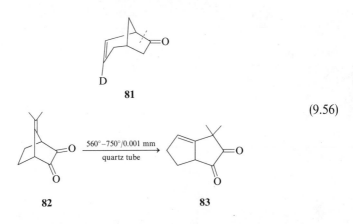

81

(9.56)

cation. Only above 750° did bisdecarbonylation to trimethylenemethane diradicals and hence hydrocarbon products become significant.

C. Vinylcyclopropane Rearrangements

The rearrangement of a vinylcyclopropane to a cyclopentene [Eq. (9.57)] is a special case of the class of 1,3 alkyl migrations outlined in Section III,B. Vinylcyclopropane rearrangements have been reviewed by Gutsche and Redmore,[77] Frey and Walsh,[78] and by Willcott *et al.*[79] Trost and Bogdanowicz[80] give a useful list of key papers. As in the previous cases, it is difficult to distinguish clearly between concerted and diradical mechanisms.

$$ \text{(9.57)} $$

Vinylcyclopropane rearrangements have recently been used extensively in the development of methods for the synthesis of five-membered rings in natural products of biological importance. A selection of typical rearrangements of synthetic interest is shown in Table 9.3. Rearrangement occurs smoothly under mild pyrolytic conditions particularly with 1-phenylthio- and 1-trimethylsilyloxycyclopropanes (entries 3, 7, and 8), whereas in cases where no stabilization of a radical center by sulfur or oxygen is possible and there is no further alkyl substitution it may be difficult to effect complete rearrangement in one pass (entry 1). The 1-phenylthio and 1-trimethylsilyloxy substituents provide the double synthetic advantage of assisting the rearrangement and generating masked potential carbonyl groups in the products.

TABLE 9.3

Preparation of Cyclopentene Derivatives by Vinylcyclopropane Rearrangement

Vinylcyclopropane	Cyclopentene	Yield (%)	Conditions	Reference
1. $(CH_3)_3SiO$ [structure]	$(CH_3)_3SiO$ [structure]	—	(1) Rearrangement incomplete at 520°/flow system	81
		73^a	(2) 360–450°; sealed tube; 0.5–3 hr	
2. $(CH_3)_3SiO$ [structure]	$(CH_3)_3SiO$ [structure]	99	360°–450°; sealed tube	81
3. $OSi(CH_3)_3$ [structure]	$OSi(CH_3)_3$ [structure]	99	330°; conditioned tube;^{b} CT 4 sec	80
4. $(CH_3)_3SiO$ [structure]	$(CH_3)_3SiO$ [structure]	50^a	450°; glass helices	82
5. O [structure]	O [structure]	76	450°; glass helices	82
6. $CO_2C_2H_5$ [structure]	CO_2R [structure] $R = H$ or C_2H_5	75	500°; N_2 flow; glass packing	83
7. C_6H_{13} SPh [structure]	C_6H_{13} SPh [structure]	93	350°; conditioned tube;^{b}	84
8. SPh [structure]	SPh [structure]	80	500°/0.2 mm; Pyrex beads	85

^{a} Yield of ketone after hydrolysis.
^{b} Surface conditioned by treatment with O,N-bis(trimethylsilyl)acetamide.

IV. CONCERTED REARRANGEMENTS OF CYCLOPROPANE DERIVATIVES

The equivalence of C=C bonds and cyclopropane C—C bonds in pericyclic reactions is commonplace, and opening of a cyclopropane ring in such reactions usually proceeds smoothly. Some examples appeared in equations in the preceding section, and we note here only three further examples chosen for their synthetic value or mechanistic interest.

Wenkert and Regodesousa[86] have developed a useful chain-lengthening sequence which produces a 1,3-diene ester from an aliphatic aldehyde [Eq. (9.58)]; the key step is a 1,5 hydrogen shift in an alkoxydialkyl cyclopropanecarboxylic ester, and elimination of the alkoxy group occurs subsequently. Some diene acid is also produced.

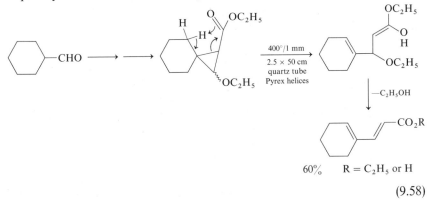

$$(9.58)$$

Some remarkable rearrangements of dimethylvinylidenecyclopropanes have been reported by Sadler and Stewart.[87] 6-Dimethylvinylidenebenzobicyclo[3.1.0]hex-2-ene (**84**) rearranges at 450° in a flow system to give the indenyldiene **85**. The formation of 2-(2',2'-dimethylvinyl)naphthalene originally found[87] is now thought to be due to an acid-catalyzed rearrangement.[88]

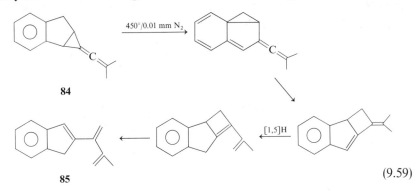

$$(9.59)$$

Under similar conditions the corresponding 1-methylbenzobicyclo[4.1.0]
heptene derivative **86** gives the acetylenic tetralin **87** in good yield. The
rearrangement of Eq. (9.60) is considered by the authors to be an example
of a concerted $\pi^2 a + \sigma^2 a + \sigma^2 s$ process.

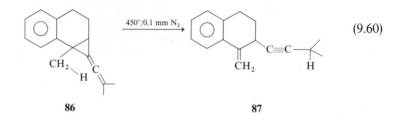

$$(9.60)$$

86   **87**

V.   SIGMATROPIC REARRANGEMENTS

A.   Introduction; Hydrogen Shifts

The theory of sigmatropic rearrangements developed by Woodward and
Hoffmann[89] has introduced order into an otherwise bewildering array of
molecular rearrangements, and has been a prime stimulus to new work in the
area. Sigmatropic rearrangements of order $[1,j]$ have been most comprehensively reviewed by Spangler,[90] and his review includes a number of examples
of rearrangements in the vapor phase at high temperatures.

We shall not attempt to deal systematically with hydrogen shifts. 1,5-
Suprafacial shifts of hydrogen are ubiquitous in appropriate systems above
$200°–300°$, and many have been mentioned in previous sections and chapters.
1,5 Shifts of hydrogen may occur readily in high-temperature flash pyrolytic
reactions even when disruption of the aromaticity of a benzene ring is required; two examples from the author's laboratory,[91,92] in which migration
was revealed by scrambling of a deuterium label, are shown in Eqs. (9.61)
and (9.62). The unlabeled form of intermediate **88** could also be generated
by pyrolysis of **89** at 550° and 0.02 mm.

Many pyrolytic reaction schemes appear to require thermally forbidden
1,3 shifts of hydrogen, but the evidence which might distinguish between

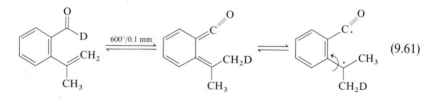

$$(9.61)$$

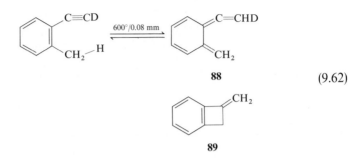

$$88 \tag{9.62}$$

$$89$$

concerted but forbidden processes requiring a high activation energy, step-wise radical processes, and surface-catalyzed processes is usually lacking. 1,7-Antarafacial shifts are occasionally found.[93]

B. 1,5 Shifts of Alkyl, Alkenyl, Aryl, Acyl, and Nitrile Groups

The occurrence of suprafacial 1,3 shifts of alkyl groups with inversion of stereochemistry at the migrating center was first established experimentally by Berson and Nelson.[94,95] Such concerted allylic shifts probably do occur in some high-temperature pyrolyses, but the stereochemical features necessary to permit concerted migration and to reveal inversion are usually not present. Thus most of the 1,3 shifts of Section III,B probably occur by step-wise formation and rearrangement of diradicals.

Sigmatropic 1,5 shifts of simple alkyl groups in cyclopentadiene rings occur very commonly. Mironov *et al.*[96] found that flow pyrolysis of 1,2,4,5,5-pentamethylcyclopentadiene (**90**) at 475° gives about 65% conversion into the isomeric pentamethyl compounds **91** and **92**, with some formation of tetramethyl compounds [Eq. (9.63)]. The loss of a methyl group in the formation of these tetramethyl compounds suggests that both concerted and radical processes may be occurring, and indeed it appears that there is a delicate balance between these processes in the migration of simple alkyl groups in cyclopentadienes.

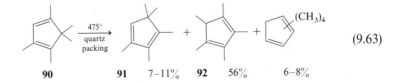

$$\tag{9.63}$$

90 **91** 7–11% **92** 56% 6–8%

Willcott and Rathburn[97] have established the occurrence of a radical-chain mechanism in the rearrangement of the deuterated trimethyl compound

93 to **94**, at a partial pressure of 10 mm of **93** in nitrogen at 1 atm. The radical-chain mechanism probably has a higher activation energy than the concerted migration but the detailed distribution of deuterium in the product suggests that both occur.[97] In the case of spirocyclic systems the cyclopropane compounds **95** and **96** undergo both radical reactions and concerted rearrangement[98,99] but the larger ring systems **97** and **98** undergo only rearrangement.[97]

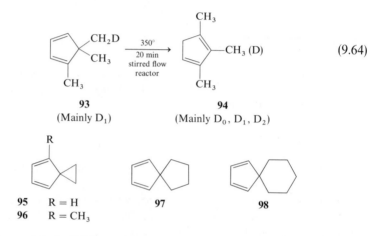

$$(9.64)$$

93
(Mainly D_1)

94
(Mainly D_0, D_1, D_2)

95   R = H
96   R = CH$_3$

97

98

Spangler and Boles[100] have found that cyclohexadienes also show competition between radical decomposition and rearrangement, as indicated by the composition of the pyrolyzate from 5,5-dimethylcyclohexa-1,3-diene [Eq. (9.65)]. They considered that rearrangement occurred via 1,5 methyl shifts. However, Schiess and Dinkel[101] conclude from their study of the pyrolysis of 5,5-diethylcyclohexa-1,3-diene [Eq. (9.66)] that both the dimethyl and diethyl compounds rearrange predominantly by a sequence of electrocyclic ring opening, hydrogen migration, and electrocyclic ring closure, as

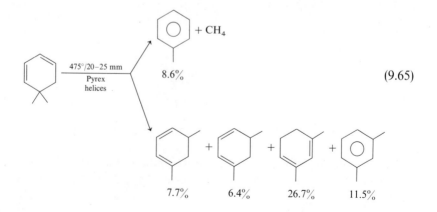

$+ CH_4$

8.6%

$$(9.65)$$

475°/20–25 mm
Pyrex
helices

7.7%   6.4%   26.7%   11.5%

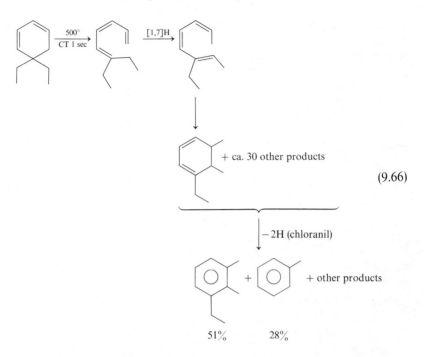

(9.66)

originally proposed by Pines and Kozlowski,[102] and that 1,5-alkyl migration occurs to the extent of less than 3%.

1,5 Migrations of hydrogen, alkyl, aryl, and other groups occur in indenes, and are to be expected whenever an indene is generated in a high-temperature pyrolytic process. Thus 2- and 3-phenylindenes are interconverted on flash pyrolysis at 700° and 0.1 mm,[103] and the 2- and 3-carbonitriles are interconverted at 800° and 0.02 mm.[104] Further examples of 1,5 migrations of nitriles will be found in Chapter 5, Section III,D. Miller and Boyer[105] have established an order of migratory aptitude (H > Ph > CH₃) for migrations in the system 3-substituted indene–isoindenes–2-substituted indene [Eq. (9.67)] from kinetic studies in solution.

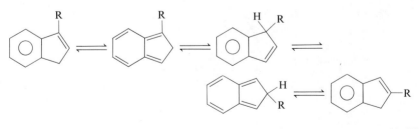

(9.67)

1,5 Shifts of bridging or spiro-bonded vinyl groups and allyl groups occur readily at relatively low temperatures, as shown in Eqs. (9.68)–(9.70) in which the migration termini are marked as (5).

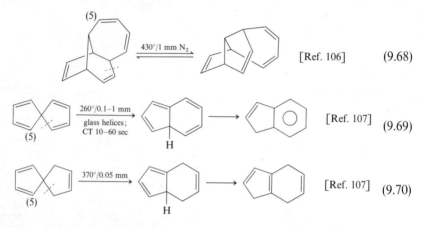

Acyl and ester groups migrate in preference to simple alkyl groups, and Schiess and Dinkel[101] have found that the cyclohexadiene **99** rearranges at 520° to give after dehydrogenation methyl *m*-toluate. Rearrangement occurs by migration of the ester group to the extent of 75%, as shown by ^{13}C-labeling, and by electrocyclic ring opening, 1,7 hydrogen migration, and ring closure to the extent of 24% [Eq. (9.71)]. The formyl group in the aldehyde corresponding to **99** migrates readily[108] at 450° and 11 mm. Field *et al.*[109] have measured rates of rearrangement in solution for many indenes hearing

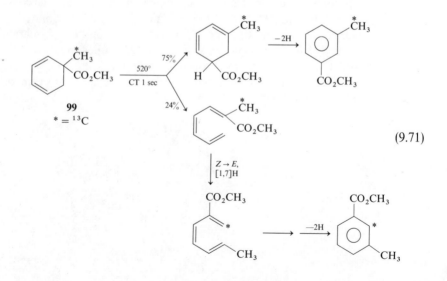

unsaturated groups, and they find that acyl groups migrate faster than most other groups and in the order $CHO > COCH_3 > COPh > CO_2CH_3$. This acceleration is attributed at least in part to secondary orbital interactions involving the π^* orbital of the carbonyl group.

C. 3,3-Sigmatropic Shifts; Claisen, Cope, and Related Rearrangements

Some examples of Claisen rearrangements of allyl aryl ethers of specialized structure, which led to intramolecular Diels–Alder reactions,have been mentioned in Section II,D. Claisen and Cope rearrangements, including many reactions in the vapor phase, have been reviewed to 1972 by Rhoads and Raulins[110] and only a few, mainly more recent, examples are discussed here. Under true flash vacuum pyrolytic conditions, at very high temperatures and with very short contact times, the Claisen rearrangement of allyl phenyl ether may not be observed; Hedaya and McNeil[111] showed that the ether can instead undergo direct homolytic fission to phenoxy and allyl radicals [Eq. (9.72)] because the low and favorable $\Delta H^{\ddagger}$ for Claisen rearrangement is offset at the high temperature by an unfavorable $T\Delta S^{\ddagger}$ term for the conformationally restrictive 3,3 transition state. The final products include diallyl, cyclopentadiene, allylcyclopentadiene, naphthalene, phenol, and benzene.

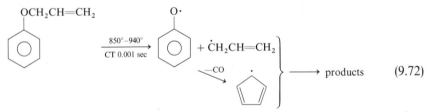

$$(9.72)$$

The first example of rearrangement shown below, that of 2-acetylmethylenecyclopropane (**100**) studied by Goldschmidt and Mauda,[112] can be classified as a formal Cope rearrangement but the authors prefer to regard it as proceeding by a diradical or dynamically continuous diradical[69] mechanism. It is of interest because the rearrangement in decalin solution at 170° gave only 10% of 2,4-dimethylfuran after 40 hr whereas flow pyrolysis at 500° gave the same furan in fair yield [Eq. (9.73)]. The isomeric methylenecyclopropane **101** rearranged to 2,4-dimethylfuran at 350°, and may possibly be an intermediate in the rearrangement of **100**.

An example of a true Cope rearrangement which is best effected by flash pyrolysis is that of the cyclohexenylpropargylmalononitrile (**102**), which when heated neat at 200° forms a mixture containing the tetralin-1,3-dicarbonitrile

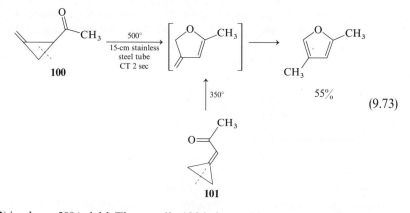

(9.73)

(**103**) in about 50% yield. The tetralin **103** is formed from the unstable product of Cope rearrangement in a sequence of about five steps including a 1,5 nitrile shift. Brown and McAllan[113] found, however, that the primary product **104** could be obtained in 65% yield by flash vacuum pyrolysis [Eq. (9.74)].

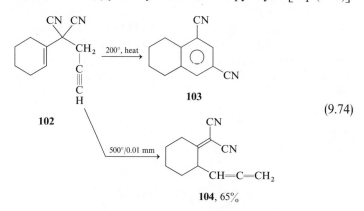

(9.74)

Japenga and co-workers [114] have identified an authentic Cope rearrangement in the pyrolysis of 3-(methylene-d_2)bicyclo[3.2.1]oct-6-ene **105**, which was transformed cleanly into the bicyclo[3.3.0]octene **106** at 450°–500°

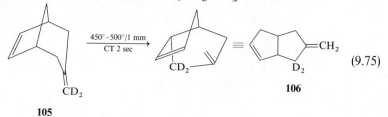

(9.75)

[Eq. (9.75)]. The corresponding ketone **107** required a temperature of 700°
for rearrangement to **108**, and in this case a nonconcerted diradical process
was suggested.[114]

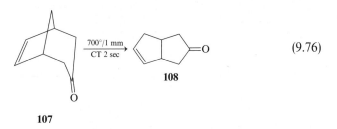

(9.76)

108

107

Karpf and Dreiding[115] have developed an interesting approach to the
synthesis of medium-ring ketones, based on a combination of 1,5 hydrogen
and 3,3 shifts in an epoxyalkyne. The simplest model system is **109**, which
underwent smooth pyrolysis at 550° to give an 80% yield of the acetylenic
and allenic ketones **110** and **111** in the ratio 5:4.

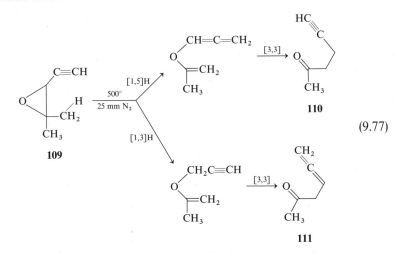

(9.77)

109

110

111

Pyrolysis of the epoxycyclohexane derivative **112** afforded a modest yield
of cyclonon-4-ynone [Eq. (9.78)] and three other products derived from

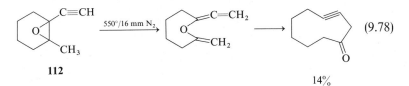

(9.78)

112

14%

epoxide ring opening with alkyl or alkynyl migration. Pyrolytic ring expansion of the ethinyl compound **113** gave a pyrolyzate containing cyclopentadec-4-ynone [Eq. (9.79)] which after hydrogenation gave cyclopentadecanone in 36% yield. The correspond 1-propynyl compound **114** gave a pyrolyzate which after hydrogenation yielded a trace of the saturated ketone, ($\pm$)-muscone.

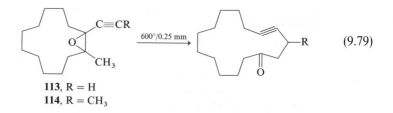

$$(9.79)$$

113, R = H
114, R = CH$_3$

Two examples of 3,3 rearrangements from those surveyed by Rhoads and Paulins[110] are mentioned here because of their general synthetic value or practical interest. Büchi and Powell[116] used the Claisen rearrangement of 3,4-dihydro-2*H*-pyranylethylenes as a new method for the synthesis of cyclohexenes starting from dimers of acrolein or methyl vinyl ketone. Both sealed-tube and flow methods were used, but at least in the case of the isopropenyl compound **115** the product **116** obtained by flow pyrolysis [Eq. (9.80)] was obtained pure after a simple distillation, whereas that from a sealed-tube reaction at 240° for 25 min contained 9% of an unidentified impurity.

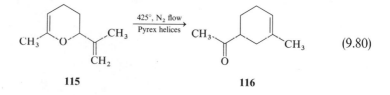

$$(9.80)$$

115   **116**

Again, in the course of the synthesis of ($\pm$)-thujopsene, Büchi and White[117] pyrolyzed β-cyclogeranyl vinyl ether on a substantial scale (0.76 mole, 135 gm) in a flow system at 320° and obtained the aldehyde **117** in high yield [Eq. (9.81)].

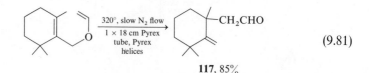

$$(9.81)$$

117, 85%

D. Rearrangements of Propargyl Esters and of Aryl Propiolates

The 3,3 rearrangement of a propargyl ester $HC\equiv CCH_2OCOR$ would be expected to form the allenic ester $RCOOCH=C=CH_2$. Trahanovsky and Mullen[118] found, however, that flash vacuum pyrolysis of propargyl esters at 580°–680° and 10^{-3}–10^{-5} mm gave instead vinyl ketones and carbon monoxide. The following results are typical:

$$HC\equiv CCH_2OCOPh \xrightarrow{660°} PhCOCH=CH_2 \quad (80\%) \tag{9.82}$$

$$HC\equiv CCH(Ph)OCOCH_3 \xrightarrow{640°-645°} CH_3COCH=CHPh \quad (79-87\%) \tag{9.83}$$

$$HC\equiv CCH_2OCOCH_3 \xrightarrow{630°} CH_3COCH=CH_2 \quad (60\%) \tag{9.84}$$

The authors propose that initial 3,3 rearrangement is followed by a 1,3 acyl shift[119] in the allenic ester. In the case of unsubstituted propargyl esters this leads to a 1-acyl-1-formylethylene, which then undergoes decarbonylation.

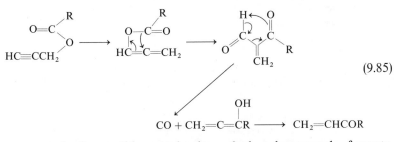

$$(9.85)$$

As expected from this mechanism 3-phenylpropargyl formate, $PhC\equiv CCH_2OCOH$, gives $PhCOCH=CH_2$ in high yield (94% on pyrolysis at 670°). The method is quite general and Trahanovsky and Emeis[120] have extended it to a number of 1-substituted and 1,1-disubstituted propargyl esters. The intermediate 1-acyl-1-formylethylene has been isolated from the pyrolysis of $HC\equiv CCH(Ph)OCOC_5H_{11}$.

Flash vacuum pyrolysis of aryl propiolates, studied by Trahanovsky and co-workers,[121] leads to an equally fascinating and synthetically useful rearrangement. Phenyl propiolate itself undergoes expansion of the phenyl ring to give 2H-cyclohepta[b]furan-2-one (**120**), and the reaction is not inhibited by the presence of o-methyl groups as in **119**. The mechanism proposed [Eq. (9.86)] again starts with a 3,3-sigmatropic rearrangement to give a transient methyleneketene **122**. This is followed by electrocyclic ring opening of **122** to give the acyclic intermediate **123** which cyclizes with the overall bond changes indicated by the arrows to give the lactone. However, a methylene-carbene mechanism also appears possible; compare Chapter 5, Section II,D and Eq. (5.24).

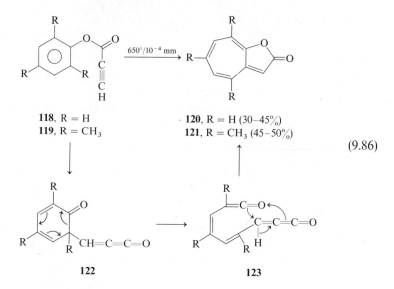

118, R = H
119, R = CH$_3$

120, R = H (30–45%)
121, R = CH$_3$ (45–50%)

122   **123**

(9.86)

E. Alkyl Migrations in Pyrroles, Indoles, and Imidazoles

1-Alkylpyrroles and indoles rearrange on flow pyrolysis at 500°–600° to give 2- and 3-alkyl compounds, and the 2- and 3-alkyl compounds are usually interconverted readily at 600°–800°. Thermal rearrangements of pyrroles are reviewed in the monograph by Jones and Bean[122] and their synthetic value is reviewed by Patterson,[123] a prolific researcher in this field.

Some examples of the rearrangements of alkyl pyrroles are shown in Eqs. (9.87)–(9.89). The kinetics of the rearrangement of 1-methylpyrrole and

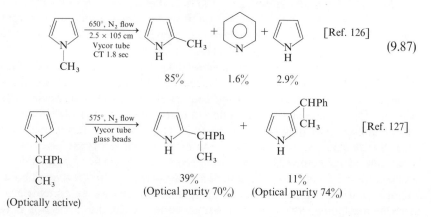

85%   1.6%   2.9%

39%   11%
(Optical purity 70%) (Optical purity 74%)

(9.88)

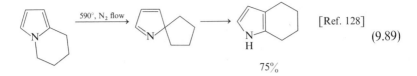

[Ref. 128]

(9.89)

75%

the pyrolysis of other 1-alkylpyrroles have been studied by Jacobsen and co-workers.[124,125] For the interconversion of 2- and 3-cyanopyrroles see Chapter 5, Section II,D, Eq. (5.63).

The rearrangement of 1-alkylpyrroles can be seen even from the few examples above to be mainly a 1,5-sigmatropic shift of the alkyl group across the azacyclopentadienyl system, occurring with substantial retention of configuration. The first intermediate must be a 2-alkyl-2*H*-pyrrole (**124**) which rapidly rearranges to the 2-alkyl-1*H*-pyrrole or, in part at high temperatures, to the 3-alkyl-1*H*-pyrrole [Eq. (9.90)]. The interconversion of the 1-, 2-, and 3-phenylindoles at 800° and 0.01 mm has been reported by Gilchrist and co-workers,[129] and that of the corresponding methyl- and benzylindoles at 300°–750° has been studied by Patterson *et al.*[130]

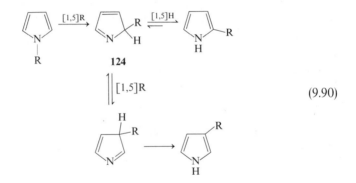

(9.90)

Competing 1,5 and 3,3 migrations can occur in the pyrolysis of 1-allylpyrroles. Patterson *et al.*[131] have found that 1-(substituted allyl)pyrroles rearrange to give the 2-substituted pyrroles without inversion of the allylic group, and to give 3-substituted pyrroles with inversion. The behavior of 1-(α-methylallyl)pyrrole at 475° is shown in Eq. (9.91); at 570°–600° the distribution of products is confused by further 1,5 and 3,3 interconversions of the products, and both 1-(α-methylallyl)- and 1-(*trans*-crotyl)pyrrole produce the same mixture of isomers.

Begg *et al.*[132] have found quite similar behavior in the alkyl imidazoles. Thus 1-methylimidazole on pyrolysis at 600° in a stream of nitrogen gave 2-methylimidazole (92%) and a little 4-methylimidazole (6%). 1-Allylimidazole isomerized to 2-allyl- (44%) and 4-allylimidazole (41%).

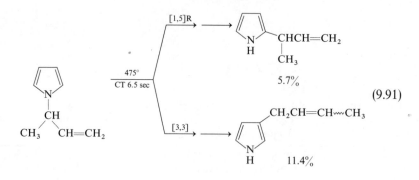

$$(9.91)$$

F.   Rearrangement of 1,5-Diynes; Ring Expansion of Fulvene to Benzene, and Related Reactions

This section deals with two groups of rearrangements which are not closely related in mechanism, but which it is convenient to introduce by considering the pyrolysis of a common starting material, 1,5-hexadiyne. Huntsman and Wristers[133] found that 1,5-hexadiyne rearranged on flow pyrolysis at 350° to give 3,4-dimethylenecyclobutene (**126**), probably by initial 3,3 rearrangement to give the bisallene **125** followed by electrocyclic ring closure.

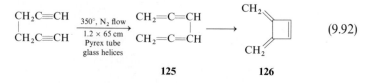

$$(9.92)$$

This rearrangement was studied over a wider temperature range by Heffernan and Jones[134] and by Kent and Jones,[135] who found that increasing proportions of fulvene, benzene, and other products were formed as the pyrolysis temperature was increased from 380° to 530°. At 380° the ratio of **126**:fulvene:benzene was 25:20:50 and at 390° this had changed to 20:5:70. This sequence of ring expansions [Eq. (9.93)] is the second type of rearrangement to be considered.

$$126 \xrightarrow{380°-390°} CH_2=\!\!\!\!\!\diagdown\!\!\!\!\!\diagup \longrightarrow \bigcirc \tag{9.93}$$

Rearrangements of alkynes and diynes have been reviewed by Huntsman,[136] and the application of the 1,5-diyne rearrangement to the preparation of bicyclic and 1,4-dehydroaromatic systems has been reviewed by Bergman.[137] Table 9.4 summarizes some further examples of the rearrangement.

TABLE 9.4

Some Products of 1,5-Diyne Rearrangement

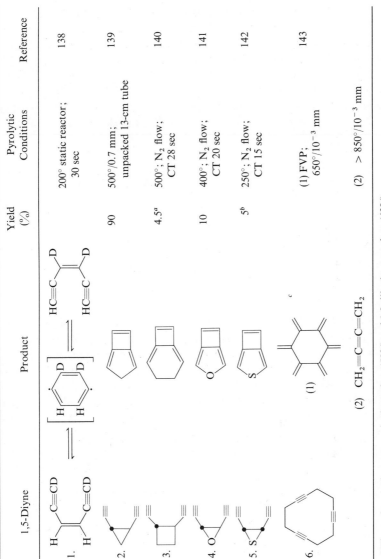

1,5-Diyne	Product	Yield (%)	Pyrolytic Conditions	Reference
1.			200° static reactor; 30 sec	138
2.		90	500°/0.7 mm; unpacked 13-cm tube	139
3.		4.5[a]	500°; N₂ flow; CT 28 sec	140
4.		10	400°; N₂ flow; CT 20 sec	141
5.		5[b]	250°; N₂ flow; CT 15 sec	142
6.	(1)[c] and (2) CH₂=C=C=CH₂		(1) FVP; 650°/10⁻³ mm (2) > 850°/10⁻³ mm	143

[a] The major products were vinylacetylene (52%) and 1,2-dihydropentalene (43%).
[b] The major products were cis- and trans-hex-3-en-1,5-diyne formed by loss of sulfur.
[c] Mechanism ambiguous, but the reaction may belong to this class of rearrangements.

The bicyclo[3.2.0]heptatriene from 1,2-diethinylcyclopropane (Table 9.4, entry 2) is of interest because at 580° it isomerizes to fulvenallene and ethinyl-cyclopentadiene,[139] products which are also formed by ring contraction of phenylcarbene or methylenecyclohexadienylidene (see Chapter 5, Sections III,B, and III,C). Indeed Wentrup et al.[144] consider a double bond isomer **127** to be a key intermediate in the ring contraction of phenylcarbene. Deuterium labeling studies by Henry and Bergman[145] have shown that up to 87% of the rearrangement follows the pathway shown in Eq. (9.94).

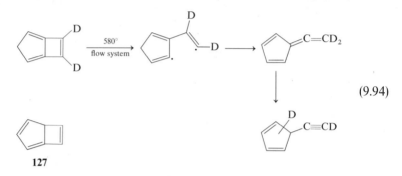

(9.94)

127

The ring expansion of 3,4-dimethylenecyclobutene to fulvene and of fulvene to benzene are reactions which do not fit readily into any well-recognized mechanistic scheme. The author has speculated, originally in the case of the rearrangement of benzofulvene to naphthalene,[146] that such reactions involve rotation of the methylene group out of the plane of the five-membered ring, ring expansion in the resulting dipolar or diradical intermediate, and rearrangement of the carbene so generated to an aromatic ring. Alternatively, the formation of the carbene might proceed through the bicyclic intermediate **128** [Eq. (9.95)]. However, there is at present no evidence to support such a mechanism, whether stepwise or concerted, and it is possible that different mechanisms operate in different cases. Table 9.5 shows a number of reactions which involve formally similar ring expansions.

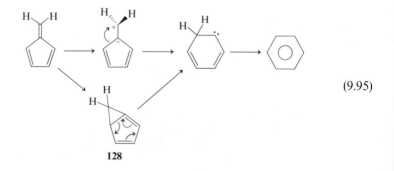

(9.95)

128

TABLE 9.5

Ring Expansions of 3,4-Dimethylenecyclobutene, Fulvene, and Related Systems

	Starting material	Product (yield)	Conditions	Reference
1.			$380°-390°$; N_2 flow	134
2.		(18%)	$700°$/low pressure of N_2	147
3.			$380°-390°$; N_2 flow	134
4.			$800°$/0.05 mm; silica packing	146
5.	[...]^a	(13%)	$820°$/0.01 mm; silica packing	148
6.	[...]^b		$820°$/0.01 mm; silica packing	148
7.	and methyl derivatives	and methyl derivatives	$350°-400°$; evacuated sealed tube; catalyzed by $CH_3 \cdot$ sources	149–150
8.			$350°-450°$; evacuated sealed tube; catalyzed by $CH_3 \cdot$ sources	151

^{a} From pyrolysis of 3-dimethylaminomethylindole (gramine).
^{b} From pyrolysis of 2-dimethylaminomethylindole.

There is a rather wide range of conditions for the reactions of Table 9.5, and there may well be an equally wide range of mechanisms. The dimethylenecyclobutenes (entires 1 and 2) have the possibility of reaction through ring-opened bisallenes which is not available to the fulvene derivatives (entries 3–6). The rearrangements of azulenes and of bifluorenylidene (entries 7 and 8) have been shown by Alder and co-workers[149–151] to be subject to radical catalysis in static systems, and almost the whole of the products can be accounted for in terms of rearrangements of intermediate radical adducts of the type shown in Eq. (9.96). There are still some anomalies in the results of labeling experiments, however, and the original papers should be consulted for details.

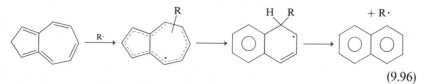

$$(9.96)$$

The formation of quinoline (entries 5 and 6) is a somewhat dubious case in that the methyleneindolenines are merely hypothetical intermediates, but it does seem likely that these and other ring expansions, such as the formation of pyridine (54%) from 1-methylpyrrole at 745°,[126] do involve methyleneindolenines or methylenepyrrolenines.

This class of rearrangements deserves further mechanistic study because, as Alder and Whittaker[149] remarked of the rearrangement of azulene considered as a possible concerted reaction, "It is not easy to propose *any* reasonable mechanism which has any precedents."

VI. MISCELLANEOUS ISOMERIZATIONS OF HETEROCYCLIC COMPOUNDS

At very high temperatures aromatic nitrogen heterocycles may undergo both fragmentation to mixtures of products including nitriles and acetylene, and isomerization. Crow and Wentrup[152] observed some isomerization of pyrazine to pyrimidine at 1000° [Eq. (9.97)] and they discussed the possibility of an intermediate or transition state of benzvalene type. Chambers and co-workers[153] found substantial isomerization of tetrafluoropyridazine to

$$\text{[pyrazine]} \xrightarrow{1000°/2\ mm} HC{\equiv}CH + HCN + \text{[pyrimidine]} + \text{[pyridazine]} + SM$$

36%   3%   0.7%   45%

$$(9.97)$$

tetrafluoropyrimidine at 745°, and they also proposed a diazabenzvalene intermediate, **129**. The yield of the pyrimidine rose to 46% at 815°, but a more complex mixture which also contained tetrafluoropyrazine was then formed.

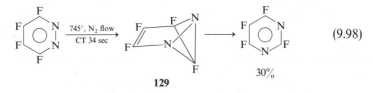

$$(9.98)$$

129

Patterson and co-workers[154] have noted the isomerization of quinoline to isoquinoline and of isoquinoline to quinoline to a small extent in one pass at 850°. Many other products are formed in trace amounts. These workers propose that rearrangement occurs mainly in diradicals of the type shown in Eq. (9.99) for the rearrangement of quinoline. However, the involvement of dihydro compounds, as in the unexpected formation of quinoline as well as the rational product isoquinoline on pyrolysis of 1-(4-pyridyl)butadiene,[25] seems equally likely [see Section II,C; Eq. (9.21)].

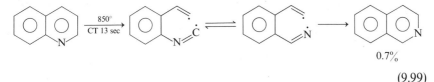

0.7%

$$(9.99)$$

Heterocycles containing an *N—O* bond are particularly prone to cleavage of this relatively weak bond. Kinstle and Darlage[155] found that 3-hydroxy-1,2-benzisoxazole (**130**) was smoothly transformed into benzoxazolone on flash pyrolysis at 440°–450°. Both acylnitrene and spiro-α-lactam intermediates were considered as intermediates, as shown in Eq. (9.100).

Aldous *et al.*[156] obtained a very complex mixture of products from pyrolysis of 3,5-diphenylisoxazole at 960°. 2,5-Diphenyloxazole was always a

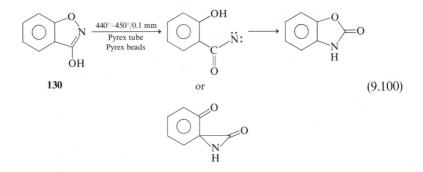

130   or $$(9.100)$$

major product, together with 2-phenylindole, diphenyl-2*H*-azirine, benz-
amide, and various bicyclic and tricyclic aromatic hydrocarbons. The major
pathways of formation of the oxazole and other products, as revealed by
[13]C-labeling and by pyrolysis of the proposed intermediates, are shown in
Eq. (9.101). A curious feature of this reaction is the elimination of carbon
monoxide from the proposed intermediate benzoylazirine **131**, which seems
without obvious precedent.

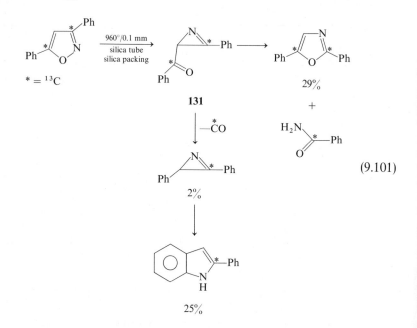

(9.101)

VII. REARRANGEMENT WITH MIGRATION OF A COORDINATED METAL ATOM

The cobalt sandwich complex **132** has recently been shown to undergo a
fascinating migration of the cyclopentadienylcobalt group on pyrolysis at
525° [Eq. (9.102)]. Fritch and Vollhardt[157] obtained the isomer **133** in good
yield by flash vacuum pyrolysis, and were also able to demonstrate the
occurrence of degenerate rearrangement in the parent system lacking the
trimethylsilyl groups by the use of starting material deuterated in the ethinyl
groups. It is suggested that migration of the cyclopentadienylcobalt group
in **132** probably occurs via the tricyclic intermediates **134–136** [R =
(CH₃)₃Si] with stepwise transfer of aromatic character across the system
[Eq. (9.103)].

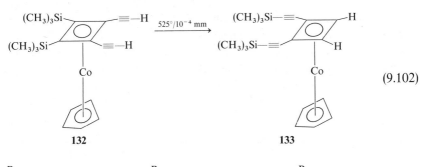

$$(9.102)$$

132   **133**

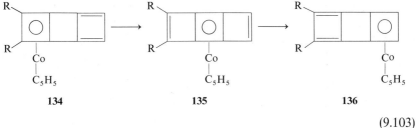

134   **135**   **136**

$$(9.103)$$

REFERENCES

1. Isomura, K., Kobayashi, S., and Taniguchi, H. (1968). *Tetrahedron Lett.*, p. 3499.
2. Wendling, L. A., and Bergman, R. G. (1976). *J. Org. Chem.* **41**, 831.
3. Demoulin, A., Gorrissen, H., Hesboin-Frisque, A. M., and Ghosez, L. (1975). *J. Am. Chem. Soc.* **97**, 4409.
4. Wendling, L. A., and Bergman, R. G. (1974). *J. Am. Chem. Soc.* **96**, 308.
5. Srinavasan, R. (1971). *Chem. Commun.*, p. 1041.
6. York, E. J., Dittmar, W., Stevenson, J. R., and Bergman, R. G. (1973). *J. Am. Chem. Soc.* **95**, 5680.
7. Frey, H. M. (1966). *In* "Advances in Physical Organic Chemistry" (V. Gold, ed.), Vol. 4, p. 147, Academic Press, New York.
8. Brown, J. M. (1973). *In* "International Review of Science. Organic Chemistry, Series One. Vol. 5, Alicyclic Compounds" (W. Parker, ed.), p. 159. Butterworth, London.
9. Becker, D., and Brodsky, N. C. (1976). *In* "International Review of Science, Series Two. Vol. 5, Alicyclic Compounds" (D. Ginsburg, ed.), p. 197. Butterworth, London.
10. Closs, G. L. (1966). *In* "Advances in Alicyclic Chemistry" (H. H. Hart and G. J. Karabatsos, eds.), Vol. 1, p. 53. Academic Press, New York.
11. Belluš, D., and Weis, C. D. (1973). *Tetrahedron Lett.*, p. 999.
12. Dowd, P., and Kang, K. (1974). *Synth. Commun.* **4**, 151.
13. Willner, I., and Rabinovitz, M. (1976). *Tetrahedron Lett.*, p. 3335.
14. Paquette, L. A., and Stowell, J. C. (1971). *J. Am. Chem. Soc.* **93**, 5735.
15. Oda, M., Kayama, Y., Miyazaki, H., and Kitahara, Y. (1975). *Angew. Chem. Int. Ed. Engl.* **14**, 418.
16. Kayama, Y., Oda, M., and Kitahara, Y. (1974). *Tetrahedron Lett.*, p. 3293.
17. Oda, M., Miyazaki, H., Kayama, Y., and Kitahara, Y. (1975). *Chem. Lett.*, p. 627.

18. Sukawa, H., Seshimoto, O., Tezuka, T., and Mukai, T. (1974). *J. Chem. Soc. Chem. Commun.*, p. 696.
19. Clifton, E. D., Flowers, W. T., and Hazeldine, R. N. (1969). *Chem. Commun.*, p. 1216.
20. Wieser, K., and Berndt, A. (1975). *Angew. Chem. Int. Ed. Engl.* **14**, 70.
21. Kitahara, Y., Oda, M., Miyakakashi, S., and Nakanishi, S. (1976). *Tetrahedron Lett.*, p. 2149.
22. Schiess, P., and Wisson, M. (1974). *Helv. Chim. Acta* **57**, 1692.
23. Schiess, P., and Fünfschilling, P. (1976). *Helv. Chim. Acta* **59**, 1745.
24. Schiess, P., and Fünfschilling, P. (1976). *Helv. Chim. Acta* **59**, 1756.
25. Yoshida, M., Sugihara, H., Tsushima, S., and Miki, T. (1969). *Chem. Commun.*, p. 1223.
26. Volkovitch, P. B., Conger, J. L., Castiello, F. A., Brodie, T. D., and Weber, W. P. (1975). *J. Am. Chem. Soc.* **97**, 901.
27. Radcliffe, M. M., and Weber, W. P. (1977). *J. Org. Chem.* **42**, 297.
28. Rosen, B. I., and Weber, W. P. (1977). *J. Org. Chem.* **42**, 47.
29. Rosen, B. I., and Weber, W. P. (1977). *Tetrahedron Lett.*, p. 151.
30. Gilchrist, T. L., Gymer, G. E., and Rees, C. W. (1975). *J. Chem. Soc. Perkin Trans. 1*, p. 1.
31. De Camp, M. R., Levin, R. H., and Jones, M. (1974). *Tetrahedron Lett.*, p. 3575.
32. Heimgartner, H., Zsindely, J., Hansen, H. J., and Schmid, H. (1973). *Helv. Chim. Acta* **56**, 2924.
33. Brown, R. F. C., and McMullen, G. L. (1974). *Aust. J. Chem.* **27**, 2385.
34. Brown, R. F. C., and Eastwood, F. W. (1980). *In* "The Chemistry of Ketenes and Allenes" (S. Patai, ed.), pp. 757–778. Wiley, New York.
35. Baxter, G. J., Brown, R. F. C., and McMullen, G. L. (1974). *Aust. J. Chem.* **27**, 2605.
36. Baxter, G. J., and Brown, R. F. C. (1975). *Aust. J. Chem.* **28**, 1551.
37. Baxter, G. J., and Brown, R. F. C. (1978). *Aust. J. Chem.* **31**, 327.
38. Cava, M. P., and Bravo, L. (1970). *Tetrahedron Lett.*, p. 4631.
39. Paquette, L. A., and Wingard, R. E. (1972). *J. Am. Chem. Soc.* **94**, 4398.
40. Paquette, L. A., and Photis, J. M. (1975). *Tetrahedron Lett.*, p. 1145.
41. von E. Doering, W., and Rosenthal, J. W. (1966). *J. Am. Chem. Soc.* **88**, 2078.
42. Paquette, L. A., Krow, G. R., and Malpass, J. R. (1969). *J. Am. Chem. Soc.* **91**, 5522.
43. Nomura, Y., Takeuchi, V., Tomoda, S., and Goldstein, M. J. (1977). *J. Chem. Soc. Chem. Commun.*, p. 545.
44. Brooke, G. M. (1974). *J. Chem. Soc. Perkin Trans. 1*, p. 233.
45. Brooke, G. M., and Hall, D. H. (1976). *J. Chem. Soc. Perkin Trans. 1*, p. 1463.
46. Trahanovsky, W. S., and Mullen, P. W. (1972). *J. Am. Chem. Soc.* **94**, 5911.
47. Riemann, J. M., and Trahanovsky, W. S. (1977). *Tetrahedron Lett.*, p. 1863.
48. Riemann, J. M., and Trahanovsky, W. S. (1977). *Tetrahedron Lett.*, p. 1867.
49. Hoffmann, H. M. R. (1969). *Angew. Chem. Int. Ed. Engl.* **8**, 556.
50. Roth, W. R., and König, J. (1965). *Justus Liebigs Ann. Chem.* **688**, 28.
51. Huntsman, W. D., Solomon, V. C., and Eros, D. (1958). *J. Am. Chem. Soc.* **80**, 5455.
52. Huntsman, W. D., and Hall, R. P. (1962). *J. Org. Chem.* **27**, 1988.
53. Bloch, R., and Bortolussi, M. (1976). *Tetrahedron Lett.*, p. 309.
54. Conia, J.-M., and Le Perchec, P. (1975). *Synthesis*, p. 1.
55. Rouessac, F., Le Perchec, P., and Conia, J.-M. (1967). *Bull. Soc. Chim. France*, p. 818.
56. Leyendecker, F., Mandville, G., and Conia, J.-M. (1970). *Bull. Soc. Chim. France*, p. 556.
57. Rosowsky, A. (1964). *In* "Heterocyclic Compounds with Three- and Four-Membered Rings" (A. Weissberger, ed.), Part 1, pp. 231–262. Wiley (Interscience), New York.
58. Hudrlik, P. F., Wan, C.-N., and Withers, G. P. (1976). *Tetrahedron Lett.*, p. 1449, and Ref. 10 therein.
59. Garin, D. L. (1969). *Can. J. Chem.* **47**, 4071.
60. Schiess, P., and Radimerski, P. (1972). *Angew. Chem. Int. Ed. Engl.* **11**, 288.

61. Alder, K., Flock, F. H., and Lessenich, H. (1957). *Chem. Ber.* **90**, 1709.
62. Turro, N. J., Morton, D. R., Hedaya, E., Kent, M. E., D'Angelo, P., and Schissel, P. (1971). *Tetrahedron Lett.*, p. 2535.
63. Khazanie, P. G., and Lee-Ruff, E. (1973). *Can. J. Chem.* **51**, 3173.
64. Eberbach, W., and Carré, J. C. (1976). *Tetrahedron Lett.*, p. 3299.
65. Crandall, J. K., Paulson, D. R., and Bunnell, C. A. (1969). *Tetranedron Lett.*, p. 4217.
66. Dolbier, W. R., Akiba, K., Riemann, J. M., Harmon, C. A., Bertrand, M., Bezaguet, A., and Santelli, M. (1971). *J. Am. Chem. Soc.* **93**, 3933.
67. Bloch, R., Leyendecker, F., and Toshima, N. (1973). *Tetrahedron Lett.*, p. 1025.
68. Jones, M., and Schwab, L. O. (1968). *J. Am. Chem. Soc.* **90**, 6549.
69. von E. Doering, W., and Sachdev, R. (1974). *J. Am. Chem. Soc.* **96**, 1168.
70. Japenga, J., Kool, M., and Klumpp, G. W. (1974). *Tetrahedron Lett.*, p. 3805.
71. Bishop, R., Parker, W., and Watt, I. (1977). *Tetrahedron Lett.*, p. 4345.
72. Subba Rao, H. N., Damodaran, N. P., and Dev, S. (1968). *Tetrahedron Lett.*, p. 2213.
73. Crandall, J. K., Arrington, J. P., and Watkins, R. J. (1967). *Chem. Commun.*, p. 1052.
74. Bertrand, M., Gil, G., Junino, A., and Mavrin, R. (1977). *Tetrahedron Lett.*, p. 1779.
75. Janusz, J. M., Gardiner, L. J., and Berson, J. A. (1977). *J. Am. Chem. Soc.* **99**, 8509.
76. Berson, J. A. (1978) private communication. See note 20 in ref. 75 above.
77. Gutsche, C. D., and Redmore, D. (1968). "Carbocyclic Ring Expansion Reactions," pp. 161–189. Academic Press, New York.
78. Frey, H. M., and Walsh, R. (1969). *Chem. Rev.* **69**, 103.
79. Willcott, M. R., Cargill, R. L., and Sears, A. B. (1972). *In* "Progress in Physical Organic Chemistry," (A. Streitwieser and R. Taft, eds.), Vol. 9, p. 25. Wiley (Interscience), New York.
80. Trost, B. M., and Bogdanowicz, M. J. (1973). *J. Am. Chem. Soc.* **95**, 5311. (See reference 18 therein.)
81. Monti, S. A., Cowherd, F. G., and McAninch, T. W. (1975). *J. Org. Chem.* **40**, 858.
82. Piers, E., Lau, C. K., and Nagakura, I. (1976). *Tetrahedron Lett.*, p. 3233.
83. Bartlett, P. D., and Sargent, G. D. (1965). *J. Am. Chem. Soc.* **87**, 1297.
84. Trost, B. M., and Keeley, D. E. (1976). *J. Am. Chem. Soc.* **98**, 248.
85. Miller, R. D. (1976). *J. Chem. Soc. Chem. Commun.*, p. 277.
86. Wenkert, E., and Regodesousa, J. (1977). *Synth. Commun.* **7**, 457.
87. Sadler, I. H., and Stewart, J. A. G. (1970). *J. Chem. Soc. Chem. Commun.*, p. 1588.
88. Sadler, I. H. (1978). Personal communication.
89. Woodward, R. B., and Hoffman, R. (1970). "The Conservation of Orbital Symmetry" pp. 114–140. Academic Press, New York.
90. Spangler, C. W. (1976). *Chem. Rev.* **76**, 187.
91. Brown, R. F. C., and Butcher, M. (1969). *Aust. J. Chem.* **22**, 1457.
92. Brown, R. F. C., Eastwood, F. W., Harrington, K. J., and McMullen, G. L. (1974). *Aust. J. Chem.* **27**, 2393.
93. Minter, D. E., and Fonken, G. J. (1977). *Tetrahedron Lett.*, p. 1717.
94. Berson, J. A., and Nelson, G. L. (1967). *J. Am. Chem. Soc.* **89**, 5503.
95. Berson, J. A. (1968). *Acc. Chem. Res.* **1**, 152.
96. Mironov, V. A., Pashegorova, V. S., Fadeeva, T. M., and Akhrem, A. A. (1968). *Tetrahedron Lett.*, p. 3997.
97. Willcott, M. R., and Rathburn, I. M. (1974). *J. Am. Chem. Soc.* **96**, 938.
98. Krekels, J. M. E., de Haan, J. W., and Kloosterziel, K. (1970). *Tetrahedron Lett.*, p. 2751.
99. Clark, R. A., Hayles, W. J., and Youngs, D. S. (1975). *J. Am. Chem. Soc.* **97**, 1966.
100. Spangler, C. W., and Boles, D. C. (1972). *J. Org. Chem.* **37**, 1020.
101. Schiess, P., and Dinkel, R. (1975). *Tetrahedron Lett.*, p. 2503.
102. Pines, H., and Kozlowski, R. (1956). *J. Am. Chem. Soc.* **78**, 3776.

103. Brown, R. F. C., Eastwood, F. W., and Jackman, G. P. (1977). *Aust. J. Chem.* **30**, 1757.
104. Wentrup, C., and Crow, W. D. (1970). *Tetrahedron* **26**, 3965.
105. Miller, L. L., and Boyer, R. F. (1971). *J. Am. Chem. Soc.* **93**, 650.
106. Paquette, L. A., Kukla, M. J., Ley, S. V., and Traynor, S. G. (1977). *J. Am. Chem. Soc.* **99**, 4756.
107. Semmelhack, M. F., Weller, H. N., and Foos, J. S. (1977). *J. Am. Chem. Soc.* **99**, 292.
108. Schiess, P., and Fünfschilling, P. (1972). *Tetrahedron Lett.*, p. 5191.
109. Field, D. J., Jones, D. W., and Kneen, G. (1976). *J. Chem. Soc. Chem. Commun.*, p. 873.
110. Rhoads, S. J., and Raulins, N. R. (1975). *In* "Organic Reactions" (W. G. Dauben, ed.), Vol. 22, p. 1. Wiley, New York.
111. Hedaya, E., and McNeil, D. W. (1967). *J. Am. Chem. Soc.* **89**, 4213.
112. Goldschmidt, Z., and Mauda, S. (1976). *Tetrahedron Lett.*, p. 4183.
113. Brown, R. F. C., and McAllan, C. G. (1977). *Aust. J. Chem.* **30**, 1747.
114. Japenga, J., Kool, M., and Klumpp, G. W. (1975). *Tetrahedron Lett.*, p. 1029.
115. Karpf, M., and Dreiding, A. S. (1977). *Helv. Chim. Acta* **60**, 3045.
116. Büchi, G., and Powell, J. E. (1970). *J. Am. Chem. Soc.* **92**, 3126.
117. Büchi, G., and White, J. D. (1964). *J. Am. Chem. Soc.* **86**, 2884.
118. Trahanovsky, W. S., and Mullen, P. W. (1972). *J. Am. Chem. Soc.* **94**, 5086.
119. Young, F. G., Frostick, F. C., Sanderson, J. J., and Hauser, C. R. (1950). *J. Am. Chem. Soc.* **72**, 3635.
120. Trahanovsky, W. S., and Emeis, S. L. (1975). *J. Am. Chem. Soc.* **97**, 3773.
121. Trahanovsky, W. S., Emeis, S. L., and Lee, A. S. (1976). *J. Org. Chem.* **41**, 4043.
122. Jones, R. A., and Bean, G. P. (1977). "The Chemistry of Pyrroles," pp. 249–252. Academic Press, New York.
123. Patterson, J. M. (1976). *Synthesis*, p. 281.
124. Jacobsen, I. A., Heady, H. H., and Dinneen, G. U. (1958). *J. Phys. Chem.* **62**, 1563.
125. Jacobsen, I. A., and Jensen, H. B. (1962). *J. Phys. Chem.* **66**, 1245.
126. Patterson, J. M., and Drenchko, P. (1962). *J. Org. Chem.* **27**, 1650.
127. Patterson, J. M., Burka, L. T., and Boyd, M. R. (1968). *J. Org. Chem.* **33**, 4033.
128. Patterson, J. M., and Soedigdo, S. (1967). *J. Org. Chem.* **32**, 2969.
129. Gilchrist, T. L., Rees, C. W., and Thomas, C. (1975). *J. Chem. Soc. Perkin Trans. 1*, p. 8.
130. Patterson, J. M., Mayer, C. F., and Smith, W. T. (1975). *J. Org. Chem.* **40**, 1511.
131. Patterson, J. M., de Haan, J. W., Boyd, M. R., and Ferry, J. D. (1972). *J. Am. Chem. Soc.* **94**, 2487.
132. Begg, C. G., Grimmett, M. R., and Wethey, P. D. (1973). *Aust. J. Chem.* **26**, 2435.
133. Huntsman W. D., and Wristers, H. J., (1963). *J. Am. Chem. Soc.* **85**, 3308.
134. Heffernan, M. L., and Jones, A. J. (1966). *Chem. Commun.*, p. 120.
135. Kent, J. E., and Jones, A. J. (1970). *Aust. J. Chem.* **23**, 1059.
136. Huntsman, W. D. (1972). *Intra-Sci. Chem. Rep.* **6**, 151.
137. Bergman, R. G. (1973). *Acc. Chem. Res.* **6**, 25.
138. Jones, R. R., and Bergman, R. G. (1972). *J. Am. Chem. Soc.* **94**, 660.
139. D'Amore, M. B., Bergman, R. G., Kent, M., and Hedaya, E. (1972). *J. Chem. Soc. Chem. Commun.*, p. 49.
140. Eisenhuth, L., and Hopf, H. (1974). *J. Am. Chem. Soc.* **96**, 5667.
141. Vollhardt, K. P. C., and Bergman, R. G. (1972). *J. Am. Chem. Soc.* **94**, 8950.
142. Vollhardt, K. P. C., and Bergman, R. G. (1973). *J. Am. Chem. Soc.* **95**, 7538.
143. Berkovich, A. J., Strauss, E. S., and Vollhardt, K. P. C. (1977). *J. Am. Chem. Soc.* **99**, 8321.
144. Wentrup, C., Wentrup-Byrne, E., and Müller, P. (1977). *J. Chem. Soc. Chem. Commun.*, p. 210.
145. Henry, T. J., and Bergman, R. G. (1972). *J. Am. Chem. Soc.* **94**, 5103.

146. Brown, R. F. C., Gream, G. E., Peters, D. E., and Solly, R. K. (1968). *Aust. J. Chem.* **21**, 2223.
147. Cava, M. P., Mitchell, M. J., and Pohl, R. J. (1963). *J. Am. Chem. Soc.* **85**, 2080.
148. Brown, R. F. C., Hooley, N., and Irvine, F. N. (1974). *Aust. J. Chem.* **27**, 671.
149. Alder, R. W., and Whittaker, G. (1975). *J. Chem. Soc. Perkin Trans. 2*, p. 714.
150. Alder, R. W., and Wilshire, C. (1975). *J. Chem. Soc. Perkin Trans. 2*, p. 1464.
151. Alder, R. W., and Whittaker, G. (1975). *J. Chem. Soc. Perkin Trans. 2*, p. 712.
152. Crow, W. D., and Wentrup, C. (1968). *Tetrahedron Lett.*, p. 3115.
153. Chambers, R. D., MacBride, J. A. H., and Musgrave, W. K. R. (1971). *J. Chem. Soc. C*, p. 3384.
154. Patterson, J. M., Issidorides, C. H., Papadopoulos, E. P., and Smith, W. T. (1970). *Tetrahedron Lett.*, p. 1247.
155. Kinstle, T. H., and Darlage, L. J. (1969). *J. Heterocycl. Chem.* **6**, 123.
156. Aldous, G. L., Bowie, J. H., and Thompson, M. J. (1976). *J. Chem. Soc. Perkin Trans. 1*, p. 16.
157. Fritch, J. R., and Vollhardt, K. P. C. (1978). *J. Am. Chem. Soc.* **100**, 3643.

Index

Q

ORGANIC CHEMISTRY

A SERIES OF MONOGRAPHS

EDITOR

HARRY H. WASSERMAN

Department of Chemistry
Yale University
New Haven, Connecticut

1. Wolfgang Kirmse. CARBENE CHEMISTRY, 1964; 2nd Edition, 1971

2. Brandes H. Smith. BRIDGED AROMATIC COMPOUNDS, 1964

3. Michael Hanack. CONFORMATION THEORY, 1965

4. Donald J. Cram. FUNDAMENTALS OF CARBANION CHEMISTRY, 1965

5. Kenneth B. Wiberg (Editor). OXIDATION IN ORGANIC CHEMISTRY, PART A, 1965; Walter S. Trahanovsky (Editor). OXIDATION IN ORGANIC CHEMISTRY, PART B, 1973; PART C, 1978

6. R. F. Hudson. STRUCTURE AND MECHANISM IN ORGANO-PHOSPHORUS CHEMISTRY, 1965

7. A. William Johnson. YLID CHEMISTRY, 1966

8. Jan Hamer (Editor). 1,4-CYCLOADDITION REACTIONS, 1967

9. Henri Ulrich. CYCLOADDITION REACTIONS OF HETEROCUMULENES, 1967

10. M. P. Cava and M. J. Mitchell. CYCLOBUTADIENE AND RELATED COMPOUNDS, 1967

11. Reinhard W. Hoffmann. DEHYDROBENZENE AND CYCLOALKYNES, 1967

12. Stanley R. Sandler and Wolf Karo. ORGANIC FUNCTIONAL GROUP PREPARATIONS, VOLUME I, 1968; VOLUME II, 1971; VOLUME III, 1972

13. Robert J. Cotter and Markus Matzner. RING-FORMING POLYMERIZATIONS, PART A, 1969; PART B, 1; B, 2, 1972

14. R. H. DeWolfe, CARBOXYLIC ORTHO ACID DERIVATIVES, 1970

15. R. Foster. ORGANIC CHARGE-TRANSFER COMPLEXES, 1969

16. James P. Snyder (Editor). NONBENZENOID AROMATICS, VOLUME I, 1969; VOLUME II, 1971

17. C. H. Rochester. ACIDITY FUNCTIONS, 1970

18. Richard J. Sundberg. THE CHEMISTRY OF INDOLES, 1970

19. A. R. Katritzky and J. M. Lagowski. CHEMISTRY OF THE HETEROCYCLIC N-OXIDES, 1970

20. Ivar Ugi (Editor). ISONITRILE CHEMISTRY, 1971

21. G. Chiurdoglu (Editor). CONFOR-
MATIONAL ANALYSIS, 1971

22. Gottfried Schill. CATENANES, ROTAX-
ANES, AND KNOTS, 1971

23. M. Liler. REACTION MECHANISMS IN
SULPHURIC ACID AND OTHER STRONG
ACID SOLUTIONS, 1971

24. J. B. Stothers. CARBON-13 NMR
SPECTROSCOPY, 1972

25. Maurice Shamma. THE ISOQUINO-
LINE ALKALOIDS: CHEMISTRY AND
PHARMACOLOGY, 1972

26. Samuel P. McManus (Editor). OR-
GANIC REACTIVE INTERMEDIATES,
1973

27. H. C. Van der Plas. RING TRANSFOR-
MATIONS OF HETEROCYCLES, VOL-
UMES 1 AND 2, 1973

28. Paul N. Rylander. ORGANIC SYNTHE-
SES WITH NOBLE CATALYSTS, 1973

29. Stanley R. Sandler and Wolf Karo.
POLYMER SYNTHESES, VOLUME I,
1974; VOLUME II, 1977; VOLUME
III, 1980

30. Robert T. Blickenstaff, Anil C.
Ghosh, and Gordon C. Wolf. TOTAL
SYNTHESIS OF STEROIDS, 1974

31. Barry M. Trost and Lawrence S.
Melvin, Jr. SULFUR YLIDES: EMERG-
ING SYNTHETIC INTERMEDIATES, 1975

32. Sidney D. Ross, Manuel Finkelstein,
and Eric J. Rudd. ANODIC OXIDATION,
1975

33. Howard Alper (Editor). TRANSITION
METAL ORGANOMETALLICS IN OR-
GANIC SYNTHESIS, VOLUME I, 1976;
VOLUME II, 1978

34. R. A. Jones and G. P. Bean. THE
CHEMISTRY OF PYRROLES, 1976

35. Alan P. Marchand and Roland E.
Lehr (Editors). PERICYCLIC REAC-
TIONS, VOLUME I, 1977; VOLUME II,
1977

36. Pierre Crabbé (Editor). PROSTAGLAN-
DIN RESEARCH, 1977

37. Eric Block. REACTIONS OF ORGANO-
SULFUR COMPOUNDS, 1978

38. Arthur Greenberg and Joel F. Lieb-
man, STRAINED ORGANIC MOLECULES,
1978

39. Philip S. Bailey. OZONATION IN OR-
GANIC CHEMISTRY, VOL. I, 1978

40. Harry H. Wasserman and Robert W.
Murray (Editors). SINGLET OXYGEN,
1979

41. Roger F. C. Brown. PYROLYTIC
METHODS IN ORGANIC CHEMISTRY:
APPLICATIONS OF FLOW AND FLASH
VACUUM PYROLYTIC TECHNIQUES,
1980

42. Paul de Mayo (Editor). REARRANGE-
MENTS IN GROUND AND EXCITED
STATES, VOLUME I, 1980; VOLUME
II, 1980; VOLUME III, 1980